Single-Layer Brazed Cubic Boron Nitride Abrasive Tools

This book systematically summarizes the state-of-the-art research in the field of single-layer brazed cubic boron nitride (CBN) abrasive tools in terms of manufacturing technology, wear mechanism, and machining performance.

The authors present manufacturing methods and related principles and explore the wear behaviour and mechanisms of single-layer brazed CBN abrasive tools, providing insights into tool manufacturing and tool life. They also clarify the scientific issues in the grinding performance of single-layer brazed CBN abrasive tools to improve machining efficiency and quality.

This book will contribute to the development of aerospace engineering and inspire academic researchers and industrial engineers in the field of ultra-high precision machining, especially grinding.

Wenfeng Ding is currently a professor of Mechanical Engineering and Doctoral Tutor at Nanjing University of Aeronautics and Astronautics, P.R. China. His research interests include grinding technology and equipment, superhard abrasive tools, machining process simulation, and control technology.

Biao Zhao is currently an associate professor of Mechanical Engineering at Nanjing University of Aeronautics and Astronautics, P.R. China. His research interests include high-efficiency and precision grinding technology, high-performance abrasive tools, grinding mechanisms, and process optimization.

Ning Qian is currently an associate professor at Nanjing University of Aeronautics and Astronautics. His research interests include high-performance grinding mechanisms and technologies.

Haonan Li is currently a professor at the School of Aerospace, University of Nottingham, Ningbo, P.R. China. His research interests include grinding technology, tooling, and machine design and build.

Jiuhua Xu is currently a professor at Nanjing University of Aeronautics and Astronautics; he is also the Changjiang scholar. His research interests include high-efficiency and precision machining technologies, intelligent manufacturing, etc.

Single-Layer Brazed Cubic Boron Nitride Abrasive Tools

Wenfeng Ding, Biao Zhao, Ning Qian,
Haonan Li, and Jiuhua Xu

CRC Press
Taylor & Francis Group
Boca Raton London New York

CRC Press is an imprint of the
Taylor & Francis Group, an **informa** business

Designed cover image: © Volodymyr Burdiak

First edition published 2024
by CRC Press
2385 NW Executive Center Drive, Suite 320, Boca Raton FL 33431

and by CRC Press
4 Park Square, Milton Park, Abingdon, Oxon, OX14 4RN

CRC Press is an imprint of Taylor & Francis Group, LLC

© 2024 Wenfeng Ding, Biao Zhao, Ning Qian, Haonan Li and Jiuhua Xu

ISBN: 978-1-032-67720-0 (hbk)
ISBN: 978-1-032-67805-4 (pbk)
ISBN: 978-1-032-67804-7 (ebk)

DOI: 10.1201/9781032678047

Typeset in Minion
by codeMantra

Contents

Preface, xv

PART I **Fabrication Technology and Mechanism**

CHAPTER 1 ▪ Introduction   3

1.1 SINGLE-LAYER BRAZED CBN ABRASIVE WEELS   3

1.2 BOOK STRUCTURE   4

REFERENCES   5

CHAPTER 2 ▪ Brazing of Monocrystalline CBN Abrasive Grains Based
            on Vacuum Furnace Heating   6

2.1 INTRODUCTION   6

2.2 BRAZING INTERFACE OF CBN GRAINS AND FILLER ALLOY
     CONTAINING TITANIUM   8

2.3 INTERFACIAL FORMATION MECHANISM DURING BRAZING
     CBN GRAINS   11

     2.3.1 Characterization of Ti-coated CBN Abrasive Grains   11

     2.3.2 Effects of Heating Temperature on Interfacial Reaction   12

     2.3.3 Compound Morphology of Interfacial Resultants   14

     2.3.4 Formation Mechanisms of the Interfacial Microstructure   16

REFERENCES   17

CHAPTER 3 ▪ Brazing of Monocrystalline CBN Abrasive Grains with
            Composite Fillers   18

3.1 INTRODUCTION   18

3.2 EFFECT OF TIX PARTICLES ON MICROSTRUCTURE
     AND STRENGTH OF COMPOSITE FILLER   20

     3.2.1 Experimental Details   20

     3.2.2 Effects of Reinforcing Particles on the Spreading Behaviour of
            Composite Fillers   22

v

3.2.3 Effects of Reinforcing Particles on the Shear Strength of the Bonding Layer 24

3.2.4 Effects of Reinforcing Particles on Bonding Layer Microstructure 26

3.2.5 Distribution of Reinforcing Particles in the Bonding Layer 28

3.2.6 Reinforcing Mechanism of the TiX Particle Modified Composite Filler 30

3.2.7 Effect of Reinforcing Particles on the Microhardness of the Bonding Layer 31

3.3 EFFECT OF TIX PARTICLES ON THE BRAZING MICROSTRUCTURE OF CBN ABRASIVE GRAINS 33

3.3.1 Experimental Details 33

3.3.2 Overall Morphology of CBN Grains Brazed with Composite Fillers 33

3.3.3 Microstructure and Microchemistry Characteristics at the Grain/Filler Interface 33

3.3.4 Morphology of Brazing Chemical Resultants 36

3.3.5 Phase Analysis of Brazing Chemical Resultants 38

3.3.6 Fracture Morphology of the Brazed Joints 39

3.3.7 Influencing Mechanism of TiX Reinforcing Particles on Brazing Reaction 40

3.4 STRENGTH ANALYSIS OF THE CBN ABRASIVE GRAINS BRAZED WITH A COMPOSITE FILLER 41

3.4.1 Measurement and Calculation of Compressive Strength of Brazed CBN Grains 41

3.4.2 Effects of TiX Particle Types on the Compressive Strength of Brazed CBN Grains 44

3.4.3 Effects of TiX Particle Contents on the Compressive Strength of Brazed Grains 46

REFERENCES 46

CHAPTER 4 ■ Brazing of Polycrystalline CBN Abrasive Grains Based on Vacuum Furnace Heating 48

4.1 INTRODUCTION 48

4.2 BRAZED POLYCRYSTALLINE CBN ABRASIVE GRAINS WITH AG-CU-TI FILLER ALLOY 48

4.2.1 Appearance and Joining Interface of Brazed Polycrystalline CBN Grains 48

4.2.2 Chemical Resultants around Brazed Polycrystalline CBN Grains 50

4.2.3 Brazing Mechanism of Polycrystalline CBN Grains 52

4.3 BRAZED POLYCRYSTALLINE CBN ABRASIVE GRAINS WITH
CU-SN-TI FILLER ALLOY   54

    4.3.1   Experimental Details   54

    4.3.2   Interfacial Microstructure of Brazed Polycrystalline CBN Grains   56

    4.3.3   Resultant Morphology of Brazed Polycrystalline CBN Grains   58

    4.3.4   Joining Performance of Brazed Polycrystalline CBN Grains   60

    4.3.5   Compressive Strength of Brazed Polycrystalline CBN Grains   60

REFERENCES   61

CHAPTER 5 ■ Brazing of CBN Abrasive Grains Based
on High-Frequency Heating   63

5.1 INTRODUCTION   63

5.2 TEMPERATURE DISTRIBUTION IN HIGH-FREQUENCY BRAZING   64

    5.2.1   Numerical Simulation Principle for Simulating Temperature
Distribution   64

    5.2.2   Geometry Modelling and Meshes for Simulating Temperature
Distribution   65

    5.2.3   Material Properties and Boundary Conditions   67

    5.2.4   Region Selections for Temperature Distribution Analysis   69

    5.2.5   Simulation Process of Temperature Distribution   70

    5.2.6   Typical Contour and Evolution of Temperature Distribution   70

    5.2.7   Effects of Induction Heating Parameters
on Temperature Distribution   73

    5.2.8   Curve Fitting of Highest Temperature Varying with Heating
Parameters   78

5.3 EXPERIMENTAL VERIFICATION OF HIGH-FREQUENCY
BRAZING OF CBN GRAINS   79

    5.3.1   Experimental Verification of the Heating Curve
and the Highest Temperature   79

    5.3.2   High-frequency Induction Brazing of CBN Grains   81

    5.3.3   Optimization of High-Frequency Heating Parameters   81

REFERENCES   82

CHAPTER 6 ■ Rhythmic Grain Distribution on the Wheel Surface   84

REFERENCES   86

PART II **Wear Behaviour and Mechanism**

CHAPTER 7 ■ Wear Behaviour and Stresses Effects of Monocrystalline
CBN Abrasive Wheels   89

  7.1 INTRODUCTION   89

      7.1.1   Wear Behaviour of Abrasive Wheels   89

      7.1.2   Stresses Effects on Grain Wear   90

  7.2 WEAR PHENOMENON OF MONOCRYSTALLINE CBN
       ABRASIVE WHEELS IN GRINDING   91

      7.2.1   Experimental Details   91

      7.2.2   Mechanical Loads of Brazed CBN Abrasive Wheels   92

      7.2.3   Thermal Loads of Brazed CBN Abrasive Wheels   94

      7.2.4   Protrusion Height of Brazed CBN Grains   95

      7.2.5   General Wear Phenomenon of Brazed CBN Abrasive Wheels   97

      7.2.6   Mild Wear Morphology of Brazed CBN Abrasive Wheels   99

      7.2.7   Severe Wear Morphology of Brazed CBN Abrasive Wheels   100

  7.3 STRESSES EFFECTS ON MONOCRYSTALLINE CBN
       GRAIN WEAR IN GRINDING   102

      7.3.1   FE Modelling   102

      7.3.2   Effect of Grain Embedding Depth on Brazing Stresses
              and Resultant Stresses   107

      7.3.3   Effect of Grain Wear on Redistributed Brazing Stresses
              and Resultant Stresses   110

      7.3.4   Effect of Grain Size on Brazing Stresses and Resultant Stresses   112

      7.3.5   Effect of Grinding Load on Resultant Stresses   112

  REFERENCES   114

CHAPTER 8 ■ Wear Behaviour of Polycrystalline CBN Abrasive Wheels   115

  8.1 INTRODUCTION   115

  8.2 WEAR PHENOMENON OF POLYCRYSTALLINE CBN ABRASIVE
       WHEELS IN GRINDING   117

      8.2.1   Experimental Details   117

      8.2.2   General Wear Phenomenon of Polycrystalline CBN
              Abrasive Wheels   118

      8.2.3   Wear Morphology of Brazed Polycrystalline CBN Grains   120

      8.2.4   Fracture Mechanism of Brazed Polycrystalline CBN Grains   123

8.3   STRESSES EFFECTS ON POLYCRYSTALLINE CBN GRAIN WEAR   124

    8.3.1   Numerical Microstructure Construction of Polycrystalline Grains   124

    8.3.2   Finite Element Model of the Brazing Stress within Polycrystalline Grains   127

    8.3.3   Materials Properties and Grinding Loads Utilized in the Finite Element Model   128

    8.3.4   Characterization of Stress Distribution in Polycrystalline CBN Grains   129

    8.3.5   Influence of Embedding Depth on Stress Distribution in PCBN Grains   130

    8.3.6   Influence of Volume Fraction on Stress Distribution within PCBN Grains   133

    8.3.7   Influence of Grinding Loads on Resultant Stress in Polycrystalline CBN Grains   134

    8.3.8   Experimental Verification of Grain Wear Topography Evolution in Grinding   136

8.4   CRACK PROPAGATION OF POLYCRYSTALLINE CBN ABRASIVE GRAINS   138

    8.4.1   Cohesive Element Model for Polycrystalline CBN Grains   138

    8.4.2   Boundary Conditions and Simulation Parameters for Polycrystalline CBN Grains   141

    8.4.3   The Effects of Bonding Strength on Grain Boundaries   142

    8.4.4   The Effects of Grain Size on Polycrystalline CBN Grain Fracture   144

    8.4.5   The Effect of the Grain Boundary Stiffness   148

REFERENCES   153

CHAPTER 9 ■ Fractal Analysis of CBN Grain Wear Morphology in Grinding   155

9.1   INTRODUCTION   155

9.2   FRACTAL ANALYSIS OF GRAIN WEAR DURING GRINDING NICKEL-BASED SUPERALLOY   156

    9.2.1   Details of the Grinding Experiment   156

    9.2.2   Basic Principles and Analysis Methods of Three-Dimensional Fractal Theory   157

    9.2.3   Reconstruction of the Polycrystalline CBN Grains Topography   160

    9.2.4   Calculation of the 3D Fractal Dimension   161

    9.2.5   Fractal Analysis of Wear Topography of Brazed PcBN Grains   163

9.3   FRACTAL ANALYSIS OF GRAIN WEAR DURING GRINDING
      TITANIUM ALLOY   164

      9.3.1   Details of the Grinding Experiment   164

      9.3.2   Comparative Fractal Analysis of Polycrystalline and
              Monocrystalline Grains   166

      9.3.3   Comparison with Different Uncut Chip Thickness   169

9.4   COMPARATIVE FRACTAL ANALYSIS OF ABRASIVE WHEEL WEAR
      DURING GRINDING   174

      9.4.1   Experimental Details   174

      9.4.2   Fractal Dimension of Typical Grain Cutting Edges Morphology   176

      9.4.3   Wear Behaviour of Monocrystalline
              and Polycrystalline CBN Grains   176

      9.4.4   Fractal Analysis of Typical Monocrystalline and Polycrystalline
              CBN Grains   178

      9.4.5   Fractal Analysis of Monocrystalline and Polycrystalline CBN
              Abrasive Wheels   179

      9.4.6   Radial Wear of Monocrystalline and Polycrystalline CBN
              Abrasive Wheels in High-Speed Grinding   180

      9.4.7   Influence of the Self-sharpening Phenomenon of Abrasive
              Wheels on Grinding Forces and Forces Ratio   181

   REFERENCES   183

CHAPTER 10 ▪ Stress Distribution Effects on Grain Wear Evolution
             in Grinding   184

   10.1   INTRODUCTION   184

   10.2   FINITE ELEMENT MODEL OF SINGLE-GRAIN WEAR EVOLUTION
          DURING GRINDING   185

          10.2.1   Modelling Framework   185

          10.2.2   Model Geometry and Governing Equations   186

          10.2.3   Abrasive Wheel Process Kinematics   188

          10.2.4   Criteria for Single-CBN Grain Wear and Workpiece Material
                   Removal   189

   10.3   WEAR EVOLUTION OF SINGLE-CBN GRAIN   190

   10.4   INFLUENCE OF INTERNAL STRESSES ON GRAIN WEAR   191

   10.5   EVOLUTION OF CUTTING EDGES DUE TO GRAIN WEAR   194

   REFERENCES   196

CHAPTER 11 ■ Grain Wear Effect on Material Removal Behaviour during Grinding   198

11.1 INTRODUCTION   198

11.2 EXPERIMENTAL DETAILS AND PROCEDURE   199

11.3 MATERIAL REMOVAL CHARACTERISTICS IN THE CASE OF GRAIN WEAR-FREE   202

11.4 CLASSIFICATION OF GRAIN WEAR BEHAVIOUR IN SINGLE-GRAIN GRINDING   204

11.5 ANALYSIS OF GRAIN WEAR BEHAVIOUR FROM THE VIEWPOINT OF THE GRINDING FORCE   206

11.6 RELATIONSHIP BETWEEN GRINDING FORCE RATIO AND NEGATIVE RAKE ANGLE   208

11.7 INFLUENCE OF GRAIN WEAR BEHAVIOUR ON THE MATERIAL REMOVAL PROCESS   212

REFERENCES   214

CHAPTER 12 ■ Undeformed Chip Thickness Nonuniformity When Considering Abrasive Wheel Wear   216

12.1 INTRODUCTION   216

12.2 EXPERIMENTAL MATERIALS AND PROCEDURE   219

12.2.1 Grinding Wheel Topology Measurement   219

12.2.2 Grinding Experiment Setup   219

12.3 WHEEL TOPOLOGY RECONSTRUCTION   220

12.3.1 Procedure for Wheel Topology Reconstruction   220

12.3.2 Normality Evaluation   221

12.3.3 Johnson Transformation   222

12.3.4 Inverse Johnson Transformation and Wheel Topology Reconstruction   225

12.4 TEXTURED WHEEL TOPOLOGY AND ACTIVE GRAINS EVOLUTION   226

12.5 MODEL DEVELOPMENT OF UNDEFORMED CHIP THICKNESS   228

12.6 RELATIONSHIP BETWEEN THE WHEEL WEAR STATUS AND THE NONUNIFORMITY OF UNDEFORMED CHIP THICKNESS   231

12.7 MODELLING OF GROUND SURFACE ROUGHNESS   231

12.8 NUMERICAL PREDICTION FOR THE WORKPIECE TOPOGRAPHY EVOLUTION   233

REFERENCES   238

PART III **Grinding Performance and Mechanism**

CHAPTER 13 ▪ Grinding Behaviour and Surface Integrity of Titanium Alloy   243

13.1 INTRODUCTION   243

13.2 EXPERIMENTAL DETAILS   244

13.3 DIMENSIONAL ACCURACY OF GROUND SPECIMENS   244

13.4 SURFACE INTEGRITY OF GROUND SPECIMENS   246

    13.4.1 Surface Roughness of the Ground Specimen   246

    13.4.2 Morphology and Microstructure of the Ground Specimen   247

    13.4.3 Microhardness of Ground Specimen   249

    13.4.4 Surface Residual Stress of Ground Specimen   249

13.5 COMPARATIVE PERFORMANCE OF BRAZED AND ELECTROPLATED CBN ABRASIVE WHEELS IN GRINDING TITANIUM ALLOY   251

REFERENCES   253

CHAPTER 14 ▪ Grinding Behaviour and Surface Integrity of Nickel-Based Superalloy   254

14.1 INTRODUCTION   254

14.2 EXPERIMENTAL DETAILS   255

14.3 GRINDABILITY OF K424 NICKEL-BASED SUPERALLOY   258

    14.3.1 Grinding Force   258

    14.3.2 Specific Grinding Energy   260

    14.3.3 Grinding Temperature   261

14.4 DIMENSIONAL ACCURACY OF THE GROUND GROOVES   262

14.5 SURFACE INTEGRITY OF THE GROUND GROOVES   264

    14.5.1 Surface Topography   264

    14.5.2 Microhardness and Microstructure Alteration of the Sub-surface   264

    14.5.3 Residual Stresses on the Ground Surface   267

14.6 COMPARATIVE PERFORMANCE OF DIFFERENT CBN ABRASIVES IN GRINDING NICKEL-BASED SUPERALLOY   268

REFERENCES   269

CHAPTER 15 ▪ Grinding Behaviour and Surface Integrity of Titanium Matrix Composites   271

15.1 INTRODUCTION   271

15.2 EXPERIMENTAL DETAILS   272

15.3 COMPARISON OF GRINDING FORCE IN HIGH-SPEED GRINDING 274

15.4 COMPARISON OF GRINDING TEMPERATURE IN HIGH-SPEED GRINDING 276

15.5 GRINDING-INDUCED SURFACE FEATURES OF TITANIUM MATRIX COMPOSITES 280

15.6 MATERIALS REMOVAL MECHANISM DURING GRINDING OF TITANIUM MATRIX COMPOSITES 282

REFERENCES 285

CHAPTER 16 ■ Speed Effect on Materials Removal during Grinding 286

16.1 INTRODUCTION 286

16.2 FINITE ELEMENT MODELLING 287

16.2.1 Materials Model 287

16.2.2 Criterion of Material Fracture 288

16.2.3 Contact Law 288

16.2.4 Two-dimensional Adaptive FE Model 289

16.3 CHIPPING STAGES IN SURFACE GRINDING 290

16.4 EFFECT OF GRINDING SPEED ON CHIP FORMATION 293

16.5 GRINDING FORCES DURING CHIP FORMATION 295

16.6 RESULTANT STRESSES UNDER DIFFERENT GRINDING SPEEDS 298

REFERENCES 301

CHAPTER 17 ■ Formation and Affecting Factors of Burrs during Grinding 303

17.1 INTRODUCTION 303

17.2 FINITE ELEMENT MODELLING 304

17.2.1 Geometric Model of Single-Grain Surface Grinding 304

17.2.2 Materials Model 305

17.2.3 Criterion of Material Fracture in Surface Grinding 307

17.2.4 Contact Law 307

17.2.5 Grinding Parameters 308

17.3 FORMATION STAGES OF EXIT-DIRECTION BURRS IN SURFACE GRINDING 308

17.3.1 Deformation and Geometry Characteristics of Exit-Direction Burrs 308

17.3.2 Formation Stages of Positive Burrs 311

17.3.3 Formation Stages of Negative Burrs 312

17.3.4 Formation Stages of Residual Burrs 313

17.4 GRINDING SPEED EFFECT ON EXIT-DIRECTION BURRS
IN SURFACE GRINDING   314

17.5 EFFECT OF UNCUT CHIP THICKNESS ON
EXIT-DIRECTION BURRS IN SURFACE GRINDING   317

17.6 EFFECT OF GRAIN RAKE ANGLE ON EXIT-DIRECTION BURRS
IN SURFACE GRINDING   320

REFERENCES   322

Preface

Advanced manufacturing can, to a large extent, define both the economic development and the defence construction of a nation. Among various manufacturing technologies, the rapid development of modern grinding technology enables not only its increasingly wide applications in the aerospace, automobile, and energy industries but also the tendency to replace cutting with grinding.

Cubic boron nitride (CBN) abrasive wheels play the key role in grinding titanium alloy and nickel-based superalloy, and therefore tool quality and performance would dominate grinding process quality, efficiency, and cost. The weak grain retention and the random grain distribution of most conventional CBN wheels with vitrified, electroplated, or resin bonds were found to be the key issues. To fix this, the High-performance Machining Technology of Difficult-to-Cut Materials Group at Nanjing University of Aeronautics and Astronautics focused on the development of single-layer brazed CBN grinding wheels. Based on the understanding of the chemical and metallurgical interaction between grains, bond, and metallic wheel hub, the strong grain holding ability and the optimized grain distribution were successfully achieved, leading to a more stable grinding process with high efficiency, long wheel life, good machined surface quality, and low cost.

This book summarizes all the relevant work relating to single-layer brazed CBN abrasive wheels and contains three key parts and 17 chapters:

- Part I Fabrication Technology and Mechanism (Chapters 1–6) studies the fabrication technology and principle of single-layered brazed CBN grinding wheels, including grinding wheel introduction, the brazing principle of the mono- and poly-crystalline CBN abrasives, composite brazing solder and CBN brazing principle, high-frequency induction brazing of CBN abrasives, and grain distribution optimization.

- Part II Wear Behaviour and Mechanism (Chapters 7–12) introduces the wear mechanism of single-layered brazed CBN grinding wheels, including the wear behaviour, fractal topography analysis, wear evaluation, stress effect, and material removal process of mono- and poly-crystalline CBN abrasives.

- Part III Grinding Performance and Mechanism (Chapters 13–17) analyses the grinding mechanism of single-layered brazed CBN grinding wheels, including grinding performances for Ti/Ni-based alloys and Ti matrix composites, the speed effect, and burring.

The authors would like to highly acknowledge the kind and strong support from many Ph.D. and master's students, whose names cannot be given here due to limited space.

We would also like to acknowledge financial sponsorship, including the National 973 Project, the National Natural Science Foundation of China, high-grade CNC machine tools and basic manufacturing equipment, national major science and technology projects, aircraft engines and gas turbines, national major science and technology projects, and civil aircraft projects.

This book is suitable for scientific and technical personnel and postgraduate students in the field of mechanical manufacturing. Due to the authors' limited knowledge and experiences, experts and readers are welcome to correct any errors or fallacies in the book.

PART I

Fabrication Technology and Mechanism

CHAPTER 1

Introduction

1.1 SINGLE-LAYER BRAZED CBN ABRASIVE WEELS

In past decades, abrasive wheels with cubic boron nitride (CBN) grains have been widely applied in place of their counterparts with conventional aluminum oxide (Al_2O_3) abrasive grains in machining difficult-to-cut materials, e.g., titanium alloys, nickel-based superalloys, and titanium matrix composites. The high performance potentials of CBN abrasive grains result from their extreme hardness, superior wear resistance, high temperature resistance, and good thermal conductivity. However, some challenges still exist to fully explore the advantages of conventional CBN abrasive grinding in the automobile and aerospace industries (Konig and Ferlemann, 1991). Single-layer CBN abrasive wheels, including electroplated wheels and brazed wheels (Yu et al., 2017; Dai et al., 2019), are usually fabricated with a single layer of abrasive grains that are bonded to a metallic wheel substrate by an electroplated nickel layer or a brazed filler layer, as schematically displayed in Figure 1.1. In comparison to the multi-layered CBN abrasive wheel types, i.e., resin-bonded wheels, vitrified-bonded wheels, and metallic-bonded wheels, single-layer CBN abrasive wheels have significant advantages, some of which include excellent form retention over long grinding times, capability of running at higher removal rates (due to high grain protrusion and large chip-storage spaces), reduction of the intricate pre-grinding wheel preparation work (i.e., periodically dressing and truing operations especially in rough grinding), and possible re-application of the wheel hub after the grains wear out (such as stripping of the abrasive layer) (Rao et al, 2021; Bredthauer et al, 2023; Pal et al, 2010; Yang et al, 2016).

In particular, the CBN abrasive wheels for high-speed grinding and high-efficiency grinding are usually subject to a special requirement regarding resistance to fracture and wear; at the same time, good damping characteristics, high rigidity, and good thermal conductivity are also desirable. Under such conditions, the abrasive wheels are normally required to be composed of a body with high mechanical strength and a comparably thin coating of CBN grains attached to the body using a high-strength adhesive. The highest cutting speed is only achievable with single-layer CBN abrasive wheels.

DOI: 10.1201/9781032678047-2

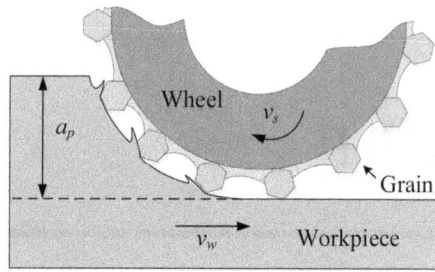

FIGURE 1.1   Schematics of a single-layer CBN abrasive wheel.

On the other hand, the electroplated CBN abrasive wheels usually exhibit random grain distribution, low mechanical bond strength, and small protrusions of the grains. Particularly in some cases of grinding metallic materials, grain pullout can occur and result in a new wheel breakdown (Shi and Malkin, 2003). Single-layer brazed CBN abrasive wheels have been reported to outperform their electroplated counterparts. A precisely controlled brazing technique can build a chemical bridge between CBN grains and metallic wheel hubs with the help of an active braze alloy. Significant advantages with the single-layer brazed abrasive wheels include higher bonding strength to grains, substantially higher exposure of grains measured from the bond level (up to 70%–80% of the total grain height), and flexibility in grain placement in any desired patterns. As such, a large inter-grain chip-accommodation space is created with a tailored grain-distribution model, which is essential to avoid premature loading in grinding operations with high material removal rates and to enhance the effectiveness of grinding fluids.

Additionally, the polycrystalline CBN abrasive grains are a typical superhard material, which is synthesized by microcrystalline CBN particles and AlN ceramic binder under the conditions of high temperature and high pressure. Compared with the traditional monocrystalline CBN abrasive grains, the unique microstructure of polycrystalline CBN grains, in theory, overcomes the drawback of anisotropy, which tends to cause cleavage in grinding due to the limited slipping planes. The polycrystalline CBN abrasive wheels have mainly worked in extreme conditions such as high temperature, friction, and impact load environments, which would lead to brittle fracture wear of the abrasive grains, resulting in a reduction in the number of useful grains to shorten the tool life. Therefore, besides the traditional CBN gains, it is also important to investigate the single-layer brazed polycrystalline CBN abrasive wheels in this book.

1.2 BOOK STRUCTURE

Chapters 1–6 explain the fabrication technologies and the different mechanisms during the production of the single-layer brazed CBN abrasive tools. The main topics include (i) the brazing interface microstructure and the chemical resultants produced with different filler materials and heating methods, (ii) the brazing reaction mechanisms, and (iii) the creation of tools with defined grain distribution based on the brazing technology.

Chapters 7–12 reveal the wear behaviours and mechanisms of the single-layer brazed CBN abrasive tools. The main topics include (i) the wear phenomena of both the mono- and poly-crystalline CBN abrasive, (ii) the grinding-induced stress influence, (iii) the fractal characteristics of the worn abrasives, and the interrelationship between abrasive wear and (iv) material removal in grinding, and (v) undeformed chip thickness.

The final five chapters (Chapters 13–17) explore the grinding performances and mechanisms of the single-layer brazed CBN abrasive tools. The main topics include the grinding behaviours for the (i) titanium alloys, (ii) nickel-based superalloys, and (iii) titanium matrix composites, and the grinding parameter effects on (iv) the material removal behaviours, and (v) the formation mechanism and affecting factors of grinding burrs.

REFERENCES

Bredthauer, M., P. Snellings, P. Mattfeld, et al. 2023. Wear-related topography changes for electroplated cBN grinding wheels and their effect on thermo-mechanical load. *Wear*, 512–513: 204543.

Dai, C., Z. Yin, W. Ding, et al. 2019. Grinding force and energy modeling of textured monolayer CBN wheels considering undeformed chip thickness nonuniformity. *International Journal of Mechanical Sciences*, 157–158: 221–230.

Konig, W., and F. Ferlemann. 1991. A new dimension for high-speed grinding. *Industrial Diamond Review*, 91(5): 237–241.

Pal, B., A. K. Chattopadhyay, and A. B. Chattopadhyay. 2010. Development and performance evaluation of monolayer brazed cBN grinding wheel on bearing steel. *International Journal of Advanced Manufacturing Technology*, 48(9–12): 935–944.

Rao, Z., G. Xiao, B. Zhao, et al. 2021. Effect of wear behaviour of single mono- and poly-crystalline cBN grains on the grinding performance of Inconel 718. *Ceramics International*, 47(12): 17049–17056.

Shi, Z., and S. Malkin. 2003. An investigation of grinding with electroplated CBN wheels. *CIRP Annals-Manufacturing Technology*, 52(1): 267–270.

Yang, X. H., J. G. Bai, W. J. Jing, et al. 2016. Strengthening of low-temperature sintered vitrified bond cBN grinding wheels by pre-oxidation of cBN abrasives. *Ceramics International*, 42: 9283–9286.

Yu, T., A. F. Bastawros, and A. Chandra. 2017. Experimental and modeling characterization of wear and life expectancy of electroplated CBN grinding wheels. *International Journal of Machine Tools and Manufacture*, 121: 70–80.

CHAPTER 2

Brazing of Monocrystalline CBN Abrasive Grains Based on Vacuum Furnace Heating

2.1 INTRODUCTION

The single-layer brazed cubic boron nitride (CBN) abrasive wheels comprise the metallic wheel hub, bonding layer, and abrasive grains, as shown in Figure 2.1. The applied metallic material of the wheel hub mainly includes AISI 1020 steel, AISI 1045 steel, alloyed steel, and hardened ball bearing steel (100Cr6); indeed, if the application does not allow for a magnetic material, aluminium or bronze/brass may also be utilized. Furthermore, the structure design and mechanical machining of the wheel hub are always carried out according to the general requirements of the machine tools and component profiles. In particular, for the profile grinding of some critical components (i.e., the aero-engine blade), the abrasive grain size must be considered when designing the specific concave and convex asperities on the circumferential surface of the metallic wheel hub. The reason is that the final profile dimension of the single-layer CBN wheel is entirely determined by both the wheel hub and

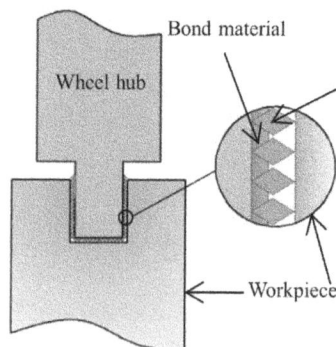

FIGURE 2.1   Schematic of the profile grinding with a single-layer brazed CBN abrasive wheel.

DOI: 10.1201/9781032678047-3

the grain size, as schematically shown in Figure 2.1. If either the wheel hub or grain size distribution exceeds the desired tolerances, the profile dimension accuracy of the ground components will not meet the desired level. For single-layer brazed abrasive wheels, preparing the wheel hub with high strength and accuracy is not as decisive for wheel performance as properly bonding the CBN grains to the wheel hub. For this reason, this section will focus on brazing techniques rather than wheel hub preparation.

Obtaining reliable joints among CBN abrasive grains, brazing filler alloy, and metallic wheel hubs is an important problem in producing single-layer brazed CBN abrasive wheels. The interfacial microstructure and chemical resultants at the joining interface are the primary concerns when investigating the brazing techniques of CBN grains. The brazing filler alloy, heating method, and heating parameters (especially brazing temperature and dwell time) significantly influence microstructure and chemical resultants.

a. Brazing filler alloy

Two methods have been applied in the brazing techniques of CBN grains: the first is the direct brazing of CBN grains using an alloy containing Ti as an active filler, such as Ag-Cu-Ti or Cu-Sn-Ti alloy. The second one is the indirect brazing of Ti-coated CBN grains using the conventional Ag-based or Ni-based filler alloy. However, good brazing results could not usually be obtained based on the second brazing method. Therefore, the direct brazing method is the dominant one to fabricate the single-layer brazed CBN abrasive wheels.

Based on the thermodynamic analysis and physical observations in experiments, the transition elements of group IV B, such as Ti or Zr, were preferred over the transition elements of group VI B, such as Cr, as activators to promote the wetting of the brazing filler alloy towards CBN grains, which are generally far more chemically stable than diamond abrasive grains. For example, Ni-Cr alloy, known for active diamond brazing, failed to show satisfactory wetting and bonding characteristics towards CBN under identical brazing conditions to those of diamond brazing. Meanwhile, wetting and bonding could not be improved by increasing the Cr content, brazing temperature, or dwell time. For this reason, direct brazing between CBN grains and metallic wheel hubs has been realized through filler alloys containing active transition elements (Ti or Zr). Mainly, Ag-Cu-Ti filler alloy has been broadly utilized owing to its high capillary effect when contacting boron nitride materials (Simhan et al., 2023) and excellent comprehensive mechanical properties (i.e., Young's modulus, thermal expansion coefficient, and yield strength). However, the cost of Ag-Cu-Ti alloy is usually high due to the high Ag content; as such, Cu-Sn-Ti alloy and Cu-Sn-Ni-Ti (or Cu-Ni-Sn-Ti) alloys are also sometimes applied for brazing CBN grains.

b. Heating method

There are generally four types of heating methods for brazing CBN grains: vacuum furnace heating (Ding et al. 2013), high-frequency induction heating (Xu et al. 2019), laser heating, and electron beam-activated heating (Huang et al. 2020; Bhaduri et al. 2010). The most popular heating methods are vacuum furnace heating and high-frequency induction heating due to their integrated advantages. For example, vacuum resistance furnace heating is advantageous for the mass production of monolayer brazed CBN wheels; however, it is difficult to control the deformation of the metallic wheel hub for large sizes (for instance, 400 mm in diameter). Compared to vacuum resistance furnace heating, the high-frequency induction heating method has different advantages, such as rapid heating rates, local heating, and easy control of wheel deformation.

c. Heating parameters

Based on the differential thermal analysis (DTA), it is known that the solidi and liquidi of both the Ag-Cu-Ti mentioned above and Cu-Sn-Ti filler alloys are usually about 780°C and 820°C, respectively. Accordingly, the brazing temperature is typically set to 880°C–940°C for brazing CBN grains. At the same time, the dwell time is chosen at 5–20 min with the vacuum furnace heating method, while it is selected at only several seconds with the high-frequency induction heating method. Research work has been made to investigate the effects of dwell time on the microstructure and mechanical properties of the brazed joints between CBN materials and medium carbon steel (CK45) hubs, in which Cusil-ABA (63 wt% Ag, 35 wt% Cu, and 2 wt% Ti) was utilized as an active filler alloy (Miab et al. 2014). The brazing temperature was kept at 920°C. With increasing the dwell time from 5 to 15 min, the interfacial reaction layers between CBN and filler become thicker and more continuous, resulting in higher joint strength.

2.2 BRAZING INTERFACE OF CBN GRAINS AND FILLER ALLOY CONTAINING TITANIUM

In the experiment, the raw materials for the brazing experiments are provided in Figure 2.2. The average size of CBN abrasive grains ranged from 150 to 180 μm. The brazing filler was $(Ag_{72}Cu_{28})_{92}Ti_8$ (wt.%) alloy powder, the composition of which has been optimized merely according to the mechanical characteristics of the brazed joint. The abrasive wheel hub used

(a)   (b)

FIGURE 2.2   Raw materials for brazing CBN grains: (a) CBN abrasive grains, (b) Ag-Cu-Ti filler powder.

was AISI 1045 steel. The eutectic composition of the Ag-Cu alloy system (melting point at 780°C) was often preferred because it was relatively ductile and could limit the resultant residual stresses produced due to the different expansion coefficients of the joined materials. The wheel hubs and the CBN grains were ultrasonically cleaned in acetone, after which the raw materials were assembled, and brazing trials were carried out in a vacuum furnace. The temperatures were 880°C, 900°C, and 920°C for dwell times of 5, 10, 15, and 20 min, respectively.

After brazing, the selected samples were cut, mounted, and polished to evaluate the interfacial microstructure and microchemistry between the grain and the brazing filler. Scanning electron microscope (SEM, JEOL-840/EDAX) coupled with an energy dispersion spectrometer (EDS) system, and X-ray diffraction (XRD, Philips PW1710, Cu Kα=0.1541 nm) were used to characterize the joints, in particular, the behaviour of active element Ti. Meanwhile, as the chemical reaction layer growth was a much more complex phenomenon than a single-reaction system due to evidence of growth competition among the different reactions, the total width of the reaction layer between the grain and the brazing alloy was experimentally measured as a function of the brazing temperature and the dwell time. It should be noted that the particular layer width reported in this investigation was regarded as a valid approximation by the average of the five values measured at the interface of grain and brazing alloy.

A representative SEM micrograph of the joint between the CBN grain and Ag-Cu-Ti filler layer brazed at 920°C for 5 min is shown in Figure 2.3a. The interface is composed of two different but associated parts. One with shallow grey is the reaction zone adjacent to the CBN grain, while the other with deep, shallow grey is the brazing filler alloy region. A typical linear distribution of elements along the white line in the brazed joints is demonstrated in Figure 2.3b. The EDS analysis detects that the content of Ti next to the grain is almost 60%, which is much higher than that in the given original filler alloy. Nevertheless, the corresponding Ti content in the region far from the grain is much lower than the value in the original brazing alloy. That is to say, the active element Ti in the brazing filler layer

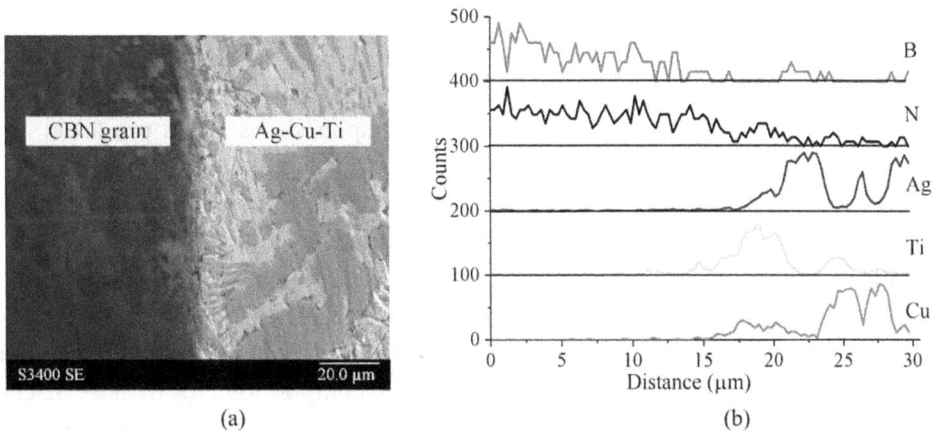

(a)   (b)

FIGURE 2.3 Microstructure and elemental diffusion across the brazing interface between CBN grain and Ag-Cu-Ti filler: (a) joining interface, (b) elemental diffusion.

FIGURE 2.4 Delamination layer and morphology of brazing resultants around a CBN grain: (a) whole morphology of brazing resultants, (b) TiN in the outer layer, (c) TiB in the middle layer, (d) TiB_2 in the inner layer.

has dramatically transferred towards the CBN grain. Moreover, in the given interfacial region, certain slopes exist among the curves of Ti, B, and N, which were confirmed to be the newly formed compounds zone.

The SEM micrograph of the brazed CBN grain, which was deeply etched, is shown in Figure 2.4. It's evident that a layer of columnar compounds has been produced on the grain surface. The brazing resultants between CBN grains and Ag-Cu-Ti filler alloy have been identified as TiB_2, TiB, and TiN, respectively. Owing to the slow heating and cooling rate and the high brazing temperature during vacuum furnace heating as well, the migration and diffusion effects, which included the directional mobility of the B and N atoms towards the brazing alloy and the Ti atoms towards the CBN grain, could be accomplished adequately. Meanwhile, the brazing temperature was above the liquidus point of the Ag-Cu-Ti filler alloy. As a result, the intermetallic compounds TiB_2 with a C32 crystal structure and TiB with a B27 crystal structure and the interstitial phase TiN with a face-centred cubic crystal structure were formed and grew in the dissolution and precipitation manner in the molten filler alloy. The shape of the compounds was similar to the ideal one in the equilibrium system. Furthermore, the preferred growth effect of the newly formed compound of the grain is prominent because of the influence of the index of the crystal plane and crystal orientation. Especially based on the thermodynamic and kinetic analyses, TiN was found to play a dominant role in controlling the growth of the resultant layer during brazing.

Furthermore, the delamination microstructure across the brazing interface of the CBN grain and Ag-Cu-Ti alloy was characterized as CBN/TiB$_2$/TiB/TiN/alloy containing Ti. Under such conditions, the different thermal expansion coefficients of the several brazing resultants have to be considered. For example, the average coefficients of thermal expansion of CBN, TiB$_2$, TiB, and TiN are 5.6×10^{-6}, 7.8×10^{-6}, 8.6×10^{-6}, and 9.35×10^{-6} K^{-1}, respectively. This gradual transition of thermal expansion coefficients reduces the residual thermal stresses at the brazing interface and minimizes potential thermal damage to brazed CBN grains.

2.3 INTERFACIAL FORMATION MECHANISM DURING BRAZING CBN GRAINS

2.3.1 Characterization of Ti-coated CBN Abrasive Grains

Because the filler alloy containing titanium is always in a liquid state during brazing, the chemical reactions of the CBN crystal and the active element Ti are so prompt that it is rather difficult to detect the characteristics of the genuine reaction process extensively. Therefore, the characteristics of the interfacial microstructure, including the formation sequence and morphology of reaction products, would be described in detail based on the solid-state interfacial diffusion reaction of Ti and CBN in the annealing process at elevated temperatures. This favours understanding the basic reaction mechanisms of Ti and CBN in brazing.

The CBN abrasive grains with Ti-deposited film (here also denoted as Ti-coated CBN grains) were used in the experiments. The average thickness of the Ti-deposited film on the surface of the CBN grain crystal was approximately 10 μm. First, DTA was conducted with 25 mg of grains from room temperature 20°C to 1,000°C at a heating rate of 10°C/min under the vacuum conditions to establish the special points of annealing temperature. Then, 8 g grains were chosen to endure the annealing operation for detecting the phase components of the interfacial reaction products at different temperatures. Especially, the Ti-coated CBN grains were analysed when the heating temperature reached a critical value, and then they were put again into the vacuum furnace instantly to continue heating for another XRD analysis at a higher particular temperature. Thus, the additional influence of uncertain factors, including different groups of grains, on the strength of XRD patterns could be effectively avoided by applying the sole group of grains to accomplish all the annealing experiments. The microstructure of the reaction products was finally extensively explored using a SEM with EDS.

Figure 2.5 shows the morphology of the Ti-coated CBN grain crystals. Though the surface Ti-deposited film of the abrasive grains could be observed through the unaided eyes, no significant Ti-coating layer would be found in the SEM image. Otherwise, the EDS spectrum pattern at point A of Figure 2.5a indicates the existence of element Ti, as demonstrated in Figure 2.6. Moreover, as displayed in the XRD pattern of Figure 2.7, the

(a)   (b)

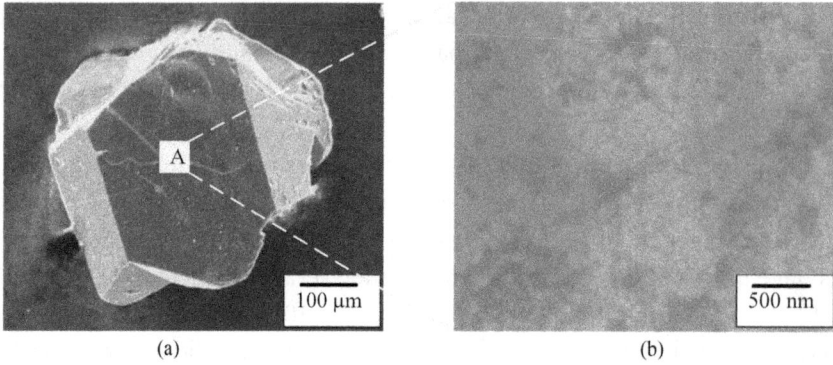

FIGURE 2.5　Micrograph of original Ti-coated CBN grains: (a) whole morphology, (b) magnified image of Point A.

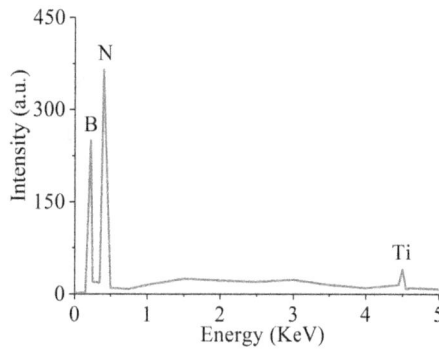

FIGURE 2.6　Energy spectrum pattern of Point A in Figure 2.5a.

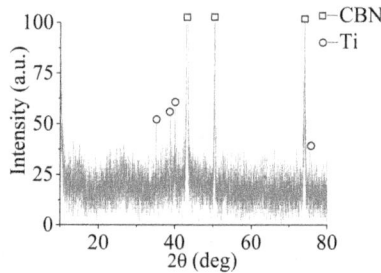

FIGURE 2.7　XRD patterns of Ti-coated CBN grains.

deposited film on the surface of the CBN grain indeed consists of the phase of pure Ti without the presence of Ti-N or Ti-B compounds, which confirms that the surface Ti film and CBN grains join together through mechanical and physical effects.

2.3.2 Effects of Heating Temperature on Interfacial Reaction

The strong interaction and high affinity between Ti and CBN allow overcoming the activation energy between the diffusion couple of Ti/CBN to form both the Ti-N and Ti-B compounds through interfacial diffusion and chemical reaction at elevated temperatures.

During 800–1,300 K (that is, 527°C–1,027°C), the possible reaction and corresponding Gibbs standard free energy between Ti and CBN could be described as follows:

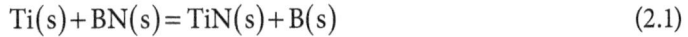

$$Ti(s) + BN(s) = TiN(s) + B(s) \tag{2.1}$$

$$\Delta GB_{TB}(1) = -75.92 - 0.0047T \tag{2.2}$$

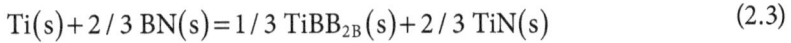

$$Ti(s) + 2/3\,BN(s) = 1/3\,TiBB_{2B}(s) + 2/3\,TiN(s) \tag{2.3}$$

$$\Delta GB_{TB}(2) = -157.20 + 0.0003T \tag{2.4}$$

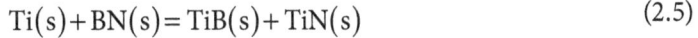

$$Ti(s) + BN(s) = TiB(s) + TiN(s) \tag{2.5}$$

$$\Delta GB_{TB}(3) = -218.25 - 0.0646T \tag{2.6}$$

According to the calculated standard Gibbs free energy shown in Figure 2.8, it is obvious that Eq. (2.5) is the most thermodynamically favourable among the three reaction equations mentioned above. It is noted that the actual interfacial reactions are not only determined by the thermodynamic factors but also impacted by many other ones, for example, the diffusion and reaction activation energy, the relevant heat-treating circumstances, etc.

For detecting the solid-state interfacial reaction of surface Ti-deposited film and CBN grains in the heat-treating process, the specimens were first discussed based on DTA analysis results, as presented in Figure 2.9. Three exothermic peaks established at 610°C, 800°C, and 860°C could be distinguished in the special plot. Moreover, different from the two strong former peaks, the last exothermic peak is extremely weak. It is inferred that the former reaction is much more powerful than the latter and hence plays a dominating role in the interfacial reaction of CBN and Ti.

Meanwhile, for determining the phase components of the special reaction products at the critical temperature such as 610°C, 800°C, and 860°C, the Ti-coated grains specimen was annealed in the vacuum furnace. Then XRD analysis was applied at several special temperatures according to the DTA curve. The results of reaction phase identification are summarized in Table 2.1. The heating temperature has a great influence on the interfacial reaction products. Three compounds' phases were identified, namely TiN, TiB_2, and TiB,

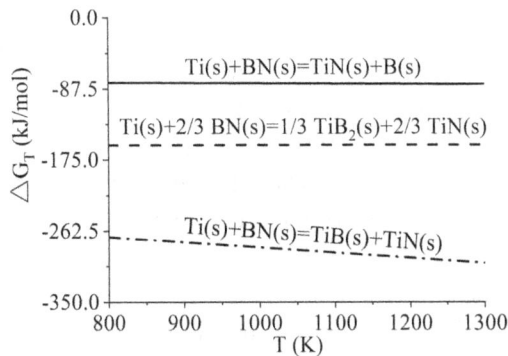

FIGURE 2.8 Change of molar Gibbs standard free energy.

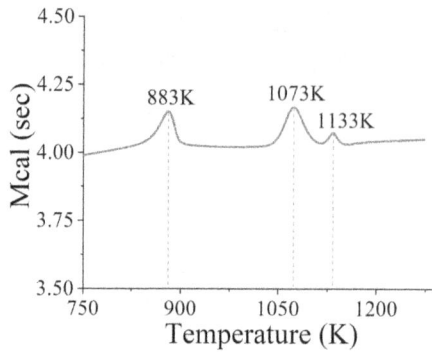

FIGURE 2.9   DTA curve of Ti-coated CBN grains.

TABLE 2.1   Phase Identification of Annealed CBN Grains with Ti-deposited Film

Processing Temperature (°C)	Original	550	620	750	820	880	950
Phases	Ti	Ti	Ti	Ti	Ti	Ti	Ti
	CBN	CBN	CBN	CBN	CBN	CBN	CBN
			TiN	TiN	TiN	TiN	TiN
					$TiBB_{2B}$	$TiBB_{2B}$	$TiBB_{2B}$
						TiB	TiB

rather than pure Ti metal at room temperature. When the temperature is below 550°C, no significant reaction product is formed between CBN and Ti. While the temperatures range from 620°C to 750°C, the production of TiN phases is preferred. TiN still exists at the subsequent higher processing temperature, and meanwhile, TiB_2 gradually increases its amount as the temperature increases above 820°C. In particular, TiB is produced at an annealing temperature of 880°C. All the interfacial reactions at different temperatures during the heat treatment indicate that the interfacial reaction is simply compatible with the DTA results shown in Figure 2.9.

2.3.3 Compound Morphology of Interfacial Resultants

According to the phase components of the reaction products on the surface of the heat-treated CBN grains, the compounds consist of TiN, TiB_2, and TiB when the processing temperature is above 880°C. For detecting the morphology of the reaction products in the interface between Ti and CBN, the grains heat-treated at 950°C for 1 h were well dealt with special measures to shuck out of the different compound layers for SEM observation, and the result is shown in Figure 2.10.

Figure 2.10a displays the top layer of network-distributed TiN. Figure 2.10b corresponds to the layer structure of different compounds on the grain surface. And Figure 2.10c shows the hypo-inner layer consisting of granular TiN and networked TiB below the external layer. In contrast, Figure 2.10d demonstrates the inner layer of columnar TiB_2, which

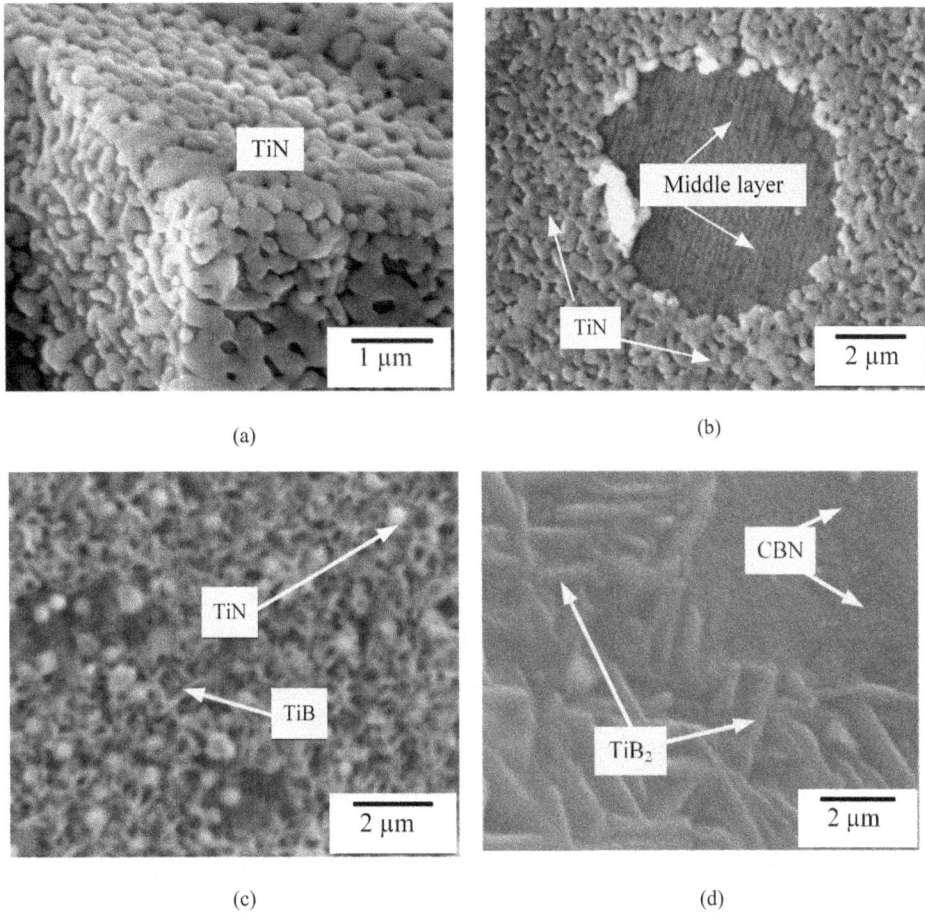

FIGURE 2.10   Compounds morphology in the different layers of Ti-coated CBN grains annealed at 950°C: (a) TiN in the outer layer, (b) Typical layer structure of compounds, (c) TiN and TiB in the hypo-inner layer, (d) TiB_2 in the inner layer.

seems connected and integrated into the CBN grains. According to the morphology of all the Ti-N and Ti-B compounds mentioned above, it's known that the reaction products of Ti and CBN are inclined to form the particular network structure, which could provide enough strength to resist the breakage of the interfacial layer and ensure the effective joining of CBN grains and the abrasive wheel hub during brazing.

Moreover, based on Figure 2.10, the surface microstructure of the Ti-coated CBN grains annealed at 950°C could be determined in a sequence of $CBN/TiB_2/TiB/(TiB+TiN)/TiN/Ti$. Thus, a schematic illustration of the atomic diffusion path from the phase and structure analysis of compounds is presented in Figure 2.11. The particular layer structure has two advantages: good transitions of chemical bonds and thermal expansion coefficients. As is well known, CBN is a strong covalent bond crystal, while a covalent bond and a metallic bond coexist in borides, and the ratio of the metallic bond increases with the ratio of Me/B. The interstitial phase TiN preliminary presents metallic bonds. So the percentage of covalent bonds decreases, but that of metallic bonds increases from $CBN \rightarrow TiB_2 \rightarrow TiB \rightarrow TiN \rightarrow$

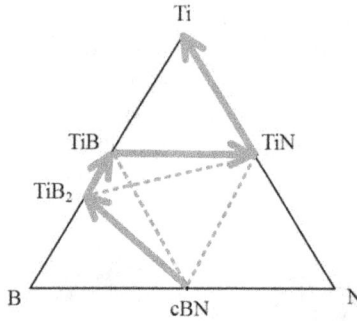

FIGURE 2.11 Atomic diffusion paths corresponding to the reaction layer of Ti-coated CBN grains annealed at 950°C.

Ti. The good transition effect of the chemical bonds provides strong adhesion between the CBN grain crystal and the surface Ti-deposited film.

2.3.4 Formation Mechanisms of the Interfacial Microstructure

The observation of the phase components and morphology of the compounds in the interfacial reaction region of CBN crystal and surface Ti-deposited film gives some information on the relevant interfacial reaction mechanism. Here, the driving force for the growth of the three phases, TiN, TiB, and TiB$_2$, is the atomic diffusion of B, N, and Ti due to the gradient of atomic composition within the different parts of the interface, as shown in Figure 2.12.

Additionally, the diffusion coefficients of B and N in the Ti would be as follows:

$$B \to Ti \ D = 4.8 \times 10^{-7} \exp(-5600/T) \ cm^2/s \tag{2.7}$$

$$N \to Ti \ D = 4.8 \times 10^{-3} \exp(-6700/T) \ cm^2/s \tag{2.8}$$

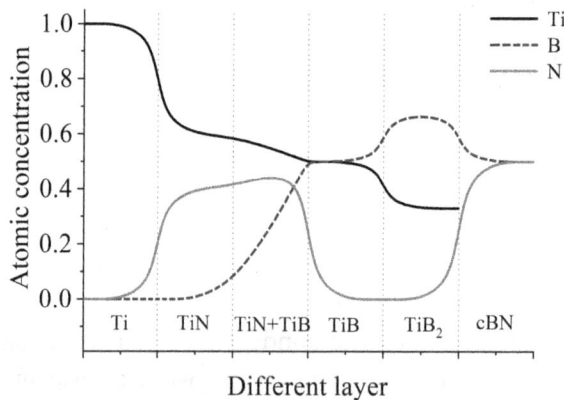

FIGURE 2.12 Schematic of atomic concentration in the different compound layers across the Ti/CBN interface.

Hence, the N atoms would always lie in the foreland of the diffusion path to form Ti nitrides in the outer layer. In contrast, the B atoms would concentrate on the position adjacent to the crystal surface to form Ti-B compounds in the inner layer because of the low diffusion speed of B atoms in the Ti lattice.

Furthermore, based on the phase rules (Li et al. 2023; Tang et al. 2002), as TiB is in equilibrium with TiN in the Ti-B-N ternary system, the freedom degree of the system, $f = 3 - 2 = 1$, in the condition of constant temperature investigated, thus a mixture layer of TiN and TiB, namely the layer of (TiN + TiB), could exist. In the mixture layer, the amount of TiN would increase with the decrease of TiB from the TiB layer to the TiN layer. On the other hand, as TiB_2 is in equilibrium with TiB in the binary system of Ti-B, the degree of freedom of the system, $f = 2 - 2 = 0$, is in the same condition. There is no mixture layer but a TiB_2/TiB interface exists between the TiB_2 and TiB layers. Therefore, the microstructure of the interfacial reaction layer after the annealing process could be determined in a sequence of TiB_2/TiB/(TiB+TiN)/TiN from the CBN grain to the growing surface from the viewpoint of thermodynamics, which has been consistent with the phase structure of CBN grains with Ti-deposited film annealed at 950°C.

REFERENCES

Bhaduri, D., and A. K. Chattopadhyay. 2010. Effect of pulsed DC CFUBM sputtered TiN coating on performance of nickel electroplated monolayer cBN wheel in grinding steel. *Surface and Coatings Technology*, 204(23): 3818–3832.

Ding, W. F., J. H. Xu, Z. Z. Chen, et al. 2013. Interface characteristics and fracture behavior of brazed polycrystalline CBN grains using Cu-Sn-Ti alloy. *Materials Science and Engineering A*, 559: 629–634.

Huang Y., G. Liang, M. Lv, et al. 2020. Nd:YAG pulsed laser brazing of cBN to steel matrix with Zr modified Ag-Cu-Ti active brazing alloy. *Diamond and Related Materials*, 104: 107732.

Li, Y., L. Yan, M. Song, et al. 2023. Integration of thermal protection and structural health monitoring for carbon fiber reinforced SiBCN composites with SiC coating enhanced interfacial performance. *Ceramics International*, 49(13): 21678–21687.

Miab, R. J., and A. M. Hadian. 2014. Effect of brazing time on microstructure and mechanical properties of cubic boron nitride/steel joints. *Ceramics International*, 40(6): 8519–8524.

Simhan, D. R., and A. Ghosh. 2023. Active brazing of cBN micro-particles with AISI 1045 steel using ceramic reinforced Ag-based fillers. *Diamond and Related Materials*, 136: 110056

Tang, W. M., Z. X. Zheng, and H. F. Ding. 2002. The interfacial stability of the coated-SiC/Fe couple. *Materials Chemistry and Physics*, 77: 236–241.

Xu, W., Y. J. Zhu, B. C. Du, et al. 2019. Residual stresses of polycrystalline CBN abrasive grits brazed with high-frequency induction heating technique. *Chinese Journal of Aeronautics*, 32(4): 1020–1029.

CHAPTER 3

Brazing of Monocrystalline CBN Abrasive Grains with Composite Fillers

3.1 INTRODUCTION

With the advancement of the correlative investigations, some problems with brazed cubic boron nitride (CBN) wheels have also been encountered. Particularly, the pure Ag-Cu-Ti filler alloy for producing a brazed CBN wheel is sometimes insufficient in strength and wear resistance, mainly when the CBN wheels are applied in deep grinding with a heavy load, because the high grinding temperature produced in the wheel-workpiece contacting zone probably leads to the softening behaviour of the Ag-Cu-Ti brazing filler layer of the brazed CBN wheels. It is therefore essential to take some effective measures to solve the problems mentioned above and explore the great potential of the brazed CBN wheels completely.

The composite filler materials, which are fabricated by introducing some reinforcing particles into the traditional filler alloy, have attracted considerable interest in brazing brittle ceramic materials to metals. For example, Wang et al. prepared the Si_3N_4 ceramic/42CrMo steel joints, which were obtained by employing TiN particles in a modified Ag-Cu-Ti filler alloy (Wang et al. 2014). With the increase in TiN particle content, more fine grains and less Ag-Cu eutectic appeared in the joint, and the interfacial reaction layers became thinner. The optimum content of TiN particles for achieving the highest bending strength (376 MPa) was determined to be 5 vol.%, at which point the joint strength was almost double that without TiN particles. Furthermore, the finite element analysis results indicated that the introduced TiN particles made it possible to lower the residual stresses in the Si_3N_4 ceramic/42CrMo steel joints.

Additionally, the microstructure evolution mechanism of the brazed joints was also revealed (Wang et al. 2015), in which the addition of TiN particles contributed to the concentration of active Ti elements. Only a small amount of Cu-Ti intermetallic compounds

DOI: 10.1201/9781032678047-4

was found in the joint brazed with the composite filler, which contained a lower content of TiN particles, whereas various Cu-Ti intermetallic phases precipitated at a higher TiN particle content. Furthermore, it was found that the addition of Si_3N_4 particles could lead to an obvious change in the interfacial microstructure of Si_3N_4/Ag-Cu-Ti+ $Si_{3}N_{4p}$/TiAl brazed joints due to its restriction effect on the diffusion of active Ti during the brazing process (Song et al. 2011). The optimal microstructure and fine resultants, which were obtained when the Si_3N_4 particles were 3 wt.%, allowed a decrease in the mismatch of thermal expansion coefficients of the brazed joints. The maximum shear strength of the joints was accordingly obtained at 115 MPa. The application of SiC particle reinforcement to Ag-Cu-Ti filler alloy during brazing Si_3N_4-TiN ceramic composite to tool steel matrix also resulted in a higher flexural strength of 395 MPa at room temperature and lower residual thermal stresses due to an increase in the allowable plastic strain in the joint (Guo et al. 2019). Meanwhile, improvements in the microstructure and strength of the brazed joints have also been reported by He et al. (2010), who joined Si_3N_4 ceramic to itself using an Ag-Cu-Ti filler alloy reinforced with SiC particles, and by Zhao et al. (2015), who brazed Ti-6Al-4V titanium alloy to Si_3N_4 ceramic using nano-Si_3N_4 reinforced Ag-Cu-Ti alloy. Based on the above analysis, it is known that the mechanical performance improvements of brazed ceramic/ceramic or ceramic/metal joints prepared by the particles (i.e., TiN, Si_3N_4, and SiC) reinforced filler materials are significant compared to the brazed joints prepared by the non-reinforced filler materials (i.e., pure Ag-Cu-Ti alloy). These improvements consist of the increased mechanical strength, microhardness, and wear resistance of the joints. The employment of particle-reinforced composite filler materials for brazing is thus a logical development. In particular, when the reinforcing particles are added to the Ag-Cu-Ti filler alloy, the active Ti elements could separate from the Ag and Cu solid solutions and then diffuse towards and enrich around the introduced particles in brazing. The formation of intermetallic compounds that react with Ti elements at the interface would be restricted to a certain extent by controlling the particle content. A thin layer of brazing resultants is accordingly obtained. The special effects resulting from the addition of reinforcing particles show remarkable benefits to the optimization of the microstructure and the overall mechanical properties of the brazed joint, which provide a new impetus to develop composite filler materials to join ceramics and metals.

TiB$_2$, TiN, and TiC particles, as reinforcing particles, have been widely believed to offer considerable advantages for advanced structural applications (i.e., brazed joints, metallic matrix composites) due to their exceptional hardness, stability at elevated temperatures, and acceptable thermal expansion coefficients. Therefore, it would be a meaningful attempt that TiB_2, TiN, and TiC particles, respectively, are added to the pure Ag-Cu-Ti alloy to join CBN superabrasive grains and AISI 1045 steel for fabricating brazed CBN wheels. Particularly, TiB_2 and TiN are also brazing resultants between CBN grains and the Ag-Cu-Ti alloy. The interfacial chemical reactions occurring in the brazed joints are accordingly expected to be further controlled with the addition of TiB_2 and TiN. Under such conditions, also optimization and utilization of such brazed CBN wheels could be made.

3.2 EFFECT OF TIX PARTICLES ON MICROSTRUCTURE AND STRENGTH OF COMPOSITE FILLER

This work aims to comparatively investigate the effects of TiX (i.e., TiB_2, TiN, or TiC, respectively) reinforcing particles on the general mechanical performance of the brazed joints between CBN superabrasive grains and AISI 1045 steel matrix. In particular, the effects of the reinforcing particles on the microstructure and strength of the Ag-Cu-Ti/TiX bonding layer are discussed. Accordingly, the optimum addition of the reinforcing particles to Ag-Cu-Ti/TiX composite filler materials could be determined.

3.2.1 Experimental Details

In the experiment, the composite filler materials were prepared by mechanically mixing Ag-Cu-Ti alloy powder and different TiX (i.e., TiB_2, TiN, and TiC, respectively) reinforcing particles. The primary physical properties of the reinforcing particles and Ag-Cu-Ti alloy applied are listed in Table 3.1. The reinforcing particles all have exceptional hardness, high bending strength, and a moderate thermal expansion coefficient. Furthermore, the morphology of these reinforcing particles is a little different, as displayed in Figure 3.1. The particle size ranges from approximately 0.2 to 8 μm.

The experimental work was generally divided into three steps, as follows: (i) the spreading tests of the molten composite filler materials, (ii) the shear strength test of brazed joints prepared with the Ag-Cu-Ti/TiX bonding layer, (iii) the metallographic microstructure detection and the microhardness evaluation of the bonding layer. Therefore, different

TABLE 3.1 Primary Physical Properties of the Reinforcing Particles and Pure Ag-Cu-Ti Filler Used in the Experiment

Items	Crystal Structure	Density (g/cm³)	Melting Point (°C)	Hardness (GPa)	Young Modulus (GPa)	Fracture Toughness (MPam$^{1/2}$)	Bending Strength (MPa)	Coefficient of Thermal Expansion (K⁻¹)	Particle Size (μm)
TiB_2	Hexagonal	4.50	2,980	30	574	6.7	750	7.80×10^{-6}	0.5–5
TiN	Cubic	5.43	2,950	21	436	6.0	431	9.35×10^{-6}	0.5–8
TiC	Cubic	4.91	3,150	40	420	-	200	6.52×10^{-6}	0.2–5
Ag-Cu-Ti	-	9.48	780–805	-	94	-	-	18.2×10^{-6}	-

(a)   (b)   (c)

FIGURE 3.1 Micrograph of the reinforcing particles: (a) TiB_2 particles, (b) TiN particles, (c) TiC particles.

TABLE 3.2   Brazing Parameters Utilized in the Current Experiments

Parameters	Heating Temperature	Holding Time	Rate of Heating & Cooling	Vacuum
Values	920°C	5 min	10°C/min	$<10^{-2}$ Pa

samples and experimental methods were applied. Additionally, it is noted that all the brazing experiments mentioned were carried out under the same parameters (Table 3.2) in a VAF-20 vacuum furnace. The parameters had been optimized based on the good wetting and reaction behaviour among the CBN grains, AISI 1045 steel matrix, and pure Ag-Cu-Ti alloy in the brazing process. That is to say, this work does not take into account the effects of the brazing parameters but merely considers the effects of the types and contents of the TiX reinforcing particles on the microstructure and mechanical strength of the brazed joints.

In the following description of the experimental procedures, the Ag-Cu-Ti/TiB$_2$ composite filler materials will be taken as an example. The same experimental procedures were also carried out with the Ag-Cu-Ti/TiN and Ag-Cu-Ti/TiC composite filler materials, respectively, in this work.

The first experimental step was the spreading test of the Ag-Cu-Ti/TiX composite filler materials. First, the AISI 1045 steel matrix with a size of 50 mm (length)×50 mm (width)×1 mm (height) was polished to a 1,200 grit finish and then cleaned ultrasonically in acetone. Afterwards, the composite filler materials of 0.025 g, which had different TiB$_2$ contents (i.e., 0, 4, 8, 12, 16, and 20 wt.%, respectively), were laid at the core region of the steel matrix according to 0.1 g/cm² (Figure 3.2). Subsequently, the composite filler materials were heated and then cooled to room temperature for solidification in the brazing process. When the brazing procedure was finished, the spreading morphology of the composite filler materials was collected with a KH-7700 type optical microscope. The spreading area was accordingly calculated. The reproducibility of the experimental results was checked three times for each composite filler material. Finally, the average values of the spreading area were obtained and analysed.

The second experimental step was the shear strength test of the bonding layer composed of the Ag-Cu-Ti/TiX composite filler materials. To prepare the samples for the shearing tests, the two matrix couples of AISI 1045 steel were first machined for each sample and then ground until the roughness of the surface to be joined was below Ra 1.6 μm. A composite filler layer with a TiB$_2$ content of 0, 4, 8, 12, 16, and 20 wt.%, respectively, was sandwiched with a thickness of 0.1 mm between the two matrix couples, as shown in Figure 3.3. Particularly, to ensure enough thickness of the solidified bonding layer to join the steel matrix couples, the density of the composite filler powder employed on the steel matrix surface was 0.25 g/cm². After brazing, the room-temperature shearing tests of the brazed joints were conducted at a cross-head speed of 1 mm/min with a CMT-SANA 5105 material test system. The average values of the shear strength of the three samples were obtained accordingly for each composite filler material.

The third experimental step was detecting the microstructure and microhardness of the Ag-Cu-Ti/TiX bonding layer. To identify the metallographic microstructure, the CBN

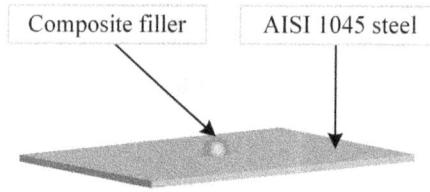

FIGURE 3.2   Schematic illustration of the spreading sample.

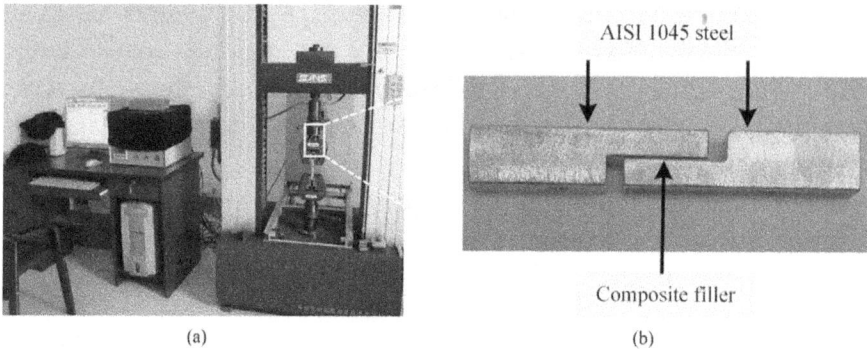

FIGURE 3.3   Strength test equipment of the brazed joints: (a) experimental setup, (b) brazed sample for shearing test.

grains were manually chosen with a size of 300–500 µm. Before the brazing experiments, all the grains and steel matrixes (AISI 1045 steel) were first cleaned ultrasonically in acetone and then air-dried. Afterward, the composite filler with 8 wt.% TiB_2 particles was sandwiched between CBN grains and the steel matrix. When the brazing process was finished, one sample was cut along the bonding interface and subsequently polished and etched with 65% nitric acid for the sake of metallographic microstructure detection of the bonding layer. Microstructure observation was carried out with a Quanta 200-type scanning electron microscope (SEM) in both secondary electron and back-scattered electron modes. Phase components of the bonding layer were identified by a Bruker D8 Advance X-ray diffraction (XRD) analyzer with Cu Kα radiation. At the same time, the other duplicated samples for each composite filler material were sectioned perpendicularly to the bonding interface and then mechanically polished using 1 µm diamond paste for the preparation of the microhardness test of the brazed joint. The microhardness measurements were made with a load of 50 g for 10 s on the HXS-1000AK Vickers hardness tester along with the joining interface at five different positions to obtain an average microhardness value.

3.2.2 Effects of Reinforcing Particles on the Spreading Behaviour of Composite Fillers

The grinding performance of brazed CBN abrasive wheels is strongly dependent on the wettability and flowability of the bonding material on the wheel matrix. Too strong flowability always has a negative influence not only on the appearance of the brazed CBN wheels but also on the interfacial joining behaviour among the CBN grains/bonding layer/wheel matrix. On the contrary, if the molten filler materials perform poorly, the brazed joint cannot be formed well, inevitably resulting in the unexpected tearing of the bonding layer or pullout of CBN grains, especially in the heavy-load grinding process. It is, therefore, necessary to evaluate the flowability of molten composite filler materials quantitatively. For this reason, the first experiment carried out was a comparison of the effects of different reinforcing particles on the spreading behaviour of the composite filler materials. The spreading area, instead of the wetting angle, was utilized. The results are comparatively displayed in Figure 3.4.

In general, the introduced reinforcing particles, i.e., TiB_2, TiN, and TiC, show a suppression effect on the spreading behaviour of the molten Ag-Cu-Ti alloy, as shown in Figure 3.4. That is to say, compared with the spreading area value of the pure Ag-Cu-Ti alloy, a significant reduction in the spreading area value of the molten composite filler materials could take place, as demonstrated in Figure 3.4e. If the spreading area of the pure Ag-Cu-Ti alloy on the AISI 1045 steel matrix is regarded as 100%, when the content of the reinforcing particles is increased from 4 to 20 wt.%, the reduction of the corresponding spreading area obtained is from 41% to 70% for Ag-Cu-Ti/TiB_2, from 73% to 87% for Ag-Cu-Ti/TiN, and from 21% to 55% for Ag-Cu-Ti/TiC. An interesting

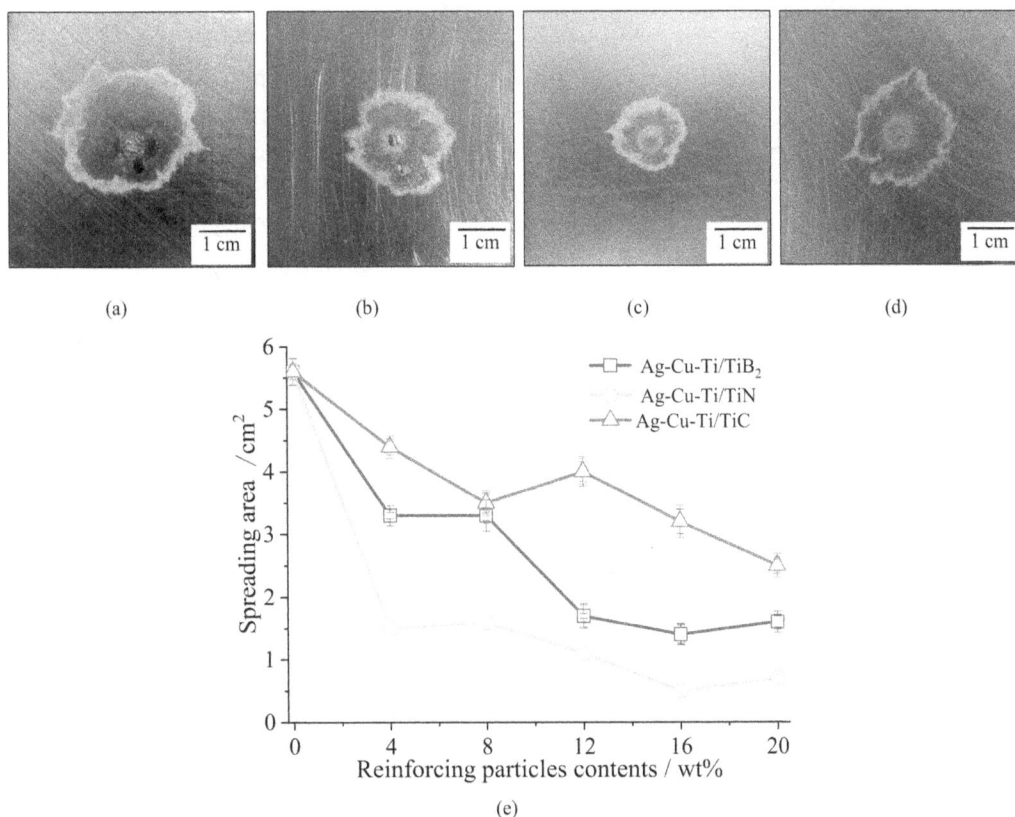

FIGURE 3.4 Effects of the reinforcing particles on the spreading morphology and area of composite filler materials: (a) Pure Ag-Cu-Ti alloy, (b) Ag-Cu-Ti/TiB_2 (8 wt.%), (c) Ag-Cu-Ti/TiN (8 wt.%), (d) Ag-Cu-Ti/TiC (8 wt.%), (e) spreading area.

phenomenon is that the spreading area of Ag-Cu-Ti/TiC composite filler material always takes the highest value, while that of Ag-Cu-Ti/TiN always takes the lowest one, whatever the content of the reinforcing particle is. In other words, the suppression effect of TiN particles on the spreading behaviour of the molten Ag-Cu-Ti alloy is more prominent than that of TiB_2 and TiC ones. Specifically, once the content of reinforcing particles increases to 8 wt.%, a slight improvement and subsequently a significant reduction of the spreading area are obtained for Ag-Cu-Ti/TiC composite filler materials, while a generally dramatic decrease in the spreading behaviour becomes obvious for Ag-Cu-Ti/TiB_2 and Ag-Cu-Ti/TiN composite filler materials.

3.2.3 Effects of Reinforcing Particles on the Shear Strength of the Bonding Layer

The effects of reinforcing particles on the shear strength of the Ag-Cu-Ti/TiX bonding layer are shown in Figure 3.5. It is found that when the contents of reinforcing particles are below 8 wt.%, the nearly linear increase of shear strength of the bonding layer prepared by Ag-Cu-Ti/TiB_2 and Ag-Cu-Ti/TiN composite fillers takes place with the increment of reinforcing particle content, which ranges from 50 to 78 MPa for Ag-Cu-Ti/TiB_2, and 50 to 128 MPa for Ag-Cu-Ti/TiN, respectively. However, the bonding layer prepared by the Ag-Cu-Ti/TiC composite filler material almost keeps a constant value of approximately 48 MPa in shear strength.

Furthermore, the strength variation of the bonding layer becomes somewhat remarkable with increasing reinforcing particle content, which ranges from 12 to 20 wt.%. Particularly, when the reinforcing particle contents of the composite filler materials are increased from 12 to 16 wt.%, the shear strength of the corresponding bonding layer varies from 54 to 124 MPa for Ag-Cu-Ti/TiB_2, 36 to 95 MPa for Ag-Cu-Ti/TiC, and 115 to 187 MPa for Ag-Cu-Ti/TiN, respectively. If the particle contents are further increased, all the shear strength of the brazed joints is drastically decreased to 60–90 MPa. As a whole, in comparison to the pure Ag-Cu-Ti alloy, both the Ag-Cu-Ti/TiB_2 and Ag-Cu-Ti/TiN composite filler materials have a favourable effect on the improvement on the shear strength

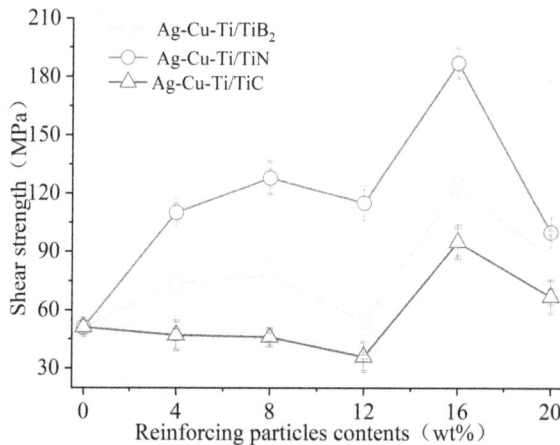

FIGURE 3.5 Effects of the reinforcing particles on the shear strength of the bonding layer.

of the bonding layer. The significantly positive effects of TiC reinforcing particles on shear strength of the bonding layer could be exhibited when the particle contents are increased to beyond 12 wt.%.

The fracture morphology of the Ag-Cu-Ti/TiX bonding layer after the shearing test is displayed in Figure 3.6. The brazed joints were mainly destroyed through the rupture of the bonding layer. As seen in Figure 3.6a, besides the thick bonding layer observed in the adjacent zone, the AISI 1045 steel matrix is also exposed in the regional zones, which is formed due to the tearing behaviour of the Ag-Cu-Ti alloy from the steel matrix. This phenomenon is produced not only due to the good spreading behaviour of the pure Ag-Cu-Ti alloy on the surface of the AISI 1045 steel matrix but also owing to the significant aggregation of the molten composite filler materials that occurred during solidification. However, in Figure 3.6b–d, a few joint fractures are observed at the joining interface of the steel matrix and the bonding layer. The main fracture patterns of ductile dimples and splitting of the bonding layer are also characterized. Also, though some reinforcing particles are pulled out of the bonding layer, the others remain with the original morphology. These fractural features suggest that the reinforcing particles are merely wetted by the Ag-Cu-Ti alloy during brazing; no chemical or metallurgical reaction happens between the reinforcing particles and

FIGURE 3.6 Fracture morphology of the Ag-Cu-Ti/TiX bonding layer after the shearing test: (a) Pure Ag-Cu-Ti alloy, (b) Ag-Cu-Ti/TiB$_2$ (16 wt.%), (c) Ag-Cu-Ti/TiN (16 wt.%), (d) Ag-Cu-Ti/TiC (16 wt.%).

the Ag-Cu-Ti alloy, which therefore permits obtaining the unique fracture morphology of the bonding layer after the shearing test.

It is worth mentioning that, when considering the effects of the addition of TiB_2, TiC, and TiN, respectively, into the Ag-Cu-Ti alloy, there is quite a difference in terms of the shear strength of the bonding layer, even though the reinforcing particle contents are equivalent. According to the present experimental work, the Ag-Cu-Ti/TiN composite filler performs best in improving the shear strength, whereas the Ag-Cu-Ti/TiC composite filler even leads to a slight reduction in comparison to the pure Ag-Cu-Ti alloy when the content of the particles is below 12 wt.%. The reason can be explained from the viewpoint of the difference in particle morphology. For example, it is apparent that the crystal outline of TiC particles is complete, but the edge is sharp. On the contrary, the TiN and TiB_2 particles have relatively rounded edges. Under such conditions, the presence of sharp particle edges contributes to the concentration of residual stresses induced by the thermal expansion mismatch between TiC particles and the Ag-Cu-Ti alloy, resulting in the formation of microcracks that weaken the brazed joints. In contrast, the rounded edges could effectively get rid of the stress concentration and thus reduce the damage effects on the bonding layer during brazing.

Furthermore, when the content of the particles is kept at 8 wt.%, both the spreading area and the shear strength of the bonding layer produce a relatively acceptable value in comparison with the results obtained at the other particle contents. The values of the corresponding spreading area and shear strength generally have a great variation when the contents of the particle are below or above 8 wt.%, which gives the impression that the content of 8 wt.% could be regarded as a turnaround point.

3.2.4 Effects of Reinforcing Particles on Bonding Layer Microstructure

The metallurgical microstructure of the bonding layer prepared by pure Ag-Cu-Ti alloy and Ag-Cu-Ti/TiX composite filler materials, respectively, is comparatively displayed in Figure 3.7. In terms of the metallurgical microstructure produced with pure Ag-Cu-Ti alloy, Ag-rich phase (α phases) with small amounts of Cu and little Ti are mainly found in the white regions, as highlighted in Figure 3.7a and b. On the contrary, the black regions are the Cu-rich ones (β phases), which always contain a rather low quantity of Ag. Furthermore, the crystals in β phases develop roughly in the form of a small dot and a narrow strip, surrounded and separated by the crystals in α phases. Accordingly, a comparison can be made on the difference in metallurgical microstructure produced by the different Ag-Cu-Ti/TiX composite filler materials.

As illustrated in Figure 3.7c and d, when the TiB_2 particles are added to the Ag-Cu-Ti alloy, concerning the microstructure shown in Figure 3.7a and b, most of the single crystals of β phases change little in width (less than 3 μm), but they have a remarkable decrease in length to 4–15 μm and have a uniform microstructure. The crystal microstructure in the α and β phases is therefore refined to a certain extent. The particular metallurgical microstructure is also observed in Figure 3.7e and f, which were prepared using the TiN particles modified in the Ag-Cu-Ti alloy. In this case, the crystals in β phases become larger and grow with the smooth edges, which leads to greater refinement

Filler material	Whole image	Regional image

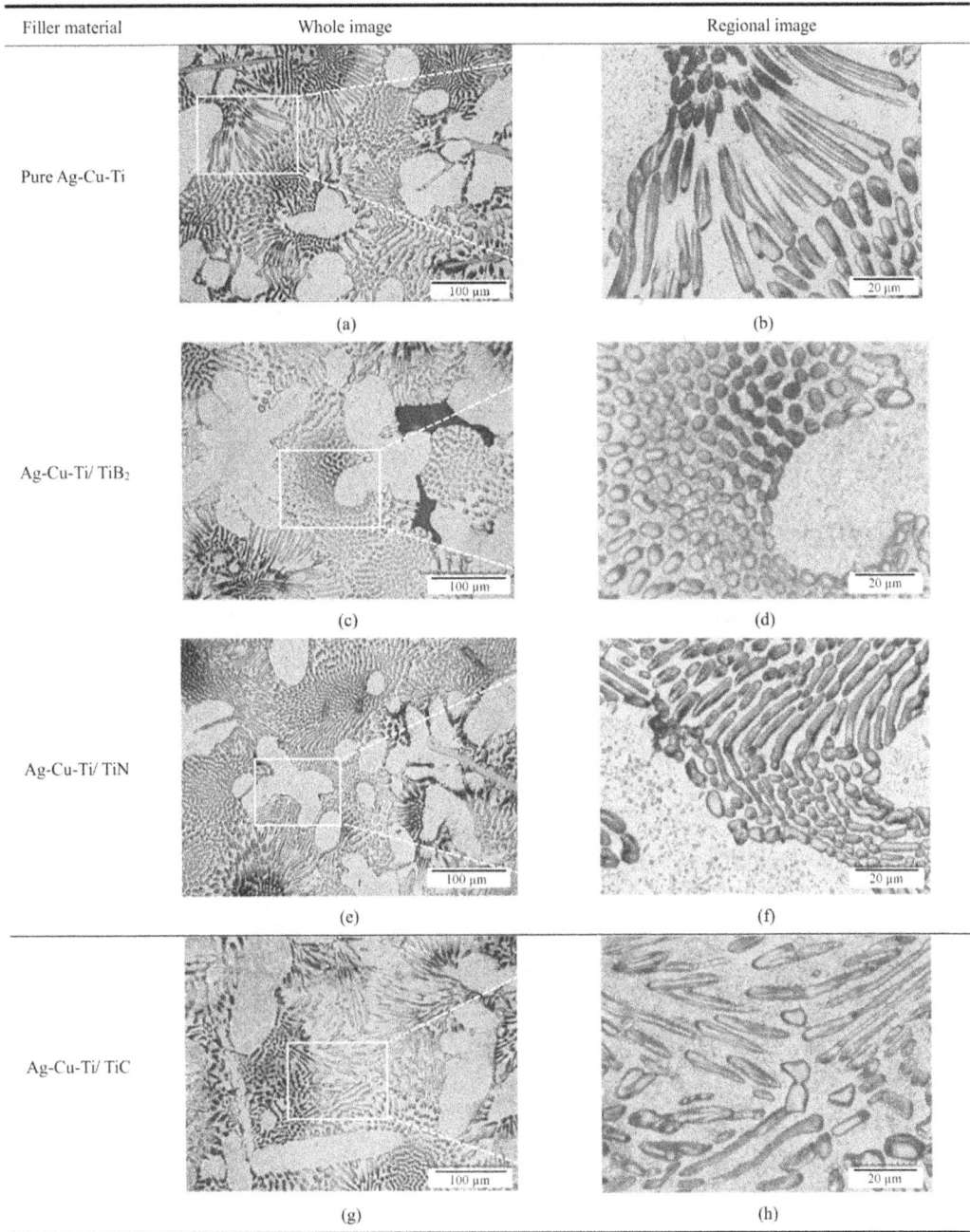

Pure Ag-Cu-Ti — (a) / (b)

Ag-Cu-Ti/ TiB₂ — (c) / (d)

Ag-Cu-Ti/ TiN — (e) / (f)

Ag-Cu-Ti/ TiC — (g) / (h)

FIGURE 3.7 The typical metallurgical microstructure of the Ag-Cu-Ti/TiX bonding layer.

of crystals in α phases in comparison to those in Figure 3.7a and d. Especially the width of the crystals (β phase) mostly has a big value of 4 μm. Also, the metallurgical microstructure of Ag-Cu-Ti/TiC composite filler materials is also characterized, as shown in Figure 3.7g and h. The needle-like crystal structure of β phases grows coarsely and sparsely.

In general, it is known from Figure 3.7 that the presence of reinforcing particles, i.e., TiB_2, TiN, and TiC, has a positive effect on the refinement of the crystal microstructure of the Ag-Cu-Ti alloy, though the crystal morphology in the metallurgical microstructure images is different. According to the literature available for the grain refinement of aluminum alloy (Zhang et al. 2022; Xiao et al. 2022; Nowak et al. 2015), a proper particle addition to the molten alloy could be a potential nucleation agent that induces heterogeneous nucleation by providing useful nucleating sites. In the case of TiX inoculation, on the one hand, owing to the small size of the reinforcing particles, it is easy for them to work as a new heterogeneous nucleus in the molten bonding layer. Thus, nucleation formation could be induced. On the other hand, the introduced particle, also serving as the second phase particle in the molten bonding layer, contributes to preventing the crystal boundary from migrating and therefore restrains the crystal's growth in α phases and β phases. All these enable the formation of fine crystals without an unacceptable rough microstructure in the current brazing practice.

3.2.5 Distribution of Reinforcing Particles in the Bonding Layer

Figure 3.8a–c display the XRD patterns of the Ag-Cu-Ti/TiX bonding layers containing different reinforcing particles. Obviously, besides the Ag-rich phase and Cu-rich phase, some intermetallic compounds such as Cu_2Ti, Cu_4Ti_3, and $AgTi_3$ were also identified. It is noted that, due to the low contents of Ti in the applied Ag-Cu-Ti alloy, the amounts of active Ti atoms are almost depleted, primarily consumed by the chemical reaction to form the above-mentioned intermetallic resultants. The additional phases (i.e., TiB_2, TiN, and TiC, respectively) were also detected demonstrably from the added reinforcing particles of the bonding layer.

To better understand the distribution characteristics of reinforcing particles in the bonding layer, the back-scattered electrons (BSE) images are displayed in Figure 3.8d–f, while Figure 3.8g–i are the SEM images of the reinforcing particles in the bonding layer etched by the nitric acid. As is well known, the imaging principle of BSE indicates that the elements with a larger atomic number reflect more electrons and appear bright on the electron micrographs, while the elements with a smaller atomic number absorb more electrons and therefore look dark (Edwin et al. 2019). The BSE images are therefore the results of atomic number contrast and contain some compositional information. Seen from Figure 3.8d–f, the bright regions are Ag-rich phases (α phases), while the grey regions are Cu-rich phases (β phases), and the black regions are TiX particles (i.e., TiB_2, TiN, and TiC, respectively).

In the case of the distribution and microstructure characteristics of TiB_2 particles in the bonding layer (Figure 3.8d and g), the reinforcing particles generally distribute uniformly. Moreover, the shape and size of the TiB_2 particles are almost the same as those of the original ones. The Ag-Cu-Ti alloy in the bonding layer tightly surrounds the TiB_2 reinforcing particles. Under such circumstances, the bonding layer could indeed be regarded as a metallic matrix composite composed of Ag-Cu-Ti alloy and TiB_2 reinforcing particles.

Figure 3.8e and h, respectively, show the BSE image and SEM microstructure of the bonding layer produced by Ag-Cu-Ti/TiN composite filler materials. According to Figure 3.8e,

Filler	XRD patterns of the bonding layer	BSE images of the bonding layer	SEM images of the bonding layer

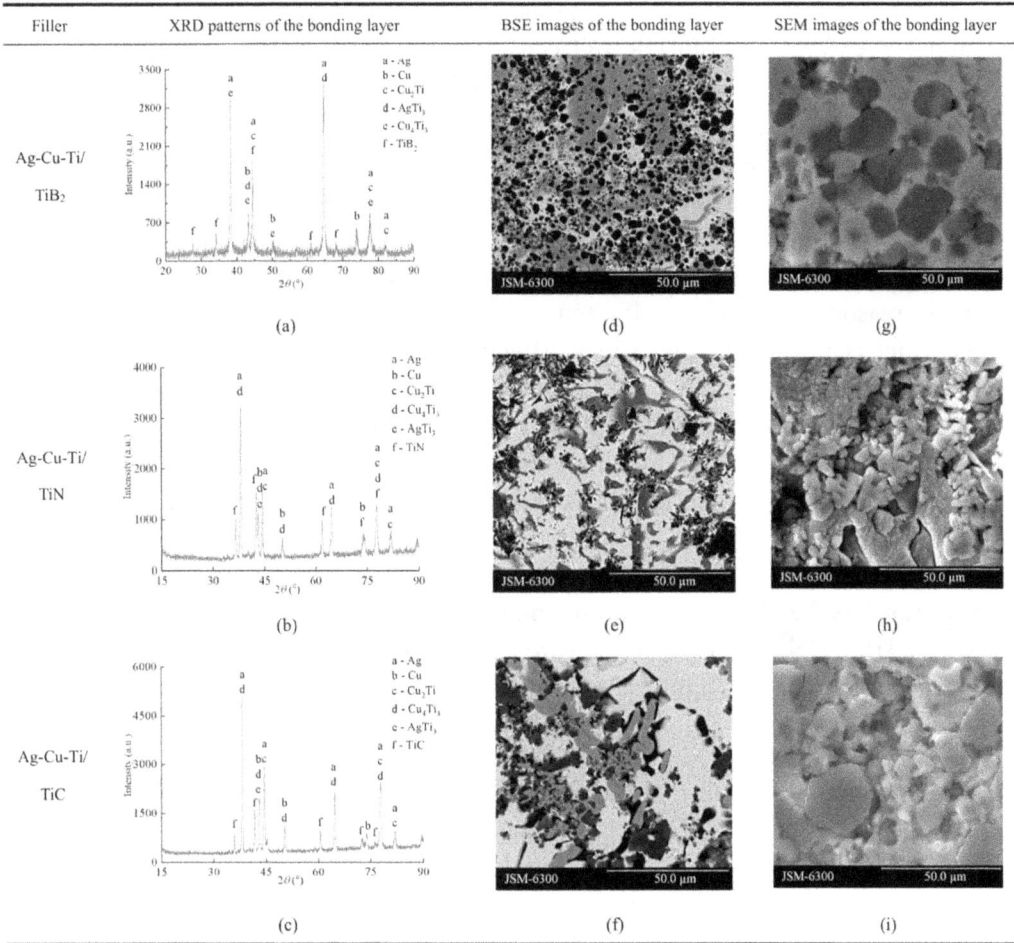

FIGURE 3.8 Reinforcing particles in the Ag-Cu-Ti/TiX bonding layer.

though the TiN particles are generally distributed evenly on the whole bonding layer, there still exists some aggregation behaviour in some regions due to the high surface energy of the TiN particles. What's more, this distribution behaviour leads to the occurrence of a phenomenon, that is, in the micro-region of aggregation, there is a small distance of 1–2 μm between the particles instead of the direct merging of the particles cluster, as demonstrated in Figure 3.8e. Thus, the TiN particles are slightly less efficient than the TiB$_2$ ones in terms of particle distribution uniformity. It is also illustrated in Figure 3.8h that the TiN particles remain in their original profile in the bonding layer.

Furthermore, in comparison with the distribution behaviour of TiB$_2$ and TiN particles in the bonding layer, TiC particles not only exist in the form of a single crystal with a length of 2–3 μm but also concentrate in some regions in the form of a crystal with a rather small size, inevitably resulting in a poor uniform distribution in the bonding layer, as shown in Figure 3.8f and i. Also, owing to the etching of nitric acid into Cu-rich phases, the TiC particles are exposed from the bonding layer, as shown in Figure 3.8i, implying that almost all TiC particles are piled up randomly in the bonding layer. However, according to Figure 3.8g and h, it is found that both TiB_2 and TiN particles are embedded deeply in the bonding layer. Under such conditions, the mechanical performance, such as the shear strength of the bonding layer and brazed joints, may be significantly affected.

As a consequence, the particles, i.e., TiB_2, TiN, and TiC, serving as the reinforcing phases in the alloy, play a key role in the microstructure variation of the bonding layer in the current investigation. Furthermore, as seen from Figure 3.8g to i, the morphology of the particles in the bonding layer has little difference compared to the original ones. The main reason is that the brazing temperature applied in the current experiments is 920°C, which is much lower than the melting point of the relevant particles, i.e., 2,980°C for TiB_2, 2,950°C for TiN, and 3,150°C for TiC. In the brazing process, the molten bonding layer is composed of the liquid phase of the Ag-Cu-Ti alloy and the solid phase of reinforcing particles. The high melting point and high thermostability of these particles result in not the chemical reaction but the wetting behaviour between the TiX reinforcing particles and the Ag-Cu-Ti alloy. The redistribution of these reinforcing particles in the liquid Ag-Cu-Ti alloy could be appropriately realized in the brazing process. In other words, it is the high melting point and high thermostability of the reinforcing particles (i.e., TiB_2, TiN, and TiC) that ensure the maintenance of the original shape, size, and properties of the TiX reinforcing particles in the molten bonding layer at elevated temperatures.

3.2.6 Reinforcing Mechanism of the TiX Particle Modified Composite Filler

It is worthwhile to note that the special particle distribution on the relatively soft Ag-Cu-Ti alloy layer could cause the formation of the dispersion strengthening effect (Wu et al. 2022; Han et al. 2022; Sarı et al. 2004), which protects the bonding layer from deformation of plastic yield. Besides, a fine and uniform microstructure contributes to the strength improvement of the bonding layer. The relationship between yield strength and crystal size for metallic material is subjected to the Hall–Petch equation (Luthra et al. 1993):

$$\sigma_s = \sigma_0 + Kd^{-\frac{1}{2}} \tag{3.1}$$

where σ_s is the yield strength, σ_0 and K are the constants on metallic material, and d is the average crystal size of the metal.

The microstructure of the bonding layer is refined when the TiX reinforcing particles are added to the composite filler materials. Then, according to Eq. (3.1), the strength of the bonding layer exhibits an increasing trend with increasing reinforcing particle content. The above analysis has already indicated that there is an apparent difference in the metallurgical microstructure of the bonding layer composed of the composite filler materials. Both the Ag-Cu-Ti/TiB_2 and Ag-Cu-Ti/TiN bonding layers have formed favourable and fine crystal microstructures. However, the crystal microstructure in the bonding layer reinforced by TiC particles grows roughly. The strength improvement induced by the particle addition is, to a certain extent, owing to the decreasing crystal size and fine microstructure.

On the other hand, the reinforcement coefficient F for the metallic composite material containing reinforcing particles can be assessed using the formula (Ibrahim et al. 1991):

$$F = \frac{\sigma_s}{\sigma_m} = \sqrt{\frac{\sqrt{3}G_m G_p b V_p^{1/2}}{\sqrt{2}d_p(1-V_p)c}} \bigg/ \sigma_m \tag{3.2}$$

where σ_s is the yield strength of composite material, σ_m is the yield strength of matrix material, G_m is the shear modulus of matrix material, G_p is the shear modulus of particle, b is the Burgers Vector, d_p is the particle diameter, V_p is the volume content of particle, and c is the constant value.

According to Eq. (3.2), there is a positive correlation between the F value of metallic composite material and the V_p of particle volume content when the other conditions are fixed. In other words, the higher the V_p of the particle is, the more effectiveness the reinforcement has. However, excessive addition of reinforcing particles may result in unexpected defects (e.g., clustering, metal-starved cavities, pores, interfacial debonding, or sites for crack initiation) in the composite material, which would lead to a decrease in the bonding strength under high particulate loading (Halbig et al. 2013). Therefore, the shear strength results suggest that only when the particle contents are controlled at a certain level, that the shear strength of the bonding layer increases with the increase in particle contents.

3.2.7 Effect of Reinforcing Particles on the Microhardness of the Bonding Layer

Hardness always plays a vital role in evaluating the mechanical properties (i.e., plastic deformation, toughness, and wear resistance) of the correlated materials. Therefore, the microhardness variation across the interface of the bonding layer and steel matrix is presented, as demonstrated in Figure 3.9. The microhardness of the joining interface, ranging

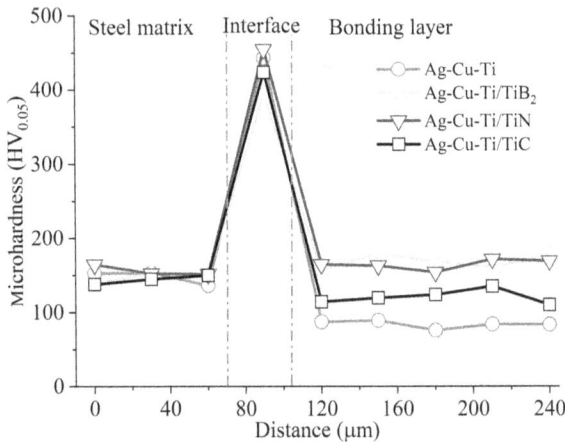

FIGURE 3.9 Microhardness variation across the interface of the joints brazed using the composite filler with 8 wt.% particle content.

from 400 to 450 $HV_{0.05}$, is significantly higher than that of the AISI 1045 steel matrix in the range of 140 to 165 $HV_{0.05}$ and that in the Ag-Cu-Ti/TiX bonding layer from 75 to 180 $HV_{0.05}$. The increase in the microhardness value of the joining interface is due to the diffusion of Fe, Cu, and Ti elements, which results in the formation of brittle intermetallic compounds with high hardness during brazing. Meanwhile, the effects of reinforcing particles on the microhardness value of the bonding layer are also evident. All the bonding layers containing reinforcing particles have a higher microhardness value than that produced by a pure Ag-Cu-Ti alloy. Based on the special distribution and microstructure characteristics of the reinforcing particles in the Ag-Cu-Ti alloy, the bonding layer composed of the Ag-Cu-Ti/TiX composite filler materials could be regarded as particle-reinforced metallic matrix materials. The reinforcing particles with high strength and modulus in the metallic matrix material normally possess a high dislocation density that enhances the mechanical strength of the metallic matrix. In the case of these reinforcing particles, the hardness and modulus are much higher in comparison to those of a pure Ag-Cu-Ti alloy, indicating that they are the potential reinforcing agents that improve the microhardness of the bonding layer. Furthermore, the microhardness of the Ag-Cu-Ti/ TiB_2 bonding layer is similar to that of the Ag-Cu-Ti/TiN bonding layer, in a range of 150–180 $HV_{0.05}$, while the microhardness of the Ag-Cu-Ti/TiC bonding layer varies at a low level of 115–130 $HV_{0.05}$, as displayed in Figure 3.9. Thus, TiB_2 and TiN particles have a more beneficial point for improving the wear resistance ability of the bonding layer than TiC particles.

Figure 3.10 shows the effects of reinforcing particles on the average microhardness values of the Ag-Cu-Ti/TiX bonding layer. It is noticed that the microhardness of the bonding layer increases with the increasing content of the reinforcing particles. Both Ag-Cu-Ti/ TiB_2 and Ag-Cu-Ti/TiN always show a higher microhardness value, and the improvement concerning Ag-Cu-Ti/TiC is much more noticeable at a low particle content (8 wt.%) where the difference in terms of microstructural features is also significant.

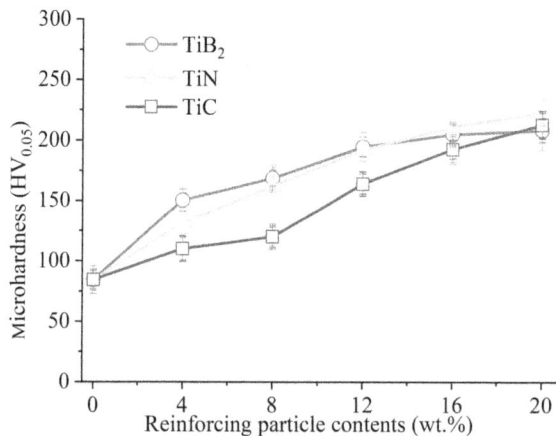

FIGURE 3.10 Effects of reinforcing particle contents on the microhardness of the Ag-Cu-Ti/TiX bonding layer.

3.3 EFFECT OF TIX PARTICLES ON THE BRAZING MICROSTRUCTURE OF CBN ABRASIVE GRAINS

3.3.1 Experimental Details

In this work, the brazing interface and compressive strength of the CBN grains brazed using different composite filler materials (i.e., Ag-Cu-Ti, Ag-Cu-Ti/TiB$_2$, Ag-Cu-Ti/TiN, and Ag-Cu-Ti/TiC) are investigated. The effects of TiX additions on the interfacial microstructure between the composite filler and CBN grains are also discussed comparatively. Meanwhile, it is also the objective to correlate the interfacial microstructure and mechanical properties of brazed CBN grains to shed light on the influencing mechanisms of the different reinforcements on the performance of brazed joints. The optimum TiX phases and contents are accordingly determined.

In the experiment, the Ag-Cu-Ti/TiX composite filler materials were fabricated by mechanically mixing Ag-Cu-Ti alloy powder and the TiX particles. The weight ratio of TiX particles to the composite filler is controlled at 0, 4, 8, 12, 16, and 20 wt.%, respectively. Particularly, based on the optimum results of the wetting and brazing behaviour of the molten composite filler on the AISI 1045 steel matrix, two typical weight fractions of the particles, i.e., 8 and 16 wt.%, are chosen in order to understand the effects of the reinforcing particles on the chemical resultants at the brazing interface.

Before brazing, the AISI 1045 steel matrix and CBN abrasive grains were first cleaned with acetone and air-dried, respectively. Then the specimens were assembled in the sequence of steel matrix/composite filler/CBN grain. Subsequently, the assembly was heated in a vacuum furnace modelled VAF-20 to the brazing temperature (920°C) with a holding time of 5 min under vacuum ($<1 \times 10^{-2}$ Pa). Both the heating and cooling rates were 10°C/min. This brazing parameter has been confirmed conclusively to braze CBN grains using pure Ag-Cu-Ti alloy powder, according to the published literature.

3.3.2 Overall Morphology of CBN Grains Brazed with Composite Fillers

Figure 3.11 shows the typical morphology of CBN grains brazed using the Ag-Cu-Ti/TiB$_2$ composite filler material and pure Ag-Cu-Ti filler alloy, respectively. No obvious defects, i.e., pores and cracks, are observed; that is to say, the interfaces generally exhibit sound joining between CBN grains and the bonding layer. However, the spreading behaviour of pure Ag-Cu-Ti alloy is so drastic that the brazed CBN grain (Figure 3.11b) is nearly surrounded, which restrains the grain protrusion; on the contrary, the Ag-Cu-Ti/TiB$_2$ composite filler permits to obtain enough protrusion of brazed CBN grain (Figure 3.11a), implying that the addition of TiB$_2$ particles could really provide the so-called suppression effect on the spreading behaviour of the molten filler materials. It is noted that the similar suppression effects induced by the other two types of TiX reinforcing particles, i.e., TiN and TiC, are also found on the brazed CBN grains.

3.3.3 Microstructure and Microchemistry Characteristics at the Grain/Filler Interface

To further characterize the joining interface between CBN grain and composite filler layer, Figure 3.12a–d displays the typical micrograph of the joints prepared with a TiX particle addition content of 8 wt.%. As can be seen, the reaction layer across the joining

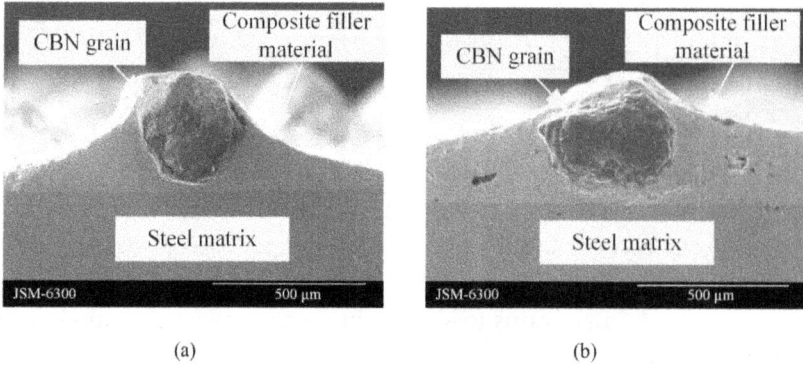

FIGURE 3.11 Typical morphology of the brazed joint of CBN grain/composite filler/steel matrix: (a) Ag-Cu-Ti/TiB$_2$ (8 wt.%), (b) pure Ag-Cu-Ti.

interface is generally continuous, which suggests that the flow of molten composite filler materials is adequate in the brazing process and that sufficient liquid-solid interaction has occurred.

Figure 3.12e–h depicts an energy dispersion spectroscopy line scan carried out across the polished section of the joining interface between CBN grains and composite fillers. The polishing has been done to keep the observed surface smooth and in a single-rough plane. The diffusion behaviour across the interface between CBN grains and composite filler is also found. For instance, the concentration of Ti atoms, as the key element in active brazing of CBN grains, gradually increases along the direction from CBN grains to composite filler materials and generally reaches a peak value at the interface, implying that many Ti atoms coming from the Ag-Cu-Ti alloy have moved towards the CBN grains.

Meanwhile, the scanning line of the B and N atoms coming from the CBN grain also has a slight slope across the joining interface. All these indicate that the diffusion and aggregation of atoms of Ti, B, and N have taken place in the brazing process at an elevated temperature of 920°C. Due to the strong atomic affinity among Ti and B and N at elevated temperatures, the formation of resultant Ti-N and Ti-B compounds could be maintained at the grain/filler interface until the dynamic balance of the related atomic diffusion is fulfilled. Moreover, the brazed joints prepared using composite filler materials with different levels of reinforcing particles show similar atomic diffusion behaviour for Ti, B, and N. An interesting phenomenon can be found in Figure 3.4 that the corresponding element content in the regions where TiX particles exist is predominantly higher than that in the neighbouring zones.

Depending on the atomic diffusion to form the integrated resultant compound layer, a chemical joining could eventually be achieved between CBN grains and composite filler materials. In other words, the formation of the interfacial microstructure is the consequence of the atomic diffusion of Ti, B, and N across the interface between the CBN grain and the bonding layer. It is worth noting that, however, as demonstrated in Figures. 3.12f–h, none of the elements of B, N, or C coming from the TiX particles have peak concentrations at the interfaces adjacent to the composite filler layer. This phenomenon may

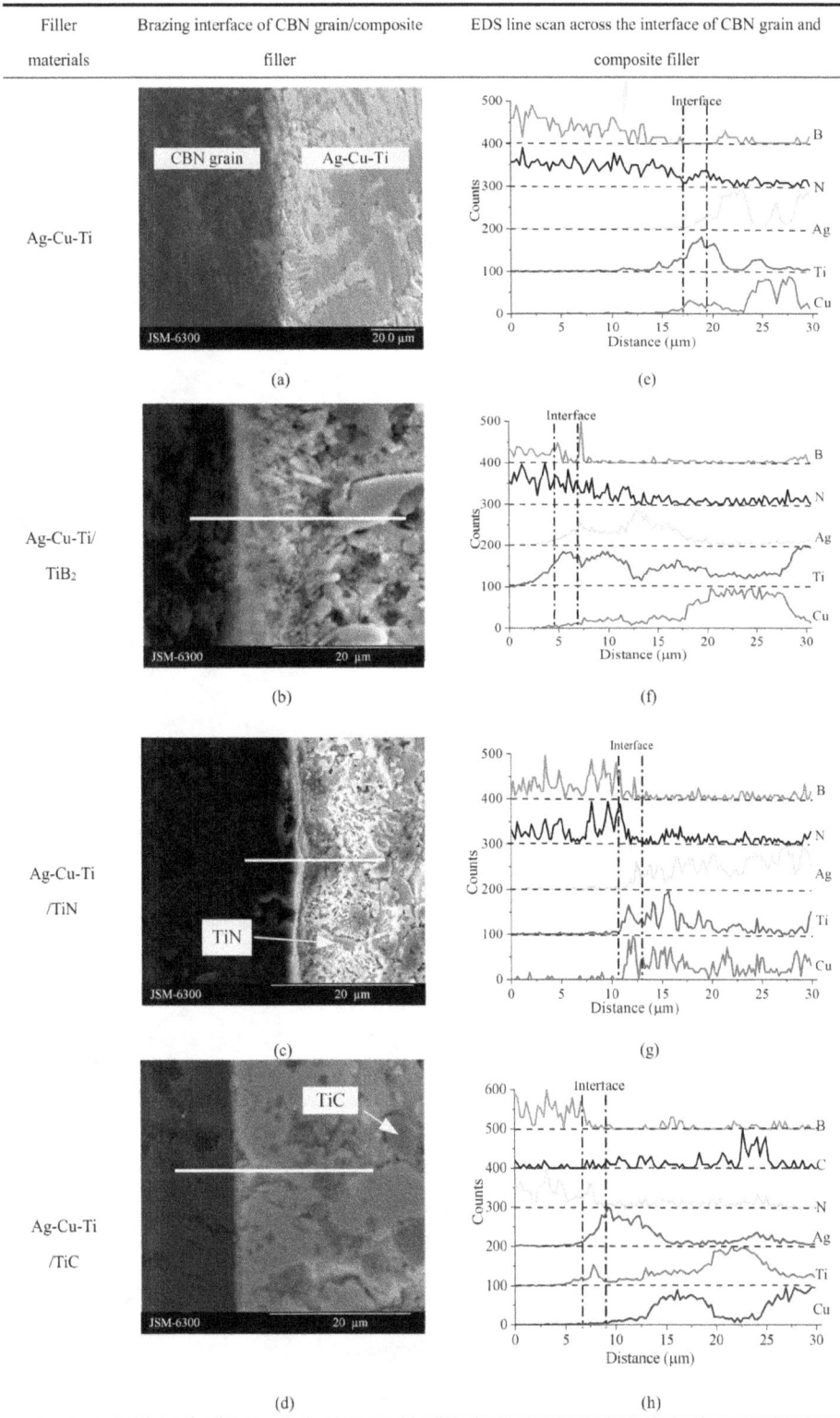

FIGURE 3.12 Typical micrograph and line-scanning element distribution across the interface between CBN grain and composite filler materials containing 8 wt.% TiX particles.

suggest that no chemical reaction but pure physical wetting behaviour happens between the reinforcing particles and the Ag-Cu-Ti alloy at an elevated temperature of 920°C in the brazing process.

3.3.4 Morphology of Brazing Chemical Resultants

A morphology collection of the brazed CBN grains is shown in Figures 3.13 and 3.14. Generally, the chemical resultants of brazed CBN grains prepared with pure Ag-Cu-Ti alloy as filler material show a considerable difference in comparison to those obtained with

FIGURE 3.13 Typical micrograph of chemical resultants of CBN grains brazed using composite filler materials containing 8 wt.% TiX particles.

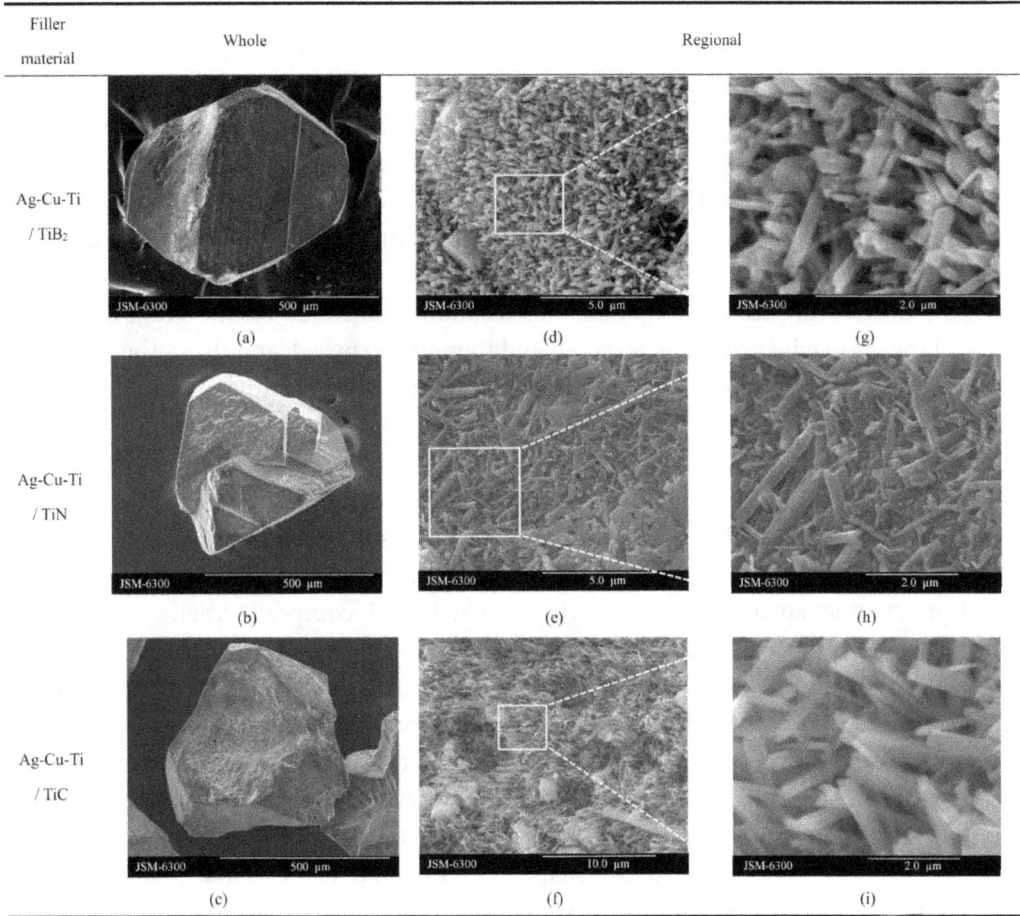

FIGURE 3.14 Typical micrograph of chemical resultants of CBN grains brazed using the composite filler materials containing 16 wt.% TiX particles.

different Ag-Cu-Ti/TiX composite filler materials. As seen in Figure 3.13a, e, and i, on the surface of CBN grain brazed with pure Ag-Cu-Ti alloy, many columnar compounds are produced, and the fragmented appearance of the chemical resultants is indicative of a very drastic chemical reaction that happened at the brazing interface. According to these images, the distribution of the chemical resultant is stochastic and sparse, and the shape is also uneven. The section width of the compounds varies from 200 nm to 1 μm, while the length is about 5–40 μm.

A similar phenomenon is also observed on the surface of the brazed CBN grain, which is prepared using the Ag-Cu-Ti composite filler materials and contains 8 wt.% TiC particles, as displayed in Figure 3.13d, h, and l. The unevenness degree in section width of the resultant compounds is slightly stronger than that shown in Figure 3.13e and i. Moreover, the brazed CBN grain is still covered with a thick layer of chemical resultants; there are also large amounts of resultants on the grain edges. As is well known, the grinding operation refers to material removal behaviour by individual grains, whose cutting edges are responsible for removing materials using chips from the workpiece. If the cutting edges of CBN grains are seriously destroyed due to the intense brazing reaction, the machining performance of brazed CBN grains would be negatively affected.

In contrast, when the TiX particle content of the composite filler is fixed at 8 wt.%, the chemical resultants on the surface of CBN grains brazed with Ag-Cu-Ti/TiB$_2$ and Ag-Cu-Ti/TiN composite filler, respectively, present a quite different morphology. Compared with Figure 3.13a and d, the morphology characteristics shown in Figure 3.13b and c can be described as follows: first, the quantity of brazing resultants around the grain surface is generally decreased; second, due to smaller resultants generated on the grain edges, the shape of the grain edges could be reserved and therefore exposed; and third, the resultants on the grain surface appear to be very fine and compact. Meanwhile, as could be found from Figure 3.13b, c, f, g, j, and k, the primary differences are as follows: (i) the brazing resultant of the CBN grain prepared with Ag-Cu-Ti/TiB$_2$ composite filler possesses a section width between 50 and 100 nm, and a length at a level of 2–5 μm, while that with Ag-Cu-Ti/TiN composite filler has a section width of about 100 nm, and a length ranging from 1 to 3 μm. (ii) the sharp edge of CBN grains brazed with Ag-Cu-Ti/TiN composite filler is much more apparent than that with Ag-Cu-Ti/TiB$_2$ composite filler.

When the TiX particle content of the composite filler materials is further increased to 16 wt.%, as shown in Figure 3.14a and b, the addition of TiB$_2$ or TiN reinforcing particles leads to a significant reduction of the total amount of brazing resultants on the CBN grain surface; however, there is not any great difference in terms of the compact degree of resultants (Figure 3.14d and e). Particularly, the columnar resultants in Figure 3.14j grow mostly in a flat shape (Figure 3.14g) when the TiB$_2$ addition content increases from 8 to 16 wt.%, while after TiN addition, the resultant shape in Figure 3.14e changes marginally concerning that in Figure 3.14g. As demonstrated in Figure 3.14c, f, and i, the addition of TiC particles to the joint results in a significant increase in resultant compact degree but a slight change in resultant quantity, and a few granular or prismatic resultants are also observed in Figure 3.14f. Based on the above phenomena, it is found that the chemical resultants at the brazing interface between CBN grain and Ag-Cu-Ti alloy are generally restrained by the added TiX (e.g., TiB$_2$, TiN, and TiC) reinforcing particles. Furthermore, it is known that the effects of TiC particles on the morphology of brazing resultants are not as prominent as those caused by TiB$_2$ or TiN particles.

3.3.5 Phase Analysis of Brazing Chemical Resultants

The phases of the brazing chemical resultants on the CBN grain surface are identified by XRD. Because the dominating phases of brazing resultants are identical after the TiX addition to the composite filler materials, this section only provides the XRD results of brazing resultants produced by the composite filler containing 8 wt.% of TiX particles, as displayed in Figure 3.15. Obviously, besides the CBN crystal phase, the brazing chemical resultants mainly consist of TiB$_2$, TiB, and TiN. Though the morphology of brazing resultants is quite different, identical phases are acquired for the resultant compounds when the CBN grains are brazed using the composite filler material containing different TiX reinforcing particles.

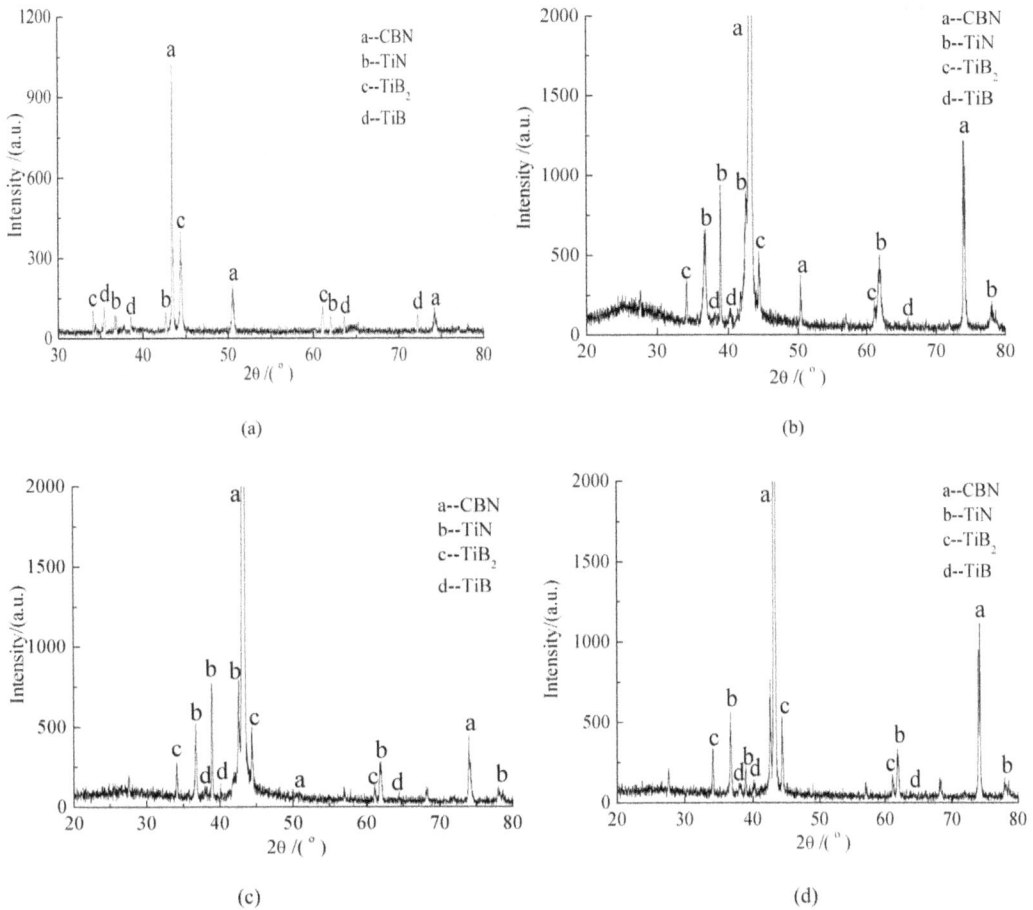

FIGURE 3.15 XRD patterns of chemical resultants on the surface of CBN grains brazed using different composite filler materials containing 8 wt.% TiX particles: (a) Brazed using pure Ag-Cu-Ti, (b) Brazed using Ag-Cu-Ti/TiB$_2$, (c) Brazed using Ag-Cu-Ti/TiN, (d) Brazed using Ag-Cu-Ti/TiC.

3.3.6 Fracture Morphology of the Brazed Joints

To examine the strength of the brazed joints visually, Figure 3.16 presents the fracture morphology of CBN/steel joints brazed using the different composite filler materials with 8 wt.% of TiX contents. Generally, after the shearing test, no grain pullout but brittle fracture of the brazed grains happens, suggesting that firm joints are achieved, though the chemical resultants at the CBN/steel brazing interface are rather different in morphology. Particularly, the identical fracture morphology of the brazed joint is also found when it is prepared by the composite filler material with other levels of TiX contents (i.e., 16 wt.%). The reason is likely because not the quantity but the compact degree of the brazing chemical resultant layer plays a vital role in ensuring the firm joining between the CBN grains and the bonding layer (that is, the composite filler layer). In the previous literature (Halbig et al. 2013), it has been indicated that a decreasing trend in joint strength resulting from Ti-scavenging interfacial reactions would be offset by the reinforcing effect of hard particulates. However, the fine but minimal amount of

Contents	Brazed using Ag-Cu-Ti/TiB$_2$ filler	Brazed using Ag-Cu-Ti/TiN filler	Brazed using Ag-Cu-Ti/TiC filler
Whole morphology	(a)	(b)	(c)
Fracture morphology	(d)	(e)	(f)

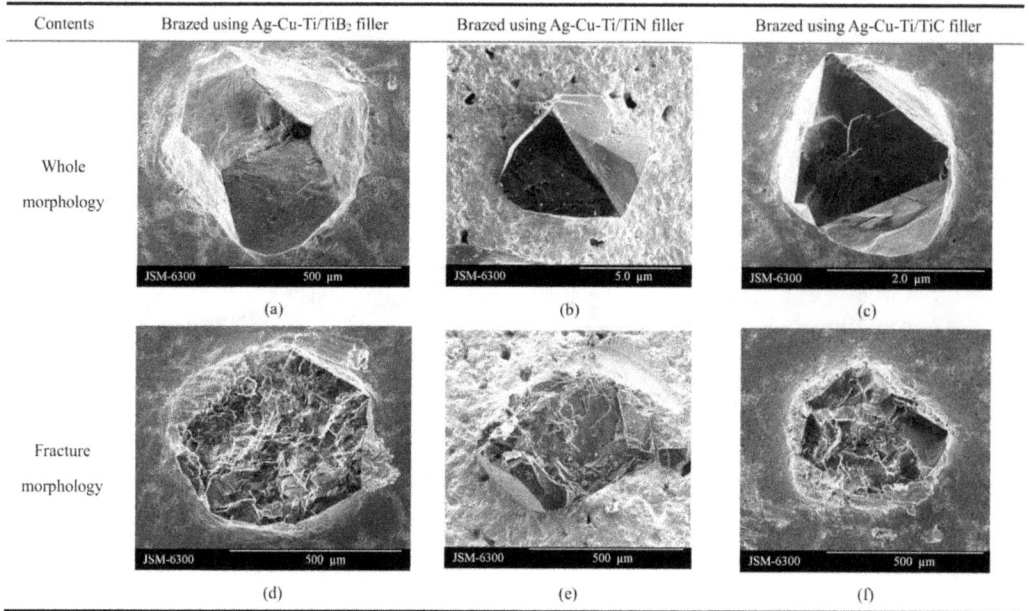

FIGURE 3.16 Fracture morphology of CBN/steel joints brazed using composite filler with TiX particle addition of 8 wt.%.

compounds may contribute to the potential weakening of the joining strength of the CBN grains because of the induced skinny resultant layer.

3.3.7 Influencing Mechanism of TiX Reinforcing Particles on Brazing Reaction

According to the experimental results mentioned above, although Ti atoms are contained in both the particles of TiB$_2$ and TiN, a significant difference exists in terms of the reaction of CBN grains and the composite filler. The TiN particles have the strongest suppression effect on the interfacial reactions, while the TiC particle has the weakest effect, and TiB$_2$ has the medium effect.

The chemical reactions that happened in the joint brazed with pure Ag-Cu-Ti alloy could also be described as follows:

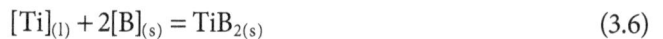

$$[Ti]_{(l)} + BN_{(s)} = TiN_{(s)} + [B]_{(s)} \tag{3.3}$$

$$[Ti]_{(l)} + [B]_{(s)} = TiB_{(s)} \tag{3.4}$$

$$3[Ti]_{(l)} + 4[B]_{(s)} = Ti_3B_{4(s)} \tag{3.5}$$

$$[Ti]_{(l)} + 2[B]_{(s)} = TiB_{2(s)} \tag{3.6}$$

The above reactions indicate that the active Ti elements act as the dominant factor in the formation of Ti-N and Ti-B compounds in molten filler materials. It is, therefore, necessary to estimate the chemical activity of Ti based on the thermodynamic analysis of the Ag-Cu-Ti system. If the Ti-N and Ti-B compounds are considered to be solutions of nitrogen and boron, respectively, in Ti, the chemical potentials of Ti in the liquid metallic phases can be written as:

$$\Delta\mu_{Ti} = \mu_{Ti}^{l} - \mu_{Ti}^{Ti\text{-}N} - \mu_{Ti}^{Ti\text{-}B} \tag{3.7}$$

Where μ_{Ti}^{l}, $\mu_{Ti}^{Ti\text{-}N}$, and $\mu_{Ti}^{Ti\text{-}B}$ are the activities of Ti in the molten filler, Ti-N, and Ti-B compounds, respectively.

In the case of Ti-N compounds, due to the large solubility of N in Ti, the Ti-N bonds, whose structure is similar to that of Ti atoms, perform as metallic bonds. Moreover, the activity of Ti in the liquid filler is much higher than that in the nitride (Wang et al. 2015), implying that the chemical potential of Ti in the filler is much higher than that in TiN. Once some TiN particles are added to the molten Ag-Cu-Ti alloy, Ti will diffuse towards and enrich around the TiN particle. For this reason, it can be inferred that the strong chemical affinity between TiN and Ti leads to a great decrease in Ti activity in the molten filler system.

In contrast, the covalent bonds in Ti-B compounds generally possess higher bond energy than the metallic bonds. With increasing Ti-B compounds, the general stability of the interatomic bond in Ti-B is enhanced whilst that of metallic bonds is reduced gradually, suggesting that the chemical potential of Ti in Ti-B compounds is relative higher in comparison with that in TiN. As a result, when TiB_2 particles are added to the molten Ag-Cu-Ti alloy, the affinity between TiB_2 and Ti becomes much lower than that between TiN and Ti. Therefore, the ability to decrease Ti activity can be distinguished according to Eq. (3.7); that is, TiN particles are much more effective than TiB_2 ones. The remarkable suppression of TiN addition on brazing chemical reactions between CBN grains and Ag-Cu-Ti filler alloy could be accordingly determined.

Due to the absence of TiC in chemical resultants between CBN and Ti interaction, the effect of TiC reinforcing particles on interfacial brazing reactions is not as significant as TiB_2 or TiN ones, especially when the TiC particle content is low. In the previous work (Miao et al. 2016), it has been illustrated that the apparent suppression of TiC on the spreading behaviour of the Ag-Cu-Ti alloy could be observed until the particle content is beyond 12 wt.%. Under such conditions, the addition of TiC particles could lead to a slight restraining effect on the interfacial brazing reactions. Nevertheless, a deeper understanding of the influencing mechanism of TiC particles on interfacial reactions needs to be further carried out from the viewpoint of thermodynamics and kinetics.

3.4 STRENGTH ANALYSIS OF THE CBN ABRASIVE GRAINS BRAZED WITH A COMPOSITE FILLER

3.4.1 Measurement and Calculation of Compressive Strength of Brazed CBN Grains

Due to the rather small size of CBN grain, the compressive strength of CBN grain can generally be characterized using compressive force. Therefore, the static compressive strength of the brazed grains was measured using ZMC-II static strength measuring equipment. The measuring range is 1–100 N, and the force error is ±1%. The loading rate is 0.5 mm/min. Figure 3.17 shows the schematic of the static compressive strength measuring test in the

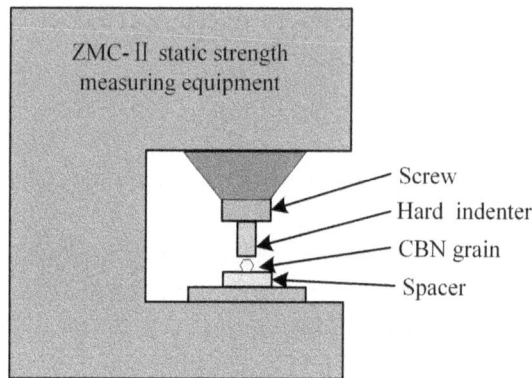

FIGURE 3.17 Static compressive strength measuring apparatus for CBN superabrasive grains.

present investigation. In particular, for each type of composite filler, three groups of grains brazed under the same conditions were selected. Each group contained 40 grains. The average compressive strength of the brazed grains was thus determined.

The detailed data processing procedures for each group of CBN grains were described as follows: One grain was first placed between the hard indenter and the spacer, and then the grain was crushed gradually until it broke up; accordingly, the compressive force was collected and recorded. Particularly if the broken particle size was less than one-third of the original grain, the force value was recalculated. When the 40 values in a group were obtained, assuming that the grain number i was corresponded to the force value X_i, the average compressive force value $\overline{X}$ of this group can be defined as:

$$\overline{X} = \frac{1}{40} \sum_{i=1}^{40} X_i \qquad (3.8)$$

Subsequently, the force values were checked carefully. If $X_i > 2\,\overline{X}$, this special force value would be rejected; under such conditions, the final value of average compressive forces could be determined by the rest grains of this group. In addition, each value of the compressive force should also meet the following requirements; otherwise, a new group of grains must be employed. The special requirements included that: (i) the number of grains whose compressive forces were lower than the average value was less than 6; (ii) the number of grains whose compressive forces were higher than the average value was more than 20.

In order to better understand the measurement and data processing procedure, Figure 3.18 shows the calculation process of the compressive strength of the CBN grains brazed using Ag-Cu-Ti/TiB$_2$ filler with a particle content of 20 wt.%. Note that the identical procedures were also carried out for the CBN grains brazed with the pure Ag-Cu-Ti, Ag-Cu-Ti/TiN, and Ag-Cu-Ti/TiC composite filler materials, respectively, in the present work. Figure 3.18a displays the original and average compressive strength values of the brazed CBN grains for Group I. Obviously, two grains were rejected because the force value was beyond the above requirements. The final average compressive strengths $\overline{X}_I$ of Group I was shown

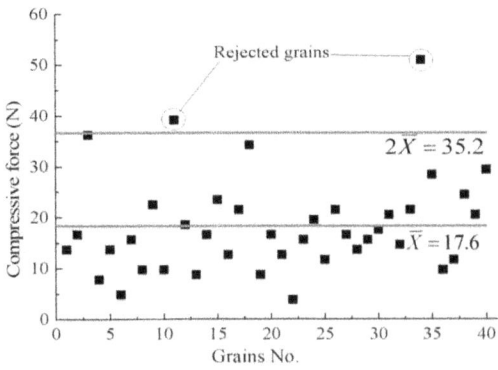

(a)

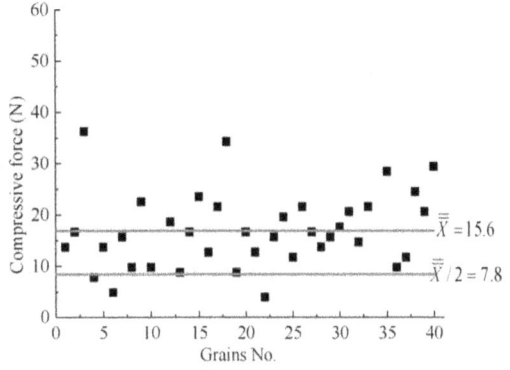

(b)

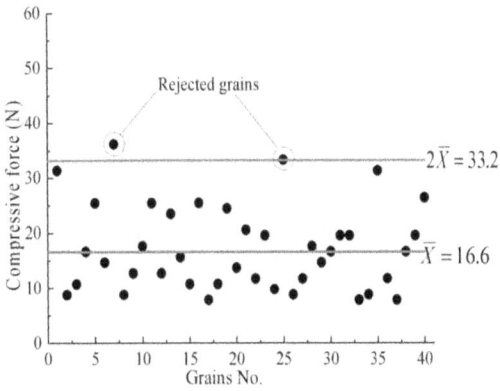

(c)

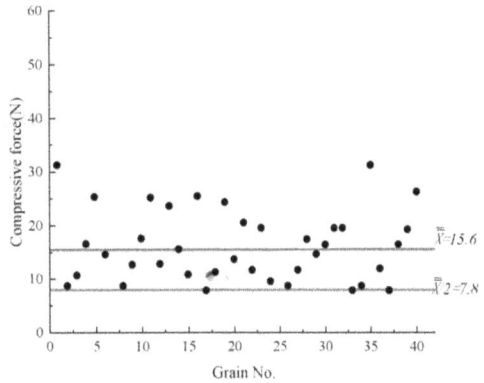

(d)

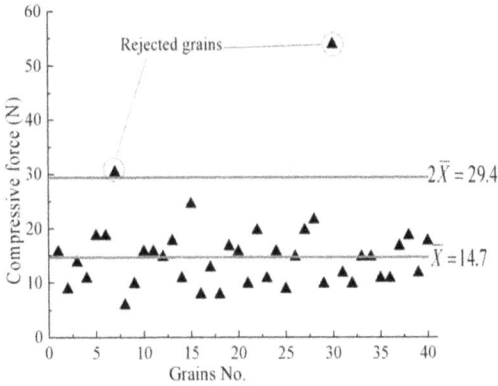

(e)

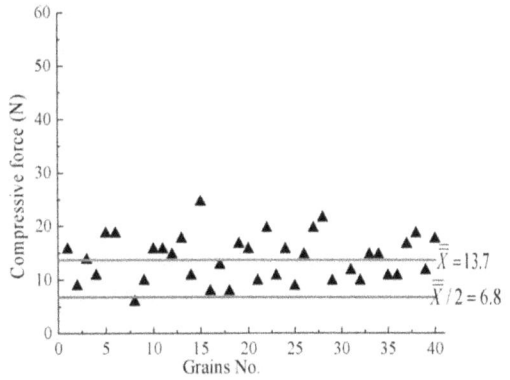

(f)

FIGURE 3.18 Calculation process of the average compressive force of CBN grains brazed with Ag-Cu-Ti/TiB$_2$ filler (20 wt.%): (a) Rejected grains in group I, (b) Final compressive force of group I, (c) Rejected grains in group II, (d) Final compressive force of group II, (e) Rejected grains in group III, (f) Final compressive force of group III.

in Figure 3.18b. As such, the corresponding average compressive strength of Group II and Group III can be calculated. Accordingly, the average compressive strength of the grains brazed using Ag-Cu-Ti filler with 20 wt.% of TiB_2 particles was eventually decided by

$$\overline{X} = \frac{1}{3}(\overline{X}_I + \overline{X}_{II} + \overline{X}_{III})$$ (3.9)

$$\overline{X} = \frac{1}{3}(15.6 + 15.6 + 13.7) = 14.9 \text{N}$$ (3.10)

3.4.2 Effects of TiX Particle Types on the Compressive Strength of Brazed CBN Grains

Figure 3.19 presents the compressive strengths of CBN grains brazed using the composite filler containing 8 wt.% TiX particles. Usually, the average compressive strength of the original CBN grains is 15.3 N. When the TiN particles are added to the composite filler material to braze CBN grains, the compressive strength of the grain, 12.7 N, is generally 17% lower than that of the grains brazed using a pure Ag-Cu-Ti alloy. Furthermore, the interesting phenomenon is that, when the Ag-Cu-Ti/TiB_2 composite filler alloy is used for brazing, the average strength of the brazed grains reaches the maximum value, 15 N, which is almost identical with that of the original grains. The compressive strength of grains brazed by Ag-Cu-Ti/TiC composite filler is 13.7 N.

The chemical resultants on the surface of CBN grains brazed using Ag-Cu-Ti/TiB_2 composite filler have a favourable growth in quantity and appearance. In the case of the grains brazed using Ag-Cu-Ti/TiN composite filler, the amount of the chemical resultant is restrained so greatly that the grain edges are exposed greatly. The compounds existing on the grain surface are not only fairly thin but also extremely small and very uniform.

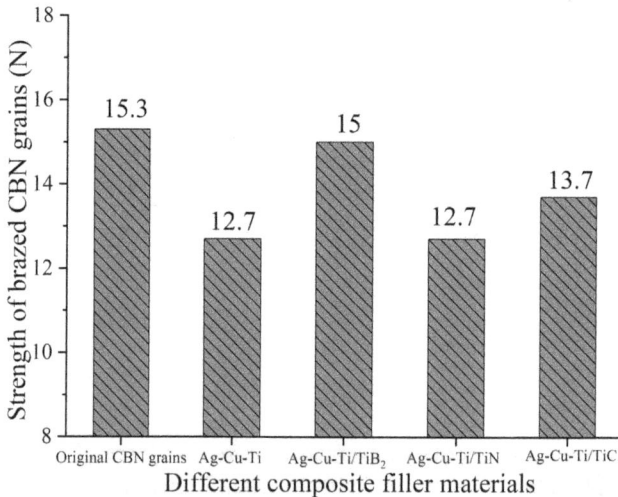

FIGURE 3.19 Average compressive strength of CBN grains brazed using different composite filler materials containing 8 wt.% of reinforcing particles.

The grains brazed either with Ag-Cu-Ti/TiC composite filler or with pure Ag-Cu-Ti have numerous chemical resultants with poor uniformity. It has been revealed that the average mechanical strength of brazed CBN grains is affected comprehensively by the enhancing effect of the chemical resultants and the reducing effect of the thermal damage in the brazing process. In general, the thermal expansion mismatch among the braze partners (i.e., CBN abrasive grains, composite filler materials, and AISI 1045 steel matrix) can bring about residual tensile stresses in the joints. The resultants generated in brazing are expected to help mitigate the mismatch in the thermal expansion coefficients. On the other hand, it is also expected for brazing resultants to compensate for the defects or damages to CBN grains caused during brazing. Through the observation of the micrograph and microstructure of the resultants on the surface of CBN grains brazed using different composite fillers, it can be stated that it keeps a good balance between the enhancing effect and the reducing effect for the grains brazed with Ag-Cu-Ti/TiB$_2$ filler alloy, therefore resulting in the highest value of compressive strength of the grains.

After the compression strength test, the typical micrographs of the broken parts of brazed CBN grains were observed, as shown in Figure 3.20. A significant difference exists in terms of the broken CBN micrograph after TiX addition. There are many cleavage lines on the broken CBN part containing TiB$_2$ particles (Figure 3.20a), and the compounds grow perpendicularly around the CBN grain surface. While in Figure 3.20b with the TiN particle, the cleavage plane with the small shoulder can be identified, a few resultants are observed on the surface of the broken CBN part. In the case of the broken CBN part with TiC particles (Figure 3.20c),

(a)   (b)   (c)

FIGURE 3.20   Typical micrograph of broken parts of brazed CBN grains after the compressive test (8 wt.% of TiX reinforcing particles): (a) Brazed using Ag-Cu-Ti/TiB$_2$, (b) Brazed using Ag-Cu-Ti/TiN, (c) Brazed using Ag-Cu-Ti/TiC.

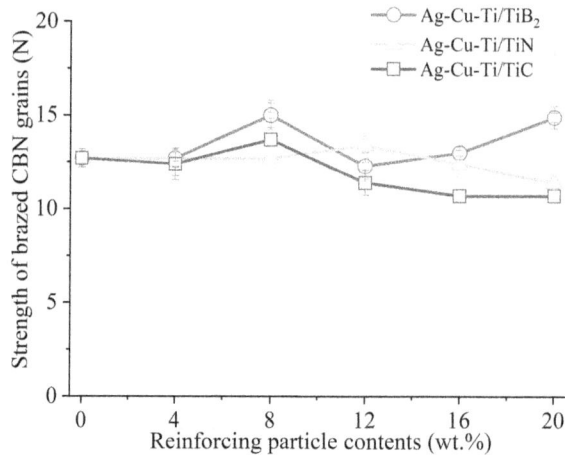

FIGURE 3.21 Effects of reinforcing particle contents on the compressive strength of brazed CBN grains.

the presence of plentiful resultants is still evidently observed. The special micrographs are strongly dependent on the formation of the resultants on the CBN grain surface, suggesting that the good chemical bridge connecting CBN grain and composite filler has already been built through the brazing reaction and that the brazed CBN grain with TiB_2 addition is favourable compared to that with TiN or TiC addition in terms of the broken micrograph.

3.4.3 Effects of TiX Particle Contents on the Compressive Strength of Brazed Grains

Figure 3.21 shows the compressive strength variation of CBN grains brazed using composite fillers with different TiX particle contents. Between 0 and 8 wt.%, there appears to be a trend towards higher particle contents, resulting in bigger compressive strengths. The increasing predominance of enhancing effects from reinforcing particles has been considered for this range of particle content. When the particle contents are further increased from 12 to 20 wt.%, the compressive strength of the grains brazed with both Ag-Cu-Ti/TiN and Ag-Cu-Ti/TiC presents a decreasing trend. In the case of TiB_2 addition, the compressive strength increases continuously, starting at 12 wt.% and reaching a value of 14.9 N at 20 wt.%. However, it is noted that excessive addition of TiX reinforcing particles is known to cause clusters, pores, and sites for crack initiation, all of which may lower the joint strength at high particle loading (Halbig et al. 2013). The results obtained here agree well with the above analysis of the resultants on the CBN grain surface. As a consequence, the optimum particle content of TiB_2 is determined as 8 wt.%, at which the maximum compressive strength can be obtained as 15 N.

REFERENCES

Edwin, R. S., M. Mushthofa, E. Gruyaert, et al. 2019. Quantitative analysis on porosity of reactive powder concrete based on automated analysis of back-scattered-electron images. *Cement and Concrete Composites*, 96: 1–10.

Guo, W., H. Zhang, K. Ma, et al. 2019. Reactive brazing of silicon nitride to Invar alloy using Ni foam and AgCuTi intermediate layers. *Ceramics International*, 45(11): 13979–13987.

Halbig, M. C., B. P. Coddingto, R. Asthana, et al. 2013. Characterization of silicon carbide joints fabricated using SiC particulate-reinforced Ag-Cu-Ti alloys. *Ceramics International*, 39: 4151–4162.

Han, L., J. Wang, Y. Chen, et al. 2022. Diffusion bonding and interface structure of advanced carbide-dispersion-strengthened Cu and oxide-dispersion-strengthened W. *Journal of Materials Processing Technology*, 302: 117508

He, Y. M., J. Zhang, C. F. Liu, et al. 2010. Microstructure and mechanical properties of Si3N4/Si3N4 joint brazed with Ag-Cu-Ti+SiCp composite filler. *Materials Science and Engineering A*, 527: 2819–2825.

Ibrahim, I. A., F. A. Mohamed, and E. J. Lavernia. 1991. Particulate reinforced metal matrix composites: A review. *Journal of Materials Science*, 26: 1137–1156.

Luthra, K. L., R. N. Singh, and M. K. Brun. 1993. Toughened silcomp composites-process and preliminary properties. *American Ceramic Society Bulletin; (United States)*, 72(7): 779–785.

Miao, Q., W. F. Ding, Y. Y. Zhu, et al. 2016. Brazing of CBN grains with Ag-Cu-Ti/TiX composite filler - The effect of TiX particles on microstructure and strength of bonding layer. *Materials and Design*, 98: 243–253.

Nowak, M., L. Bolzoni, and N. Hari Babu. 2015. Grain refinement of Al-Si alloys by Nb-B inoculation. Part I: Concept development and effect on binary alloys. *Materials and Design*, 66: 366–375.

Sarı, U., S. Agan, I. Aksoy, et al. 2004. Dispersion-strengthening effect of Cu-based Mn, Al, and Zn rich alloys. *Journal of Materials Processing Technology*, 150(3): 208–214.

Song, X. G., J. Cao, Y. F. Wang, et al. 2011. Effect of Si3N4-particles addition in Ag-Cu-Ti filler alloy on Si3N4/TiAl brazed joint. *Materials Science and Engineering A*, 528: 5135–5140.

Wang, T. P., T. Ivas, C. Leinenbach, et al. 2015. Microstructural characterization of of Si3N4/42CrMo joints brazed with Ag-Cu-Ti+TiNp composite filler. *Journal of Alloys and Compounds*, 651: 623–630.

Wang, T. P., Z. Zhang, C. F. Liu, et al. 2014. Microstructure and mechanical properties of Si3N4/42CrMo joints brazed with TiNp modified active filler. *Ceramics International*, 40: 6881–6890.

Wu, Z., X. Jiang, Y. Li, et al. 2022. Microstructures and properties of graphene nanoplatelets reinforced Cu/Ti3SiC2/C nanocomposites with efficient dispersion and strengthening achieved by high-pressure torsion. *Materials Characterization*, 193: 112308

Xiao, F., M. Wu, Y. Wang, et al. 2022. Effect of trace boron on grain refinement of commercially pure aluminum by Al-5Ti-1B. *Transactions of Nonferrous Metals Society of China*, 32(4): 1061–1069.

Zhang, L., H. Jiang, J. He, et al. 2022. Improved grain refinement in aluminium alloys by re-precipitated TiB2 particles. *Materials Letters*, 312: 131657.

Zhao, Y. X., M. R. Wang, J. Cao, et al. 2015. Brazing TC4 alloy to Si3N4 ceramic using nano-reinforced AgCu composite filler. *Materials and Design*, 76: 40–46.

CHAPTER 4

Brazing of Polycrystalline CBN Abrasive Grains Based on Vacuum Furnace Heating

4.1 INTRODUCTION

Polycrystalline cubic boron nitride (PCBN) abrasive grains are composed of microcrystalline CBN particles and AlN ceramic binder, which are sintered at high temperature and high pressure. The grains may be engineered to continually wear through microfracture. This tends to produce a continual resharpening action with a low overall rate of wear. Therefore, PCBN abrasive grains overcome some of the disadvantages of monocrystalline CBN grains, such as the fracture along the well-defined crystal planes on a large scale under a compressive load. The excellent mechanical properties make that polycrystalline CBN abrasive grains have a promising future in the grinding process of difficult-to-cut materials.

Considering the great potential of the PCBN abrasive grains in terms of their continual resharpening action, single-layer brazed PCBN abrasive wheels have been suggested as candidates for brazed monocrystalline CBN abrasive wheels. Different from the monocrystalline CBN grains, the PCBN abrasive grains are composed of microcrystalline CBN particles and AlN binder material. Therefore, not only CBN but also AlN may react with the filler alloy in the brazing process. To address this issue, a detailed investigation is performed on the multiple chemical reactions and the interfacial microstructure between PCBN abrasive grains and the Ag-Cu-Ti filler alloy. Meanwhile, the compressive strength of the brazed grains is evaluated.

4.2 BRAZED POLYCRYSTALLINE CBN ABRASIVE GRAINS WITH AG-CU-TI FILLER ALLOY

4.2.1 Appearance and Joining Interface of Brazed Polycrystalline CBN Grains

The size of the PCBN abrasive grains applied here was between 300 and 400 μm. In particular, the contents of CBN and AlN were about 80 and 20 wt.%, respectively. $(Ag_{72}Cu_{28})_{95}Ti_5$ (wt.%)

DOI: 10.1201/9781032678047-5

filler alloy was utilized. AISI 1045 steel was chosen as the abrasive tool hub, which was cut into pieces with a diameter of 15 mm and a height of 5 mm. The specimens were prepared in a sandwich structure of tool matrix/filler alloy/abrasive grains. The brazing temperatures were 890°C, 900°C, and 920°C, respectively, which was the frequently used heating temperature for the Ag-Cu-Ti filler alloy. The holding time was 8 min. During the brazing process, the vacuum in the heating furnace modelled VAF-20 was not worse than 1×10^{-2} Pa.

The typical morphology of the brazed samples processed at 900°C is displayed in Figure 4.1. Good wetting behaviour of the molten Ag-Cu-Ti alloy has taken place on the abrasive grains. Similar morphology has also been exhibited on the samples brazed at 890°C and 920°C, respectively. The microstructure and the correspondent elemental distribution across the joining interface of polycrystalline CBN abrasive grains and the Ag-Cu-Ti filler

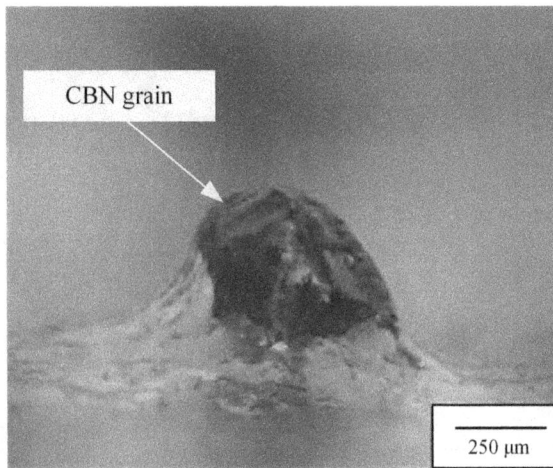

FIGURE 4.1   Morphology of the sample brazed at 900°C.

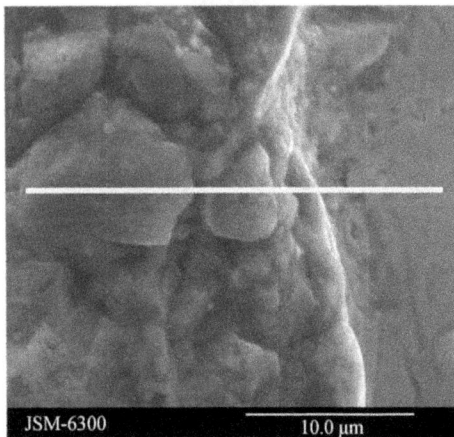

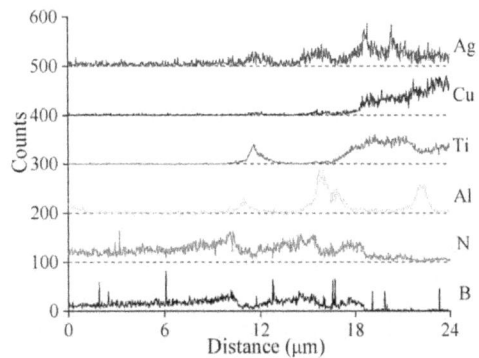

(a)

(b)

FIGURE 4.2   Joining interface and elemental distribution between polycrystalline CBN abrasive grains and Ag-Cu-Ti filler alloy: (a) joining interface, (b) elemental distribution.

layer are displayed in Figure 4.2. According to Figure 4.2b, not only the Ti atoms but also the Al ones have preferentially penetrated towards the joining interface in the exact opposite direction. At the same time, the B and N atoms coming from microcrystalline CBN particles have been gathered at the joining interface. The atomic diffusion across the interface has implied the possibility of chemical reactions among the elements Ti, Al, B, and N.

4.2.2 Chemical Resultants around Brazed Polycrystalline CBN Grains

To detect the morphology of the chemical resultants at the joining interface of the abrasive grains and the filler layer, the brazed samples were particularly prepared by covering the polycrystalline CBN abrasive grains with Ag-Cu-Ti filler alloy powder. Then, the brazed grains were separated from the samples in the etching solution of nitride acid and hydrofluoric acid. Thus, the reaction products remained on the surface of the brazed

FIGURE 4.3 Chemical resultants around the polycrystalline CBN abrasive grains brazed at 900°C: (a) single grain, (b) outer compounds, (c) inner compounds, (d) magnified inner compounds.

polycrystalline CBN abrasive grains. Figure 4.3a–c show the representative morphology of the chemical resultants around polycrystalline CBN abrasive grains brazed at 900°C. Compact columnar compounds have been formed. This reveals that the brazing reaction of abrasive grains and the Ag-Cu-Ti alloy is enough. It is mainly due to the fact that titanium preferentially penetrates the grain boundaries, resulting in finger-like compound intrusions into the filler layer, which covers the grain in a resultant shell. Figure 4.3d displays more details of the inner parts of the reaction zone than Figure 4.3c. At the very interface around the microcrystalline CBN particle, a continuous resultant layer is observed. A particular orientation relationship exists at the interface of the CBN particle and the chemical resultants. This is always correlated with the minimum lattice mismatch.

The X-ray diffraction technique is employed as a supplementary method of phase identification of the reaction products around the brazed polycrystalline CBN abrasive grains. The result is demonstrated in Figure 4.4. Since the thickness of the resultant layer is much smaller than the size of the abrasive grains, not only the phases of the reaction products but also those of CBN and AlN are detected.

As seen in Figure 4.4, the compounds mainly consist of TiB_2, TiB, TiN, and Ti_3AlN. In particular, the formation of the Ti-borides and some Ti-nitrides is mainly determined by the reaction of the CBN particles and Ti in the Ag-Cu-Ti filler alloy. Ti_3AlN and other Ti-nitrides are produced dependent on the chemical reaction between the AlN binder material and the Ag-Cu-Ti filler alloy. The joining of the composite grits and the filler layer is indeed due to the brazing resultants across the grit-filler interface.

The typical scanning electron microscope (SEM) images of the chemical resultants on the surface of polycrystalline CBN abrasive grains brazed at 890°C are shown in Figure 4.5. Though some columnar compounds have grown up around the grains, the quantity and shape of the reaction products are smaller than those of the compounds formed at the brazing temperature of 900°C. This indicates that the brazing reaction at 890°C is much weaker than that at 900°C. Figure 4.6 displays the micrograph of the reaction products around the polycrystalline CBN abrasive grains brazed at 920°C. It is found that the interfacial reaction at this time is more drastic than that at 900°C, which destroys the sharp

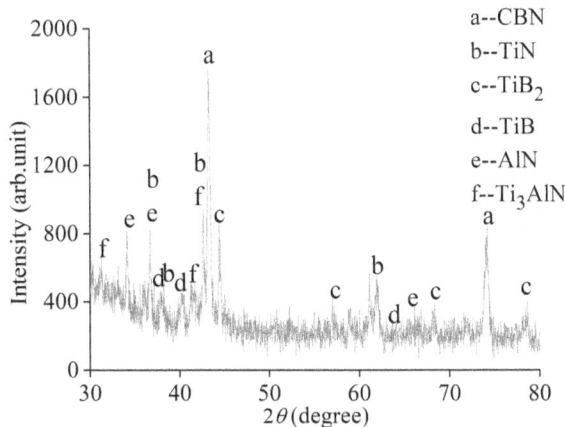

FIGURE 4.4   XRD patterns of the polycrystalline CBN abrasive grains brazed at 900°C.

FIGURE 4.5 Chemical resultants around the polycrystalline CBN grains brazed at 890°C: (a) brazed grain, (b) chemical resultants.

FIGURE 4.6 Chemical resultants around the polycrystalline CBN grain brazed at 920°C: (a) brazed grain, (b) chemical resultants.

grain edges and perhaps degrades the machining performance of the brazed polycrystalline CBN abrasive tools.

4.2.3 Brazing Mechanism of Polycrystalline CBN Grains

The brazing of polycrystalline CBN abrasive grains and Ag-Cu-Ti filler alloy is mainly dependent upon the chemical reaction of CBN-Ti and AlN-Ti couples. On one hand, the chemical composition of the molten Ag-Cu-Ti filler alloy has an important influence on the activity of the Ti element in terms of thermodynamics. According to the reported data of the ternary Ag-Cu-Ti system (Paulasto et al. 1995; Liao et al. 2022), the relationship between the Ti activity a_{Ti} and the chemical composition of the Ag-Cu-Ti alloy at 1,223 K (or 950°C) is given by (Qu et al. 2003):

$$\ln a_{Ti} = 3.102(1 - x_{Ti})^2 + 1.505(x_{Cu})^2 - 1.251 x_{Cu}(1 - x_{Ti}) + \ln x_{Ti} \tag{4.1}$$

where the parameters x_{Cu} and x_{Ti} represent the contents of Cu and Ti in the molten Ag-Cu-Ti alloy, respectively.

At the same time, the activity of the Ti element at different temperatures may also be described as

$$\ln a_2 = \frac{T_1}{T_2} \ln a_1 + \frac{T_2 - T_1}{T_2} \ln x_1 \tag{4.2}$$

where T_1 is 1,223 K, and T_2 is the calculated absolute temperature.

Combined with Eqs. (4.1) and (4.2), the activity of the Ti element in $(Ag_{72}Cu_{28})_{95}Ti_5$ (wt.%) alloy at different temperatures is obtained, as displayed in Figure 4.7. As could be seen, the temperature affects Ti activity, which is one of the reasons for the difference between the brazing reaction and the correspondent resultants. Note that the kinetic factor is another reason for the brazing reaction.

On the other hand, the previous investigation has stated that, because of the high chemical affinity of titanium to nitrogen, the reaction leading to the formation of TiN is the dominant one among all the corresponding reactions of the couple between Ti and cubic BN, as follows (Ding et al. 2006):

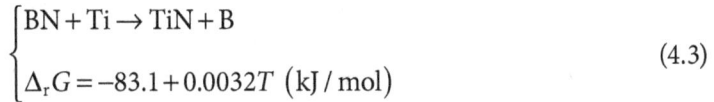

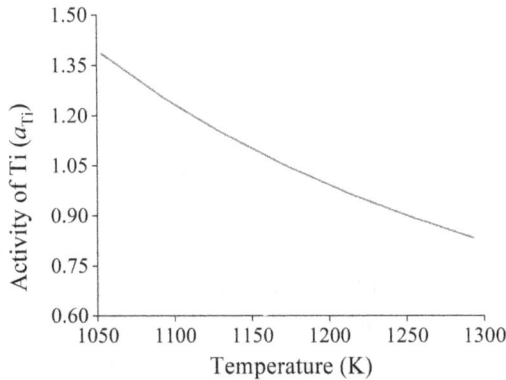

$$\begin{cases} BN + Ti \rightarrow TiN + B \\ \Delta_r G = -83.1 + 0.0032T \ (kJ/mol) \end{cases} \tag{4.3}$$

FIGURE 4.7   Activity of Ti in $(Ag_{72}Cu_{28})_{95}Ti_5$ (wt.%) filler alloy.

Gibbs free energy of this reaction, $\Delta_r G$, is negative (approximately 79.35 kJ/mol at 900°C). Therefore, the TiN crystals have been nucleated first around the surface of microcrystalline CBN particles and grown into the molten Ag-Cu-Ti alloy. Then, TiB_2 and TiB were formed in the subsequent reactions.

Also, considering the reaction couple of AlN binder and reactive Ti, the ternary phases may be formed, such as a cubic TiN phase (NaCl type), a cubic Ti_3AlN phase (perovskite type, τ phase) with the average composition $Ti_3Al_{0.8}N_{0.8}$, and a hexagonal Ti_3Al phase (Ni_3Sn type) (Zhang et al. 2023; Li et al. 2021; Qian et al. 2023). The presence of two layers has been confirmed at the brazing interface of AlN ceramics and 63Ag-35.3Cu-1.7Ti (wt.%) at 900°C. The resultant layer adjacent to AlN is TiN, with some intergranular webbing of element Cu and isolated Cu and Ag. The second layer is identified as a compound of the $(TiCuAl)_6N$ type. Consequently, it is also feasible to form Ti_3AlN and Ti-nitrides at the brazing interface of polycrystalline CBN grains and Ag-Cu-Ti filler alloy.

4.3 BRAZED POLYCRYSTALLINE CBN ABRASIVE GRAINS WITH CU-SN-TI FILLER ALLOY

4.3.1 Experimental Details

Cu-Sn-Ti alloy is also gaining importance as an active brazing alloy for joining various covalent-bonded materials to metallic materials due to its high melting point, good mechanical properties, and low cost. It has been thought of as an effective alternative to the Ag-Cu-Ti alloy for single-layer brazed abrasive tools (Ding et al. 2013). Some research has been done, for example, on the thermodynamic assessment of the Cu-Sn-Ti ternary system, the microstructural development of the Cu-Sn-Ti alloys on graphite, the surface and transport properties of the Cu-Sn-Ti alloys, and the spreading behaviour of the Cu-Sn-Ti alloy on vitreous carbon and alumina.

The present Investigation intends to research the brazing behaviour of Cu-Sn-Ti alloy on polycrystalline CBN grain. The commercially obtained pre-alloyed powders with chemical contents of Cu 72 wt.%, Sn 18 wt.%, and Ti 10 wt.% were utilized in the present investigation. Figure 4.8a displays an SEM micrograph of Cu-Sn-Ti alloy powder. The compositional analysis based on energy-dispersive X-ray spectrometry (EDS) showed that the spherical particles (Point A) were mainly composed of Cu and Sn, while others (Point B) mainly consisted of Ti. Figure 4.8b presents the differential thermal analysis (DTA) curve, which indicated that the solidi and liquidi of the Cu-Sn-Ti alloy were 790°C and 810°C, respectively, according to the two endothermic peaks. Polycrystalline CBN grains having particle sizes ranging from 355 to 425 μm were adopted. The size of microcrystalline CBN particles within the grains was 1–3 μm, as displayed in Figure 4.9.

The brazing process was carried out in the vacuum furnace modelled VAF-20 under a high vacuum (better than 10^{-2} Pa). The specimens were heat-treated to the brazing temperature (880°C, 900°C, 920°C), isothermally soaked for 8 min at the brazing temperature, and then cooled down to room temperature at a rate of 10°C/min. Figure 4.10 displays the morphology of polycrystalline CBN after brazing at 900°C for 8 min. It is obvious that good wetting behaviour of the molten Cu-Sn-Ti alloy has occurred on the polycrystalline CBN grains.

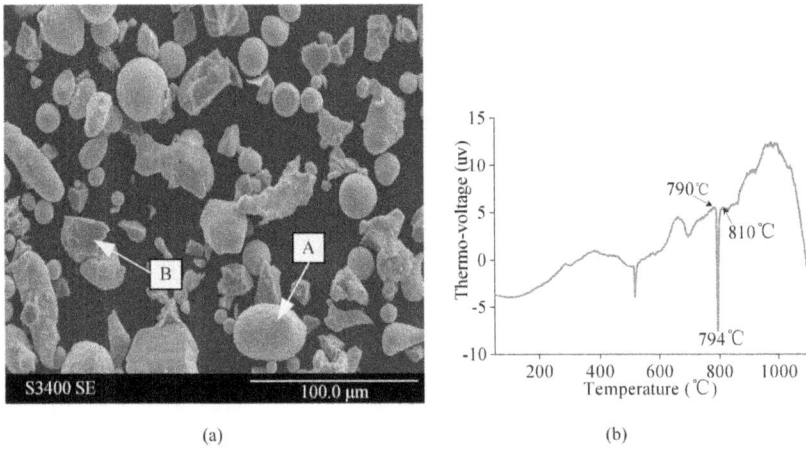

(a)   (b)

FIGURE 4.8   Morphology and DTA analysis of Cu-Sn-Ti brazing alloy powder: (a) morphology, (b) DTA analysis.

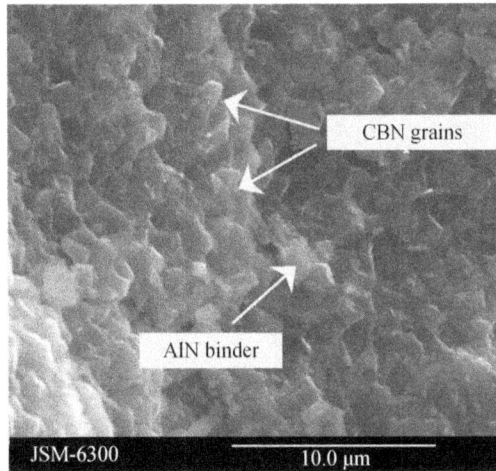

FIGURE 4.9   Microstructure of polycrystalline CBN grain.

FIGURE 4.10   Typical morphology of brazed polycrystalline CBN specimen with Cu-Sn-Ti alloy.

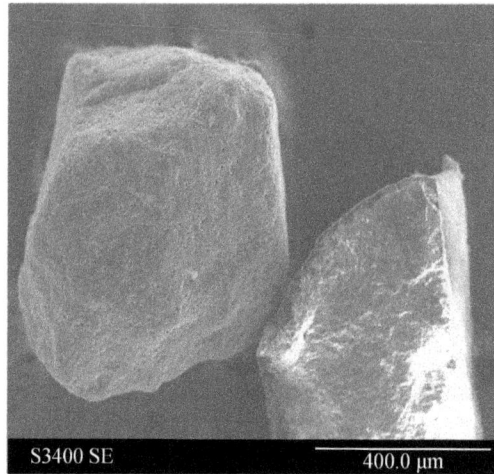

FIGURE 4.11   Brazed polycrystalline CBN grains after etching.

Some brazed specimens were etched in strong acid to separate the brazed polycrystalline CBN grains, as displayed in Figure 4.11. The static compressive strength of the brazed grains was measured using ZMC-II static strength measuring equipment. The measuring range was 1–100 N, and the force error was ±1%. The loading rate was 0.5 mm/min. Forty grains were measured for each brazing parameter. The arithmetic mean value P was calculated. In this experiment, the static compressive strength σ (MPa) of the grains was determined by the following relation:

$$\sigma = 1.37 P / d^2 \qquad (4.4)$$

where P is the mean value of the maximum compressive load (N) and d is the diameter of the polycrystalline CBN grain (mm).

4.3.2 Interfacial Microstructure of Brazed Polycrystalline CBN Grains

The SEM micrograph in Figure 4.12a shows an overview of the whole brazing gap of a specimen brazed at 900°C for 8 min. The joining interface is microstructurally sound, well-bonded, and devoid of imperfections such as voids or cracks. The energy-dispersive X-ray spectrometry elemental linear distribution of the important elements of the joint is displayed in Figure 4.12b. The curves connecting the individual elements are construed as suggesting a trend. Preferential segregation of interphases at the alloy/steel interface and grain/alloy interface is noted in the joint. At interface I (the interface between the steel matrix and Cu-Sn-Ti alloy), a titanium-containing intermetallic interlayer, such as (Cu, Sn) matrix and (Fe, Ti) matrix, is developed. In the reaction between the interlayers, a brazing alloy with different intermetallic phases may be formed. Interface II (the interface between Cu-Sn-Ti alloy and polycrystalline CBN grain) is enriched in Ti and also displays a small percentage of Al, N, and B. The higher Ti concentration in the vicinity of the polycrystalline CBN grain surface implies that Ti from the Cu-Sn-Ti alloy has segregated. This phenomenon is consistent with the high chemical affinity of Ti towards Al, N, and B, which

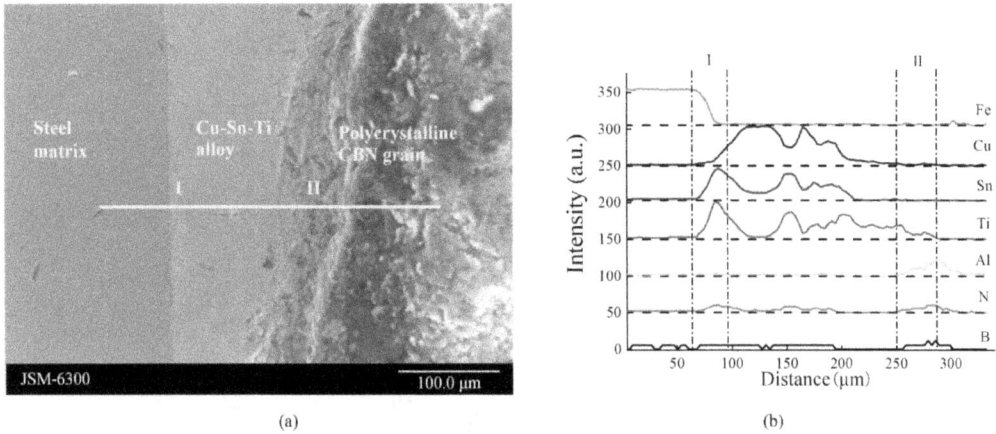

FIGURE 4.12 Joining interface of steel matrix/Cu-Sn-Ti alloy/polycrystalline CBN grain: (a) interfacial microstructure, (b) elemental distribution curves.

is known to cause beneficial near-interfacial changes that promote wetting and joining. During brazing, the redistribution of Ti through diffusion within the joint is a function of the brazing temperature and holding time. Because some of the nitrides and borides that formed may not be stable, the active metallic element Ti of the molten Cu-Sn-Ti alloy diffuses faster and forms the compounds Ti-Al, Ti-B, and Ti-N.

Figure 4.13 shows the X-ray diffraction patterns of the brazed polycrystalline CBN grains. The compounds mainly contain CBN, AlN, TiN, TiB_2, TiB, and $TiAl_3$. The phases of CBN and AlN originate from the polycrystalline CBN grains. Other phases of TiN, TiB_2, TiB, and $TiAl_3$ are generated in the brazing process, which conforms to the chemical combination developed between polycrystalline CBN grains and Cu-Sn-Ti alloy. The chemical reaction between polycrystalline CBN grains and Cu-Sn-Ti alloy in the brazing process can be summarized as follows:

For the reaction of AlN binder material and Cu-Sn-Ti alloy, at the early stage of brazing, active Ti reacts with AlN to form a Ti solid solution containing elemental Al and N, as described by:

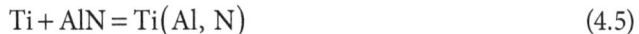

$$Ti + AlN = Ti(Al, N) \tag{4.5}$$

Subsequently, at the AlN/Ti interface, TiN and Ti_3AlN are formed adjacent to AlN, as described by:

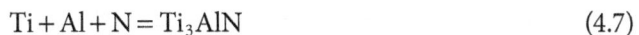

$$Ti + N = TiN \tag{4.6}$$

$$Ti + Al + N = Ti_3AlN \tag{4.7}$$

Further, Al diffuses through TiN and Ti_3AlN to Ti to form Ti_3Al intermetallic compounds adjacent to Ti. A noticeable increase in thickness by increasing holding time and brazing temperature takes place for the Ti_3Al layer. Previous research has revealed that, during brazing AlN ceramic material and Ti metal, the activation energy for the Ti_3Al growth

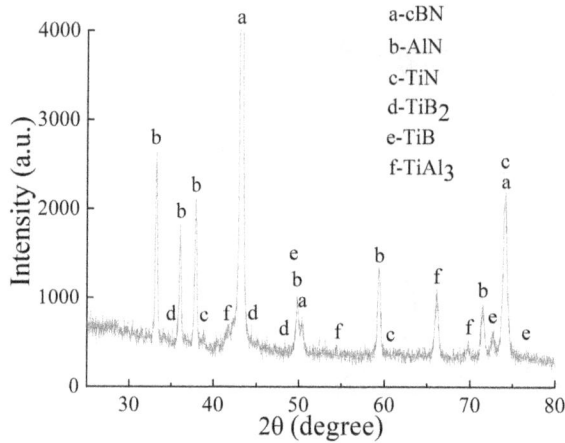

FIGURE 4.13   XRD patterns of the resultants of polycrystalline CBN grains brazed at 900°C for 8 min.

(146 kJ/mol) is comparable to the activation energy for Ti diffusion (169 kJ/mol) and that of Al diffusion (115 kJ/mol) in titanium. In particular, the presence of a diffusion barrier consisting of Ti-nitrides (TiN or Ti_3AlN) at the AlN/Ti interface retards Al diffusion into the Ti central layer of the Ti/AlN joining interface.

The reaction between CBN particle material and Ti in the Cu-Sn-Ti alloy is similar to the reaction between CBN and Ti in the Ag-Cu-Ti alloy.

4.3.3 Resultant Morphology of Brazed Polycrystalline CBN Grains

Figure 4.14 shows the SEM micrograph of the surviving particles of the brazed polycrystalline CBN grains after the compressive tests. No microcrystalline CBN particle is exposed on the outer grain surface but in the inner. The reason is that the polycrystalline CBN grain is entirely covered by a continuous resultant layer, whose cross-section image is observed at the edge of the fractured grain. According to the X-ray diffraction pattern analysis, it is known that the continuous resultant layer mainly consists of the compounds TiN, TiB_2, TiB, and $TiAl_3$.

As seen in Figure 4.14a, when the brazing temperature is 880°C, the thickness of the resultant layer with a flat lamellar structure is 2–3 μm. When the brazing temperature is elevated to 900°C, the corresponding thickness is increased to 3–4 μm. Under such circumstances, the resultants grow rather compactly around the grain so that the boundary is not obvious among them, as displayed in Figure 4.14b. When the brazing temperature is increased further to 920°C, the thickness of the resultant layer reaches 4–5 μm (Figure 4.14c). Moreover, the compounds grow promiscuously, which corresponds to the drastic chemical reaction at 920°C. The coarse resultants may weaken the joining strength between the brazed polycrystalline CBN grain and the Cu-Sn-Ti alloy. Therefore, when brazed at 900°C, the patterns and thickness of the resultant layer are a great benefit to the good joint of polycrystalline CBN grains/Cu-Sn-Ti alloy/steel matrix. This is due to the variation in the mean value of the static compressive strength of the grains brazed at different temperatures. In the present investigation, because of the tiny size

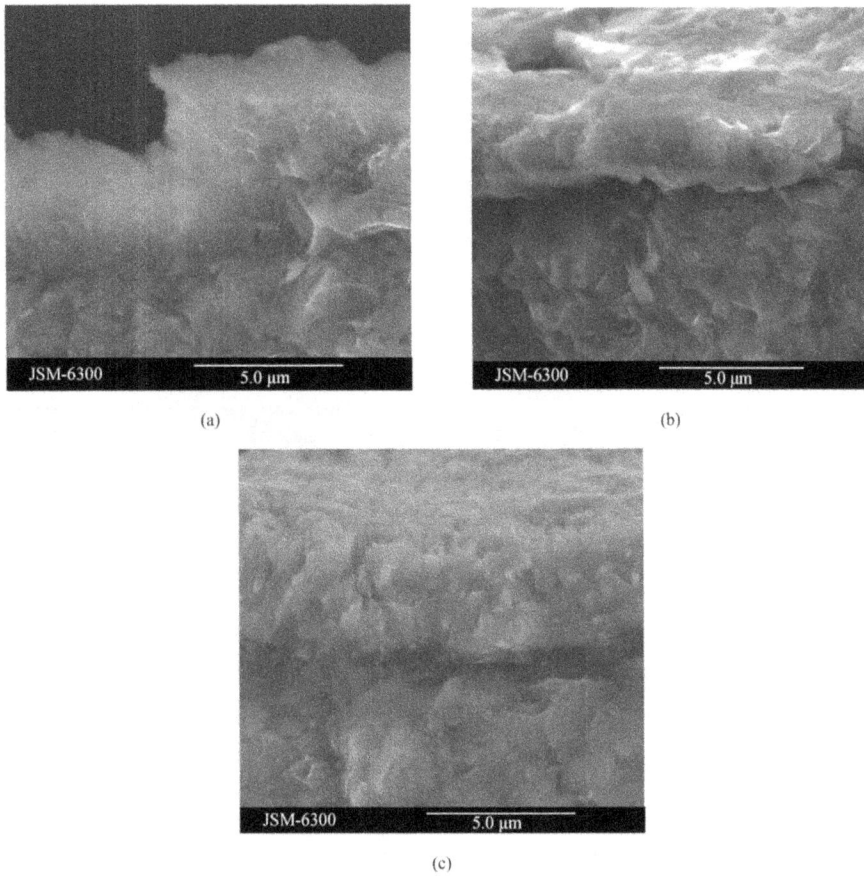

FIGURE 4.14 Resultants morphology of brazed polycrystalline CBN grains: (a) 880°C, (b) 900°C, (c) 920°C.

and compact degree of the brazing resultants, it is rather difficult to distinguish the different compounds within the resultant layers.

The mechanical strength of the brazed grains is affected comprehensively by the enhancing effect of the chemical resultants and the reducing effect of the thermal damage in the brazing process. In particular, the thermal damage is caused by the difference in thermal expansion coefficients of the microcrystalline CBN particles and the AlN binder of the polycrystalline CBN grains. On the one hand, the resultants formed around the brazed grain fill up the cracks, craters, and other deficiencies of the original grains to a certain degree. As a consequence, the mechanical strength of the polycrystalline grains brazed at 880°C is reinforced and shows an upward tendency against the original grains. The mean value of the grain compressive strength brazed at 900°C reaches the highest value because of the enhancing effect of the chemical resultants. On the other hand, when the grains are brazed at 920°C, the thermal damage plays a dominantly negative role in the fracture strength of the brazed polycrystalline grains. Therefore, the mean value of the grain fracture strength has a decreasing tendency.

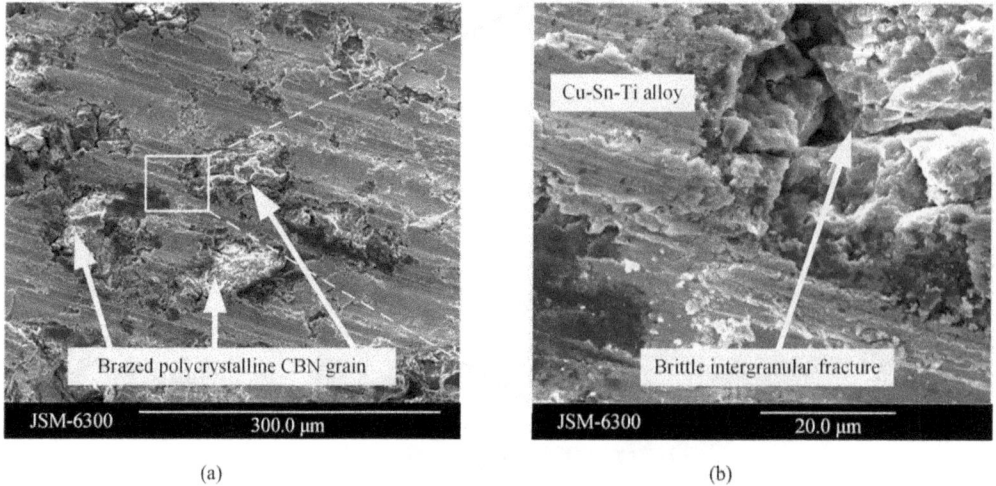

FIGURE 4.15 Fracture morphology of the brazed polycrystalline CBN grain during grinding: (a) whole, (b) regional.

4.3.4 Joining Performance of Brazed Polycrystalline CBN Grains

To evaluate the joining performance of polycrystalline CBN grains/Cu-Sn-Ti alloy/steel matrix, a grinding test of a brazed polycrystalline CBN abrasive tool was conducted with a fixed normal load until the grain failed. The SEM micrographs of the fracture surface of the grain brazed at 900°C for 8 min are presented in Figure 4.15. According to the literature (Buhl et al. 2012), four different fracture behaviours and corresponding mechanisms may have an effect on the brazed joints. They are partially ductile shearing of the brazing alloy, brittle fracture in the intermetallic interlayer, failure in the brazing resultant layer, and the partial destruction of the polycrystalline grain as well. As seen in Figure 4.15, the fracture happens in the brittle intermetallic interlayer and directly at the grain/alloy interface. Additionally, the brazed polycrystalline CBN grain is also partly destroyed. For the brazing temperatures of 880°C and 920°C, the fracture surface of the brazed grains looks similar to that in Figure 4.15a. This phenomenon suggests that the joining strength between the grains and the Cu-Sn-Ti alloy is stronger than the strength of the brazing alloy. As a consequence, the requirement for high joining strength in polycrystalline CBN grains is met. In particular, the brittle intergranular fracture along the CBN/CBN crystal particle boundary is the main fracture pattern of brazed polycrystalline CBN grains, as shown in Figure 4.15b.

4.3.5 Compressive Strength of Brazed Polycrystalline CBN Grains

The static compressive tests have been conducted to measure the grain strength in the present investigation. Figure 4.16a displays the mean strength value of four groups of polycrystalline CBN grains brazed at different heating temperatures. Forty grains were chosen at random to test for every group. The brazing temperature has an impact on the mean strength of polycrystalline CBN grains to a certain extent. For the original grains, the mean value of the compressive strength is 770.6 MPa. When the brazing temperature is

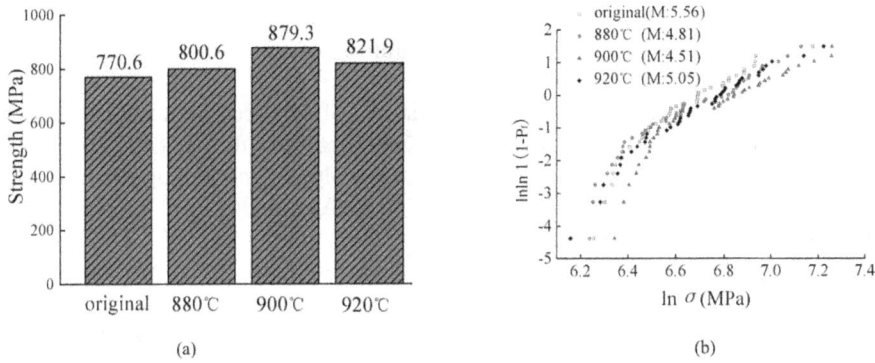

FIGURE 4.16 Static compressive strength of brazed polycrystalline CBN grains: (a) comparison of compressive strength, (b) Weibull distribution of compressive strength.

880°C, the mean strength of the polycrystalline CBN grains is 800.6 MPa. The mean value of the compressive strength increases while elevating the brazing temperature to 900°C, at which the maximum value (879.3 MPa) is obtained. The mean strength decreases to 821.9 MPa when the brazing temperature is elevated to 920°C. Generally, the strength of brazed polycrystalline CBN grains is higher than that of the original grains.

On the other hand, the polycrystalline CBN grains are brittle with surface cracks, pores, and grain boundaries. Thus, the Weibull weak link theory of fracture mechanics is always applied to investigate the strength distribution (Weibull et al. 1951). The Weibull strength distribution determined from the static compressive tests for the four groups of polycrystalline CBN grains is plotted in Figure 4.16b. The Weibull modulus M is also listed. Obviously, the mean compressive strength and the moduli M corresponding to the four groups of polycrystalline CBN grains appear to be dependent on the brazing temperature. According to the relevant literature (Ichida et al. 2010), the larger the Weibull modulus M, the more drastic the grain strength distribution is. Therefore, it is known that the difference in Weibull modulus among the four groups of polycrystalline CBN grains indicates that the strength distribution of the original grains is the most stable, while the fluctuation of the strength distribution of the grains brazed at 900°C is the strongest.

REFERENCES

Buhl, S., C. Leinenbach, and R. Spolenak. 2012. Microstructure, residual stresses and shear strength of diamond-steel-joints brazed with a Cu-Sn-based active filler alloy. *International Journal of Refractory Metals & Hard Materials*, 30(1): 16–24.

Ding, W. F., J. H. Xu, Z. Z. Chen, et al. 2013. Interface characteristics and fracture behavior of brazed polycrystalline CBN grains using Cu-Sn-Ti alloy. *Materials Science and Engineering: A*, 559: 629–634.

Ding, W. F., J. H. Xu, M. Shen, et al. 2006. Thermodynamic and kinetic analysis of interfacial reaction between CBN and titanium activated Ag-Cu alloy. *Materials Science and Technology*, 22(1): 105–109.

Ichida, Y., M. Fujimoto, Y. Inoue, et al. 2010. Development of a high performance vitrified grinding wheel using ultrafine-crystalline cBN abrasive grains. *Journal of Advanced Mechanical Design, Systems, and Manufacturing*, 4: 1005–1014.

Li, Z., W. Fu, S. Hu, et al. 2021. First-principles calculations of the effects of the interface microstructure on the wettability of a Cu-Ti/AlN system. *Ceramics International*, 47(13): 18592–18601.

Liao, X., Z. Liu, R. Liu, and D. Mu. 2022. Microstructure evolution and joining strength of diamond brazed on Ti-6Al-4V substrates using Ti-free eutectic Ag-Cu filler alloy. *Diamond and Related Materials*, 127: 109198.

Paulasto, M., F. J. J. Loo, and J. K. Kivilahti. 1995. Thermodynamic and experimental study of Ti-Ag-Cu alloy. *Journal of Alloys and Compounds*, 220: 136–141.

Qian, H., Y. Han, K. Zhang, et al. 2023. The dependence of microstructure and mechanical properties on substrate heat treatment in AlN ceramics/AgCuTi/316 stainless steel brazed joints. *Vacuum*, 213: 112094.

Qu, S. Y., Z. D. Zou, and X. H. Wang. 2003. Thermodynamic analysis of a Ag-Cu-Ti active brazing alloy. *Transactions of the China Welding Institution*, 24(4): 13–16.

Weibull, W. 1951. A statistical distribution function of wide applicability. *Journal of Applied Mechanics*, 18: 293–297.

Zhang, J., J. Xu, J. Huang, et al. 2023. High shear strength Kovar/AlN joints brazed with AgCuTi/Cu/AgCuTi sandwich composite filler. *Materials Science and Engineering: A*, 862: 144435.

CHAPTER 5

Brazing of CBN Abrasive Grains Based on High-Frequency Heating

5.1 INTRODUCTION

As for the fabrication techniques of single-layer brazed cubic boron nitride (CBN) abrasive wheels, different thermal sources have their own characteristics and advantages. For example, the high-frequency induction brazing technique has the great advantages of a rapid heating rate and local heating. Particularly, it is more suitable to fabricate the brazed CBN wheels with a large size (i.e., 400 mm in diameter) due to the fact that the large abrasive wheel could not be brazed in the vacuum resistance furnace because of the limited chamber size and significant deformation of the metallic wheel substrate, and the quality control of the laser brazing is complex and also very difficult. It is noted that, however, there still exist some problems when fabricating the single-layer brazed CBN abrasive wheels based on high-frequency induction heating techniques, the reason for which is the insufficient understanding of the heating temperature distribution. Under such conditions, the temperature control and heating parameter optimization could not be conducted well, which resulted in the bad joining among the metallic wheel substrate, Ag-Cu-Ti filler alloy, and CBN abrasive grains (Li et al. 2015). Therefore, it is necessary to deeply understand the temperature distribution and influencing factors during high-frequency induction brazing of CBN grains (Xu et al. 2017).

Experimental detection is still an important method to study the temperature distribution in induction heating. For instance, Codrington et al. (2009) developed the high temperature measuring equipment with which the thermocouples are used to measure the heating temperature and provide a feedback signal for PID control of the induction heater. Jang et al. (2007) applied the thermocouples and infrared thermal imaging systems, respectively, to measure the temperatures at the inside and outside surfaces of a hollow steel cylinder during the induction heating process. Franco et al. (2012) proposed an infrared sensor and detected real-time temperature evolution. However, all the reported

DOI: 10.1201/9781032678047-6

experimental techniques could merely measure the temperature at one point or a rather local zone within the induction-heating zone and cannot characterize the temperature distribution within the whole heating zone. In fact, besides the experimental method, the finite element (FE) simulation method has also been broadly applied to investigate quantitatively the temperature distribution in the current heating analysis. Chen et al. (2004) established an FE model of a 2-inch SiC growth system and investigated the effects of induction heating parameters on temperature distribution. Mei et al. (2011) proposed a heat-generating rate model of slabs in induction heating from a hot rolling plant, and the temperature evolution of the slab was solved by the developed FE code.

For the reasons mentioned above, a FE simulation is used to characterize and understand the temperature distribution in the high-frequency heating system of brazing CBN grains. The effects of some key influencing factors on the resultant temperature are investigated. At the same time, experimental measurement is also carried out to testify the simulation results.

5.2 TEMPERATURE DISTRIBUTION IN HIGH-FREQUENCY BRAZING

5.2.1 Numerical Simulation Principle for Simulating Temperature Distribution

Since the properties of heated materials are temperature-dependent, the induction heating process may be considered the coupling of a steady-state alternating current (AC) electromagnetic field and a transient temperature field. The surface of the heated material is in the AC electromagnetic field, and the induced eddy current field is accordingly produced. The material's surface is heated by the internal heat source, which comes from the induced eddy current field. Under such conditions, the simulation analysis should include both the electromagnetic field and the temperature field, in which the electromagnetic field is governed by Maxwell's equations, and its differential form could be described as follows:

$$
\left\{
\begin{aligned}
&\nabla \times \vec{H} = \vec{J} + \frac{\partial \vec{D}}{\partial t} \\
&\nabla \times \vec{E} = -\frac{\partial \vec{B}}{\partial t} \\
&\nabla \cdot \vec{B} = 0 \\
&\nabla \cdot \vec{D} = \rho_e
\end{aligned}
\right.
\tag{5.1}
$$

where $\vec{E}$ is the electric field intensity vector (V/m or N/C); $\vec{H}$ is the magnetic field intensity vector (A/m); $\vec{A}$ is the conduction current density vector (A/m²); $\vec{D}$ is the electric flux density vector (C/m²); $\vec{B}$ is the magnetic flux density vector (T or N/(A•m)); ρ_e is the electric charge density (C/m³).

The field variables have relational expressions as follows:

$$
\left\{
\begin{aligned}
&\vec{J} = \sigma \vec{E} \\
&\vec{B} = \mu \vec{H}
\end{aligned}
\right.
\tag{5.2}
$$

where μ is the relative magnetic permeability (H/m); σ is the electrical conductivity (S/m).

To solve Eq. (5.1), magnetic vector potential $\vec{A}$ and an electric scalar potential φ are brought into play as follows:

$$\vec{B} = \nabla \times \vec{A} \tag{5.3}$$

$$\vec{E} = -\nabla\varphi - \frac{\partial \vec{A}}{\partial t} \tag{5.4}$$

The differential form of Maxwell's equations is solved by a comprehensive analysis of the above equations. Considering the following:

$$\nabla \times \left(\frac{\nabla \times \vec{A}}{\mu} \right) + \sigma \left(\frac{\partial \vec{A}}{\partial t} + \nabla\varphi \right) = 0 \tag{5.5}$$

where $\vec{A}$ is the magnetic vector potential (Wb/m); φ is the electric scalar potential (V).

The basic equations of the transient temperature field are specified as follows:

$$\vec{\varphi} = -\lambda \nabla \vec{T} \tag{5.6}$$

$$\nabla \cdot \vec{\varphi} + C_v \frac{\partial T}{\partial t} = q_v \tag{5.7}$$

where $\vec{\varphi}$ is the heat flux density vector (W/m²).

The final temperature field equation is solved, given as follows (Zhu et al. 2013; Eastwood et al. 2015):

$$\nabla\left(-\lambda \nabla \vec{T}\right) + C_v \frac{\partial T}{\partial t} = q_v = \rho\left|\vec{J}\right|^2 \tag{5.8}$$

where T is the temperature on the heated material surface; C_v is the volume heat capacity (J/(m³·K)); q_v is the internal heat source density (A/m²).

In the present investigation, the FE simulation processes in a time increment step are clarified as follows: first, solve the eddy current field based on the initial temperature condition; second, use the eddy current field to calculate the heat and determine the temperature field based on thermal analysis; third, update the heated workpiece material properties with temperature and recalculate the temperature in the next time increment step.

5.2.2 Geometry Modelling and Meshes for Simulating Temperature Distribution

The two-dimensional (2D) model of a high-frequency induction heating system applied in this work is displayed in Figure 5.1a. The single-layer brazed CBN abrasive wheel is composed of a commercial AISI 1045 steel substrate, Ag-Cu-Ti filler alloy, and CBN grains. Particularly, the filler alloy and CBN grains spread uniformly on the top surface of the metallic wheel substrate, as shown in Figure 5.1b. The inductor above the CBN wheel

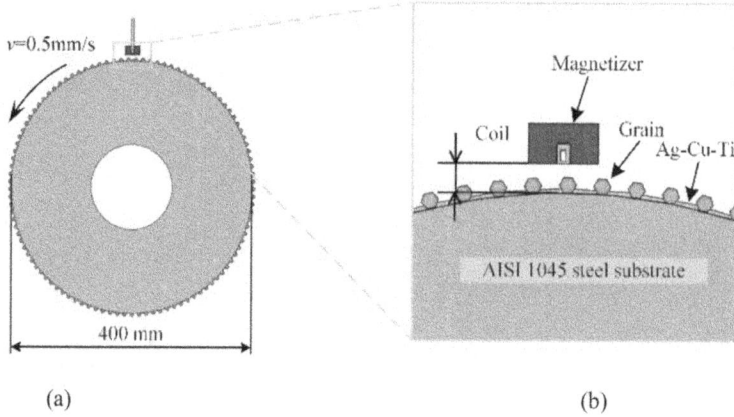

FIGURE 5.1 Geometry model of high-frequency induction heating system: (a) whole geometry model, (b) the inductor and heated wheel surface.

consists of a trough-type magnetizer and a coil. The magnetizer material is Ferrotron 559H. The coil is made from a rectangular copper tube with good electrical conductivity, and cooling water flows through it to keep the coil at a low temperature.

The external and internal diameters of the wheel substrate are 400 and 127 mm respectively; meanwhile, the width of the outer wheel circumference, that is, the width of the heated top surface of the wheel substrate, is 12 mm. The cross-section diagram of the inductor is schematically displayed in Figure 5.1b, in which the coil is 3×2 mm and the magnetizer is 9×12 mm. The thickness of the coil is 0.5 mm.

Considering the FE model established in two dimensions, the electromagnetic edge effect is not considered in the current simulation work. Additionally, because the eddy current only occurs at the surface of the heated part of the metallic wheel substrate, the heat produced in the induction heating only exists on the metallic wheel substrate, while the heat transfers from the metallic wheel substrate to the CBN grains and Ag-Cu-Ti filler alloy, which makes the filler alloy melt smoothly. Therefore, CBN grains and Ag-Cu-Ti filler alloy have no influence on the temperature distribution during induction heating. Under such conditions, a simplification of the induction heating model without CBN grains and Ag-Cu-Ti alloy could be made, like the FE model shown in Figure 5.2.

Figure 5.2a shows the meshes of the whole FE model, which is mainly meshed with the triangular elements. The outer regions including the wheel substrate and the inductor are infinite box and inner air, and the internal and external diameters are 1,000 and 1,200 mm, respectively. The local heating regions are illustrated in Figure 5.2b. The red section is the coil with a hollow structure, while the yellow section covering the wheel is compressible air. The mesh of the heated wheel substrate surface contains two quadrangular elements in the skin depth. The heating zone is meshed by a thin grid with the smallest line element of 0.2 mm. The skin depth is 0.55 mm in the case of a heating temperature of 1,000°C during induction heating. When the heating gap is 2 mm, 145,011 nodes and 70,422 triangular elements are contained in the whole mesh.

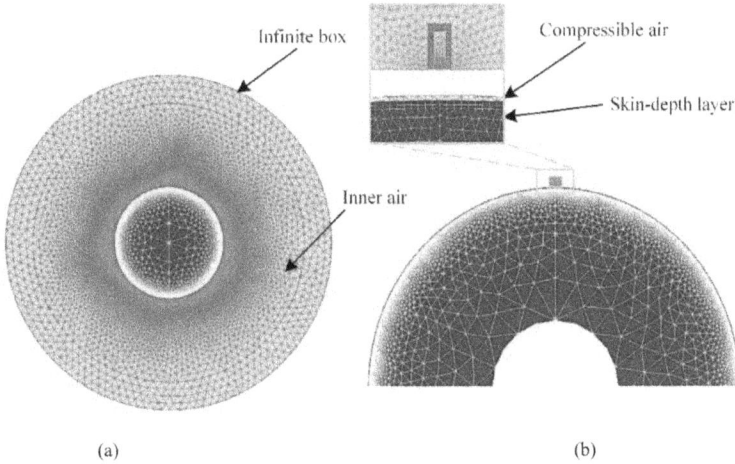

FIGURE 5.2   Meshing of the finite element model: (a) meshing of the whole model, (b) meshing of the coil and wheel top surface.

During the induction brazing process, the melting of the Ag-Cu-Ti filler alloy is completely dependent on the large quantities of heat and the elevated temperatures of the metallic wheel substrate. Previous literature has reported that (Ding et al. 2010) when the heating temperature at the surface layer of AISI 1045 steel substrate arrives at 880°C–940°C (especially at nearly 920°C), a good interfacial reaction could take place, and strong joining is therefore formed among CBN grains, Ag-Cu-Ti alloy, and AISI 1045 steel in the vacuum-furnace brazing process. For this reason, it is necessary to control the heating temperature of the wheel substrate surface, which ranges from 880°C to 940°C in the high-frequency induction brazing process.

5.2.3 Material Properties and Boundary Conditions

The applied material properties (i.e., relative permeability, electrical resistivity, thermal conductivity, and volumetric heat capacity) are taken from the published literature (Sun et al. 2019, 2023; Shen et al. 2014; Norouzifard et al. 2014) or provided by the manufacturers. The magnetic non-linearity and the dependence of physical properties on temperature are considered in the current FE analysis.

The material properties of AISI 1045 steel are shown in Figure 5.3. Particularly, some properties of the materials have been simplified for FE stimulation. For example, the electrical resistivity and thermal conductivity of AISI 1045 steel are provided as linear functions with independent variables of temperature, respectively, as follows:

$$\rho(T)= \rho_0(1+\alpha_1 T) \tag{5.9}$$

$$\lambda(T)= \lambda_0(1+\alpha_2 T) \tag{5.10}$$

where $\rho_0=1.3\times10^{-7}$ Ω·m is the electrical resistivity of AISI 1045 steel at 0°C; $\alpha_1=8.8\times10^{-3}$ is the slope of the straight line expressed by Eq. (5.9); $\lambda_0=446$ W/(m·K) is the thermal

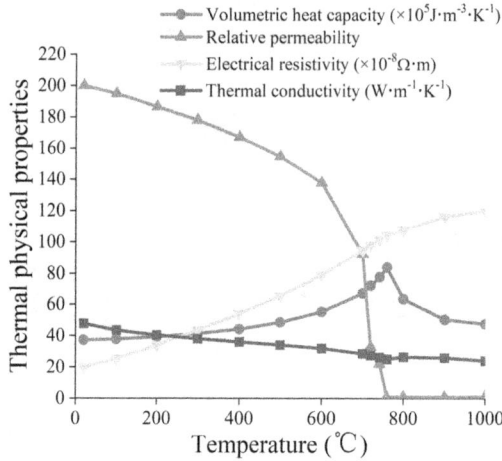

FIGURE 5.3 Material properties of AISI 1045 steel.

conductivity of the wheel substrate at 0°C; $\alpha_2 = -5.3 \times 10^{-4}$ is the slope of the straight line expressed by Eq. (5.10).

In FLUX FE analysis software, the volumetric heat capacity of AISI 1045 steel can be achieved by the curve fitting as the superposition of the Gauss function and exponential function:

$$C_v = C_{vi} + (C_{v0} - C_{vi})e^{-\frac{T}{\tau}} + E \cdot \text{Gauss}(T) \tag{5.11}$$

$$\text{Gauss}(T) = \frac{1}{\sigma_0 \sqrt{2\pi}} e^{-\frac{1}{2}\left(\frac{T-T_c}{\sigma_0}\right)^2} \tag{5.12}$$

where $C_{v0} = 3.5 \times 10^6$ J/(m^3·K) and $C_{vi} = 5.2 \times 10^6$ J/(m^3·K) indicate the volumetric heat capacity of AISI 1045 steel at 0°C and ∞°C, respectively; $E = 6.7 \times 10^8$ J/m^3 is the energy of phase transition of AISI 1045 steel. $\sigma_0 = 85$ is the Gaussian standard deviation, $T_c = 760$°C is the temperature of the phase transition, and $\tau = 400$ is the temperature coefficient.

In FLUX software, the magnetization curve (also named the B-H curve) of AISI 1045 steel can be described by a series of straight lines of different slopes starting from the origin, as illustrated in Figure 5.4. The B-H curves are expressed as follows:

$$B(H) = \mu_0 H + \mu_0(\mu_{r0} - 1)H \cdot \text{COEF}(T) \tag{5.13}$$

$$\text{COEF}(T) = \begin{cases} 1 - e^{\left(\frac{T-T_C}{C_0}\right)}, & T < T_1 \\ e^{10\left(\frac{T_2-T}{C_0}\right)}, & T > T_1 \end{cases} \tag{5.14}$$

where $\mu_0 = 4\pi \times 10^{-7}$ H/m is the permeability of vacuum, $\mu_{r0} = 200$ is the initial relative permeability of AISI 1045 steel; COEF(T) is the temperature coefficient, in which $T_c = 760$°C is the temperature of phase transition, $C_0 = 50$ is the temperature constant, and COEF(T_1) $= 0.1$.

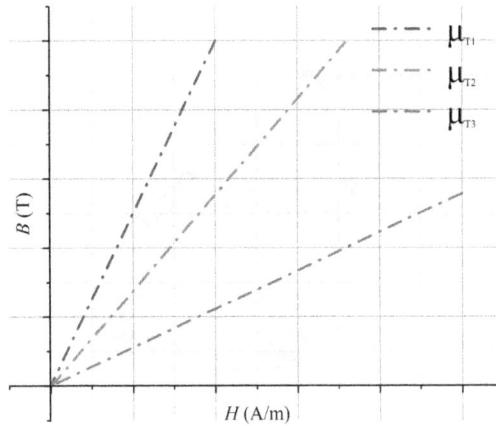

FIGURE 5.4   Diagrammatic B-H curves of AISI 1045 steel.

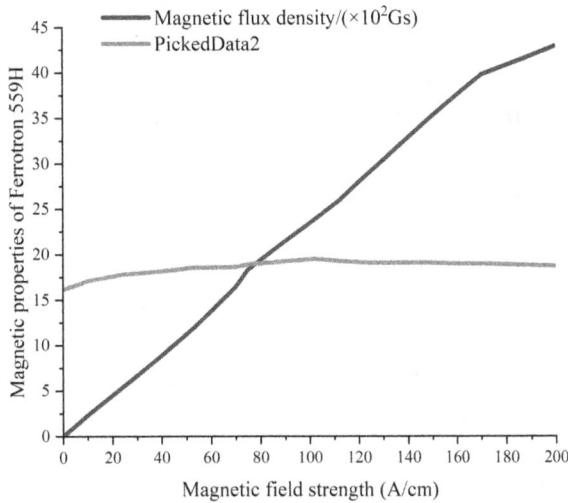

FIGURE 5.5   Magnetic properties of Ferrotron 559H.

The magnetizer is made up of Ferrotron 559H. Its magnetic properties (including flux density and the relative permeability) variation with the magnetic field strength is shown in Figure 5.5 (Jang et al. 2007), and the relative permeability can be approximated as a constant (Li et al. 2013).

The thermal radiation and cross-ventilation are simulated for analysis; the environmental temperature is defined as 20°C. The convective heat transfer coefficient and heat radiation coefficient between the brazing surface of the wheel and surrounding air are defined as 20 W/(m²·K) and 0.5 W/(m²·K⁴), respectively.

5.2.4 Region Selections for Temperature Distribution Analysis

The temperature distribution obtained in induction heating is significantly influenced by four parameters, as follows: current frequency, current magnitude, heating gap, and scanning speed. Figure 5.6 displays the region selection for temperature distribution analysis

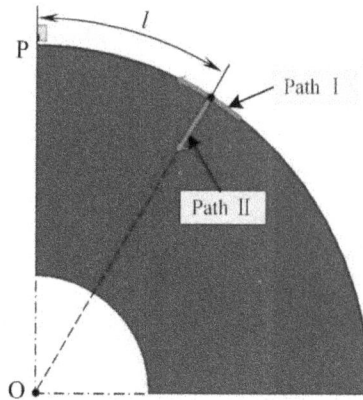

FIGURE 5.6   Point A and two reference paths.

in a quarter of the metallic wheel substrate. Point P is on the outer circumference of the wheel substrate below the inductor at the beginning of heating. Point A is also on the outer wheel circumference, and the arc length l of PA is 90 mm. Two paths, denoted by I and II, are chosen in two different directions, as shown in Figure 5.6. Path I is in the anticlockwise direction with an arc length of 40 mm, and the midpoint of Path I is Point A. Path II is also 40 mm in length and starts from Point A along the radius direction of the wheel.

5.2.5 Simulation Process of Temperature Distribution

During the high-frequency induction brazing process, the temperature distribution characteristics at the effective heating zone are very important to form good joining among CBN grains, Ag-Cu-Ti filler alloy, and metallic wheel substrate (i.e., AISI 1045 steel). Therefore, this work will primarily focus on investigating the temperature variation obtained at Point A within the heating region. Based on the high-frequency induction heating technique, the brazing process could be described as follows:

i. Adjust the distance (that is, the heating gap) between the inductor and the heating surface of the metallic wheel substrate.

ii. Load the electric current with a particular frequency and constant magnitude through the coil and transfer the cooling water passing through the coil.

iii. Rotate the wheel in an anticlockwise direction at a constant speed; in this case, the resultant heat and the elevated temperature of 880°C–940°C could enable that the Ag-Cu-Ti filler alloy melts sufficiently to form good brazed joints between CBN grains and AISI 1045 wheel substrate. The Ag-Cu-Ti filler alloy melts merely within a narrow zone owing to the heating resulted from the heated wheel substrate of AISI 1045 steel.

5.2.6 Typical Contour and Evolution of Temperature Distribution

A typical contour of temperature distribution in the heating zone of the metallic wheel substrate is displayed in Figure 5.7a. The heating gap between the inductor and wheel top

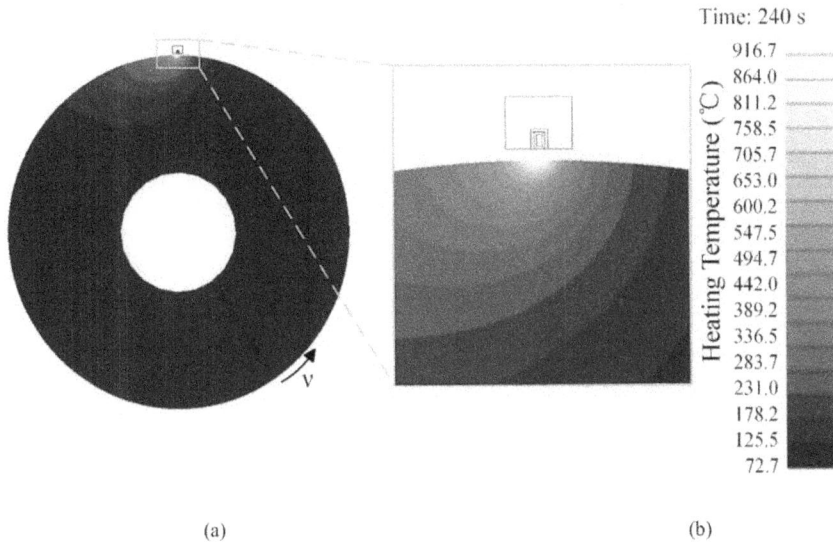

(a)　　　　　　　　　　　　　　(b)

FIGURE 5.7　Finite element simulation results of the typical temperature field: (a) contour maps of the typical temperature field at the heating zone, (b) magnified view of the heating region.

surface is 2 mm, and the wheel is rotated in an anticlockwise direction at a peripheral speed of 0.5 mm/s. Meanwhile, the AC is loaded into the coil with a magnitude of 20 A and a high frequency of 1 MHz. The highest temperature obtained is 916°C when the wheel substrate is heated for 240 s. The curve shown in Figure 5.8a is obtained by calculating the highest temperature of the outer circumference of the wheel substrate during the heating process. The heating procedure could be divided into two stages: the first is an initial temperature-rising stage (0–40 s), in which the temperature rises rapidly from 20°C to 880°C with the decreasing heating rate; the second is an effective heating stage (after 40 s), in which the highest temperature of wheel substrate increases from 880°C, and reaches the temperature of 916°C at 180 s. The highest temperature stays stable at 916°C after 180 s, as shown by the smooth part of the curve in Figure 5.8a.

Figure 5.8b shows the change curve of the resultant heating temperature at Point A, which rotates following the wheel substrate during induction heating. The highest temperature is 916°C at 180.4 s, while the displacement of Point A is 90.2 mm. Point A is 0.2 mm beyond the central coil in the scanning direction and near the highest temperature point. Additionally, the highest temperature appears mainly in the heating region below the inductor, as shown in the contour maps of the heating region in Figure 5.8b.

The temperature distribution along Paths I and II at 180 s is demonstrated in Figure 5.9, while the temperature of Point A below the inductor is 914°C and the highest temperature is 916°C at the peak. As shown in Figure 5.9a, Point A is near the curve peak, and the temperature decreases quickly from the peak to the two ends of Path I. At the same time, the temperature gradient in the positive direction of Path I is slightly smaller compared with that in its negative direction. The lowest temperature is 880°C at the 2 mm interval from 19.2 to 21.2 mm and reduced by merely 4% compared with the highest value; however,

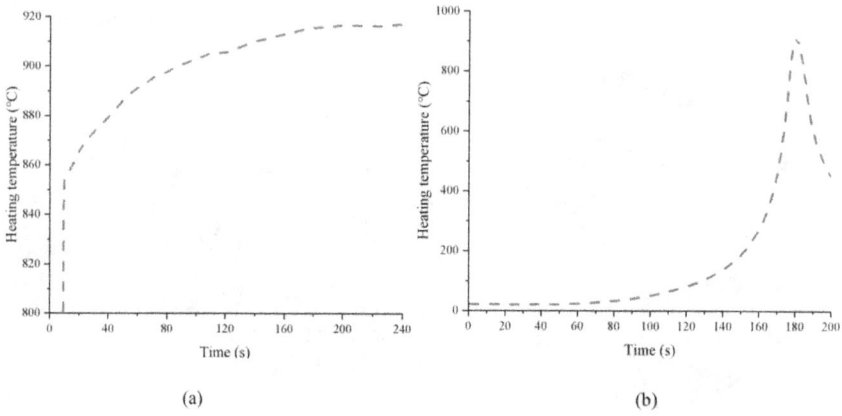

FIGURE 5.8 Change curves of temperature during induction heating: (a) time-varying highest temperature curve in the wheel substrate, (b) change curve of temperature at Point A.

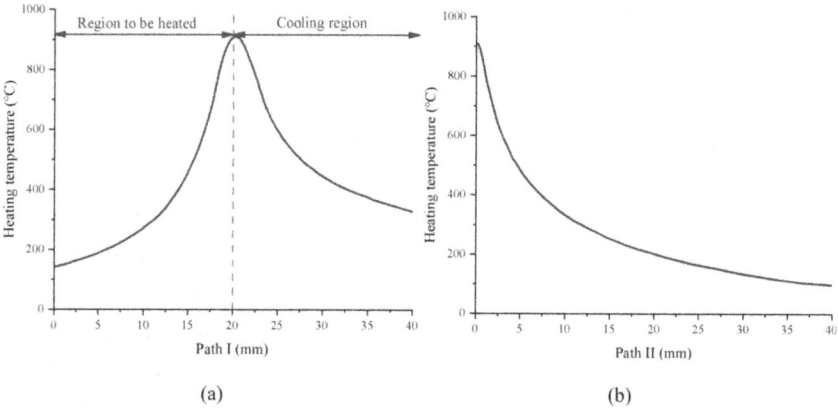

FIGURE 5.9 Temperature distribution along two paths: (a) along Path I, (b) along Path II.

the lowest temperature is 466°C at the 10 mm interval from 15 to 25 mm and reduced by about 49% compared with the highest one. Now it can be inferred that the effective heating region mainly concentrates at the 2 mm interval below the inductor.

The temperature distribution along Path II is affected significantly by the skin effect during high-frequency induction heating. The eddy current is always generated on the top surface of the wheel substrate and greatly focuses on the skin depth. Therefore, the region heated by the eddy current is shallow and focuses on the top surface of the metallic wheel substrate. Figure 5.9b shows the temperature distribution along Path II. The highest temperature, i.e., 914°C, is obtained at Point A. Along the positive direction of Path II, the temperature decreases rapidly from 0 (that is, the starting Point A) to 10 mm, while the temperature gradient becomes smaller gradually from 10 to 40 mm. The temperature is 864°C at 0.55 mm (equal to the skin depth), which is decreased by 6% compared to that at Point A. The temperature is 776°C at 1.1 mm (twice the skin depth), which is decreased by 15% compared to that at Point A. The temperature is 652°C at 2.2 mm (four times the skin depth), which is decreased by 29% compared with that at Point A. Based on the above analysis, it is known that the depth of the dominating heating zone is equal to the skin depth on the top surface of the wheel substrate.

5.2.7 Effects of Induction Heating Parameters on Temperature Distribution

5.2.7.1 Current Frequency

To analyse the effects of the current frequency on the resultant temperature distribution, the other induction heating parameters are chosen as follows: the heating gap of 2 mm, the current magnitude of 20 A, and the peripheral scanning speed of 0.5 mm/s. The temperature curves of Point A varying with the heating time are demonstrated in Figure 5.10a. The resultant temperature of Point A rises with an increasing heating rate, during which Point A is rotated following the wheel substrate and approaches the inductor gradually before 180 s. The temperature of Point A is 914°C when it is below the inductor at 180 s, while it reaches the highest at 916°C at 180.4 s. Subsequently, Point A rotates away from the coil, and the temperature at Point A decreases gradually with a decreasing heating rate. Meanwhile, the heating rate and the highest temperature of Point A increase with a higher frequency. When the heating time is 180 s, the highest temperature values of Point A with different current frequencies are illustrated in Figure 5.10b. The temperature is 914°C with a current frequency of 1 MHz, which is decreased by 17% in comparison to the temperature of 1,096°C with a current frequency of 1.5 MHz and increased by 8% compared with the temperature of 844°C with a current frequency of 500 kHz.

Figure 5.10c and d display the effects of the current frequency on the temperature distribution along Paths I and II, respectively. The heating zone concentrates on the centre more violently with a higher current frequency, which is more beneficial for increasing the heating rate and efficiency during the induction heating process. Considering that the temperature of 880°C–940°C is essential for brazing CBN grains, the current frequency is optimized at 1 MHz.

5.2.7.2 Heating Gap

The effects of the heating gap on the temperature distribution in induction heating are displayed in Figure 5.11a. In this case, the peripheral scanning speed is 0.5 mm/s; the current is 20 A in magnitude and 1 MHz in frequency. It is significant that the heating rate and the highest temperature obtained at Point A are increased with a decrease in the heating gap. Figure 5.11b shows that the temperature of Point A is 819°C when the heating gap is 3 mm, which is decreased by 11% compared to 914°C in the case of the heating gap of 2 mm. The induction heating on the surface of the metallic wheel substrate is consumed by the surrounding environment using thermal radiation. The heat loss increases when a higher temperature is obtained. Therefore, the highest temperature at Point A is inversely proportional to the heating gap.

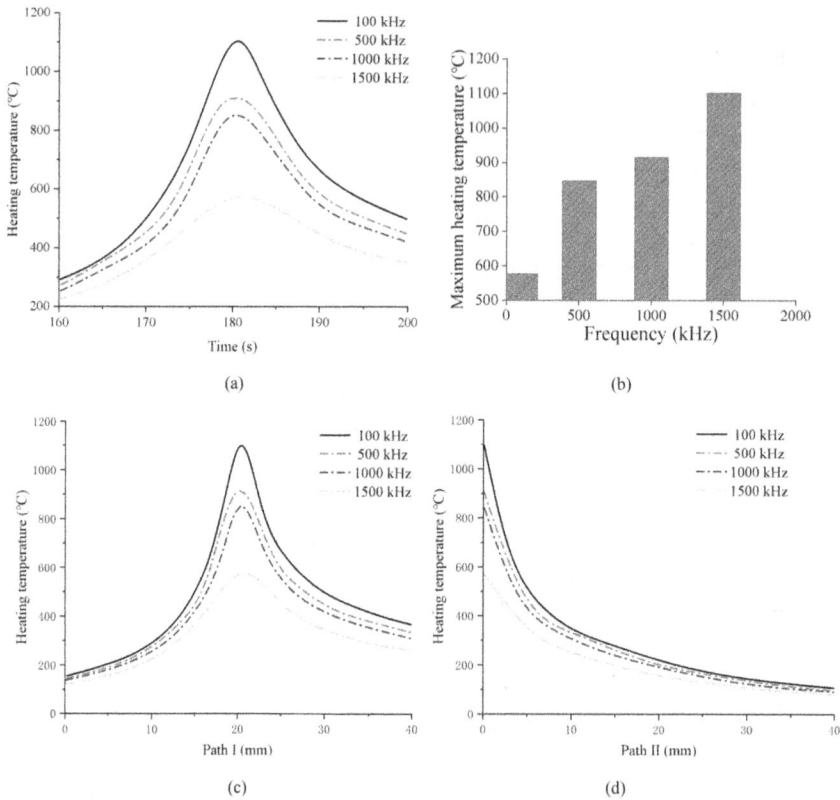

FIGURE 5.10 Variation of temperature distribution versus current frequency: (a) variation of heating curve, (b) variation of highest temperature, (c) temperature distribution along Path I, (d) temperature distribution along Path II.

Figure 5.11c and d demonstrate the temperature distribution along Path I and Path II, respectively, versus the heating gap. When the heating gap value becomes smaller, the heating zone tends to concentrate on the central region of induction heating, and the rate and efficiency of induction heating become larger as well. Furthermore, the minimum heating gap is also required during induction brazing when taking into account the particle size of CBN grains (i.e., 150–400 μm) and the filler layer thickness (i.e., 60–160 μm). As a result, a heating gap of 2 mm is selected as the optimal parameter.

5.2.7.3 Current Magnitude

On the one hand, when the current magnitude increases, the inducted alternating magnetic field becomes stronger, which enlarges the eddy current generated on the metallic wheel substrate. On the other hand, a larger induction heating current results in a higher heating power and heating rate. Figure 5.12a shows the effects of the current magnitude on the heating temperature of Point A. The current frequency is 1 MHz with a heating gap of 2 mm, and the peripheral scanning speed is 0.5 mm/s. Figure 5.12a displays that both

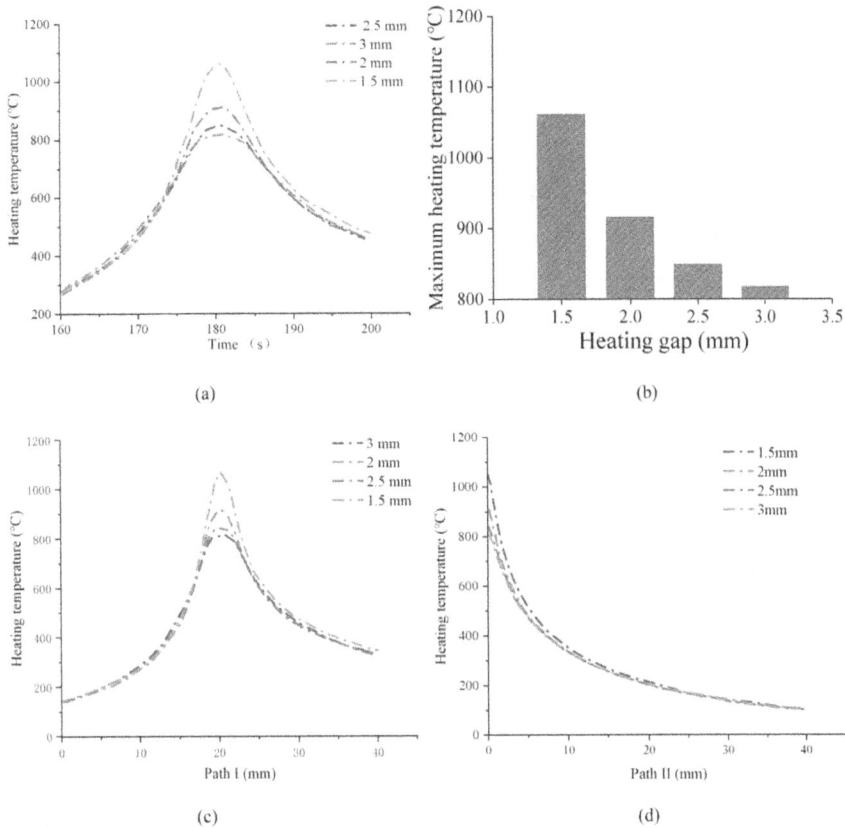

FIGURE 5.11 Variation of temperature distribution versus heating gap: (a) variation of temperature curve; (b) variation of highest temperature; (c) temperature distribution along Path I; (d) temperature distribution along Path II.

the highest temperature at Point A and the heating rate increase with the larger current magnitude. Figure 5.12b demonstrates the highest temperature of Point A versus different current magnitudes. The heating temperature is 849°C in the case of 18 A current, which is decreased by about 7% compared with 914°C in the case of 20 A current. Figure 5.12c and d both display that, when the current magnitude increases, the effective heating zone becomes thinner in the central area of induction heating, and the heating rate and efficiency get higher as well.

5.2.7.4 Scanning Speed

The heating curves of Point A at different scanning speeds are shown in Figure 5.13a, in which 20 A current, 1 MHz frequency, and a 2 mm heating gap are applied. The heating time is 360.5, 180.4, 90.2, and 45.2 s, respectively, when the temperatures of Point A reach the peak value at the different peripheral scanning speeds from 0.25 to 2 mm/s. According to the above facts, the distances of Point A relative to the inductor could be calculated at the heating time mentioned above, and the distances are 0.125, 0.2, 0.2, and 0.4 mm, respectively, which are all much smaller than the inductor size of 12 mm and can be ignored.

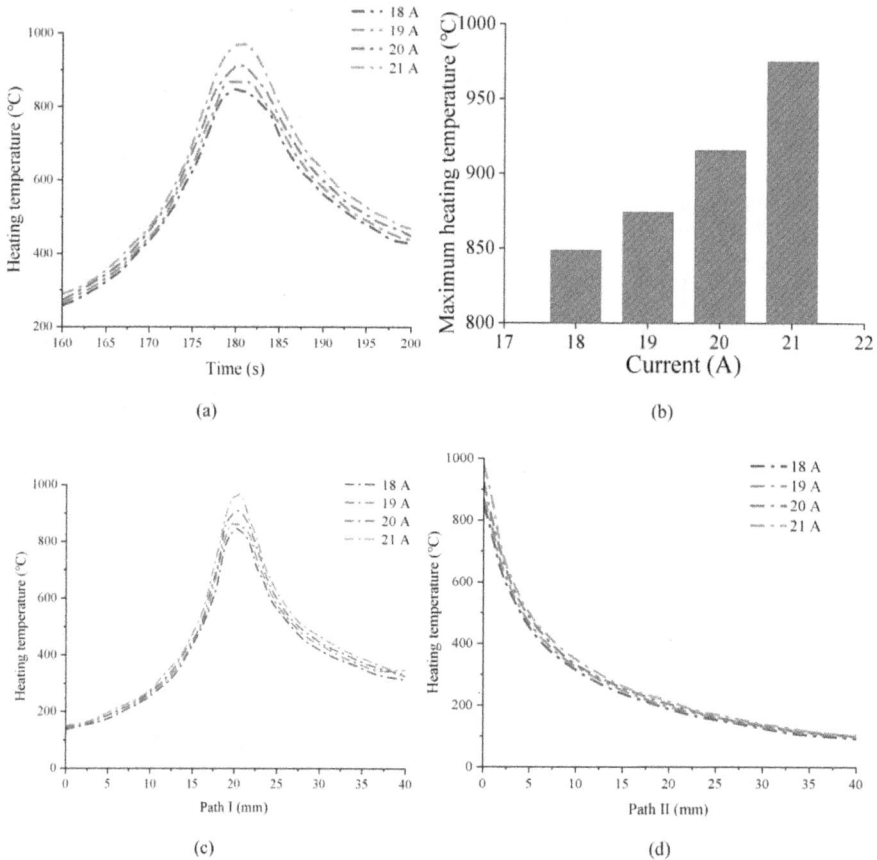

FIGURE 5.12 Variation of temperature distribution versus current magnitude: (a) variation of heating curve, (b) variation of highest temperature, (c) temperature distribution along Path I, and (d) temperature distribution along Path II.

This implies that the scanning speed has little influence on the position where Point A reaches the highest temperature. However, the highest temperature at Point A is highly affected by the peripheral scanning speeds. As shown in Figure 5.13b, the heating temperature is 977°C in the case of the peripheral scanning speed of 0.25 mm/s, which is decreased by 6% compared to 916°C in the case of the peripheral scanning speed of 0.5 mm/s. The heating temperature is 861°C in the case of the peripheral scanning speed of 2 mm/s, which is decreased by 6% compared to 916°C in the case of the peripheral scanning speed of 0.5 mm/s. The higher peripheral scanning speed contributes to the smaller heating temperature at the same position in the metallic wheel substrate.

When Point A is rotated below the coil at 180 s, the temperature distributions along Paths I and II at different scanning speeds are provided in Figure 5.13c and d, respectively. Point A is near the peak of the curve along Path I, as shown in Figure 5.13c. The region to

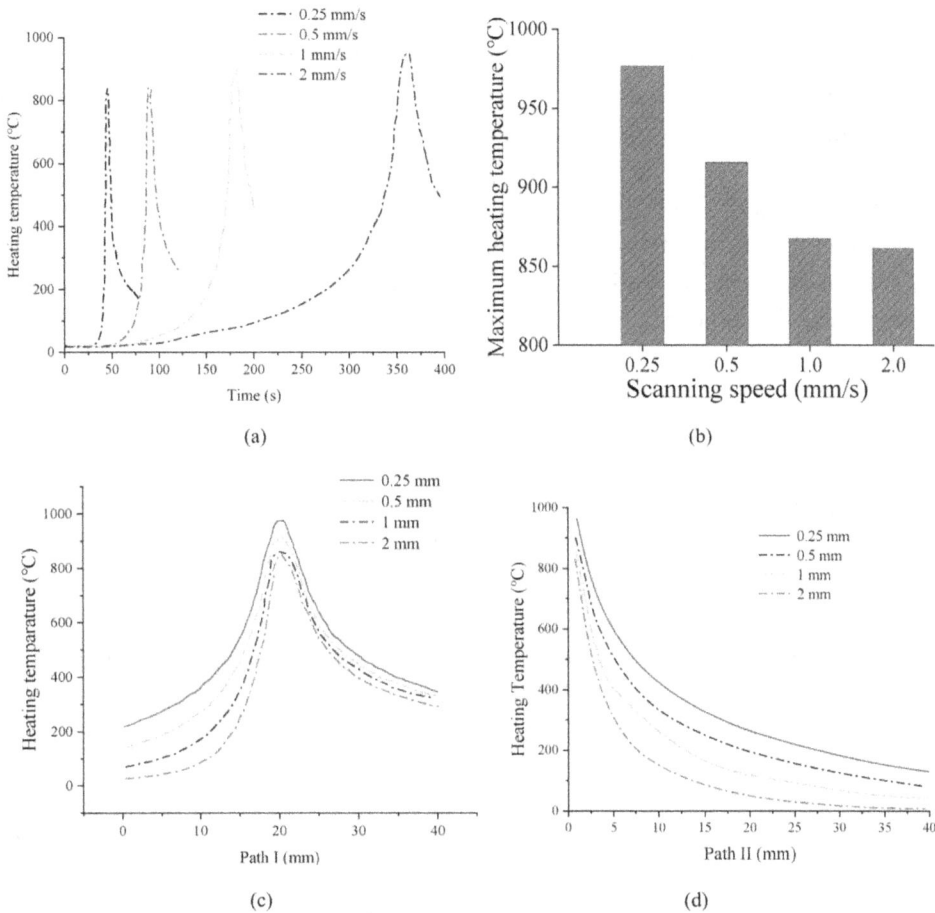

FIGURE 5.13 Variation of temperature distribution versus peripheral scanning speed: (a) variation of heating curve, (b) variation of highest temperature, (c) temperature distribution along Path I, and (d) temperature distribution along Path II.

be heated is on the left of the peak. It is obvious that the temperature distribution becomes more concentrated at higher scanning speeds. The region on the right of the peak is the cooling region, in which the average temperature gradients of the curves are 31°C/mm, 29°C/mm, 27°C/mm, and 28°C/mm at different peripheral scanning speeds from 0.25 to 2 mm/s. It is obvious that the gradient values at different peripheral scanning speeds are approximately identical. Thus, the scanning speed nearly does not influence the temperature distribution within the cooling part of the wheel substrate. Figure 5.13d displays the temperature distributions of Point A along Path II, and the curves are similar to those on the left part along Path I from 0 to 20 mm. That is to say, the induction heating at higher scanning speed would produce a higher temperature gradient, which makes the temperature distribution near Point A much more inhomogeneous. When the scanning speed is lower, more time is provided to transfer the heat produced in the narrow central region below the coil to the surrounding regions in the wheel substrate. As such, the preheating effect of the region to be heated in the steel wheel substrate is more sufficient, and the temperature gradient becomes smaller at the lower scanning speed.

5.2.8 Curve Fitting of Highest Temperature Varying with Heating Parameters

The highest temperature curves varying with four heating parameters (i.e., induction current I, current frequency f, heating gap h, and scanning speed v) are fitted with the least squares method. The empirical formula is expressed as follows:

$$T(I,f,h,v) = A \cdot (I^{b_1} + c_1)(f^{b_2} + c_2)(h^{b_3} + c_3)(v^{b_4} + c_4) \quad (5.15)$$

where T is the highest temperature (°C). The parameters I, f, h, and v represent the induction current, current frequency, heating gap, and scanning speed, respectively. $A = -9.48 \times 10^{-10}$ is the coefficient in the empirical formula. The other coefficients are listed in Table 5.1.

When the heating parameters are changed, the highest temperature values obtained in both the simulation and the empirical formula are compared, as displayed in Figure 5.14. It is obvious that the variation tendencies of the highest temperatures in both the simulation and the empirical formula are generally consistent. The calculation formula for the deviation is given as follows:

$$\delta = (T_s - T_{fit}) / T_s \times 100\% \quad (5.16)$$

where T_s means the highest temperature obtained in the simulation, and T_{fit} represents the highest temperature calculated by the empirical formula. Based on Eq. (5.16), the deviation of the highest temperatures between the simulation and empirical formula is analyzed. The highest temperature deviations, ranging within ±1%, account for 81.3%. When the frequencies are 100, 500, and 1,500 KHz, the deviations are larger than 1%, and those values are 4.8%, 7.3%, and 7.6%, respectively. That is to say, the fitting results correspond well to the simulation results, and the fitting curves are reasonable in this work.

TABLE 5.1 Values of Curve-fitting Parameters

Fitting Parameters		Value
I	b_1	8.46
	c_1	1.21×10^{12}
f	b_2	−0.405
	c_2	−0.249
h	b_3	−2.60
	c_3	1.30
v	b_4	−0.245
	c_4	1.58

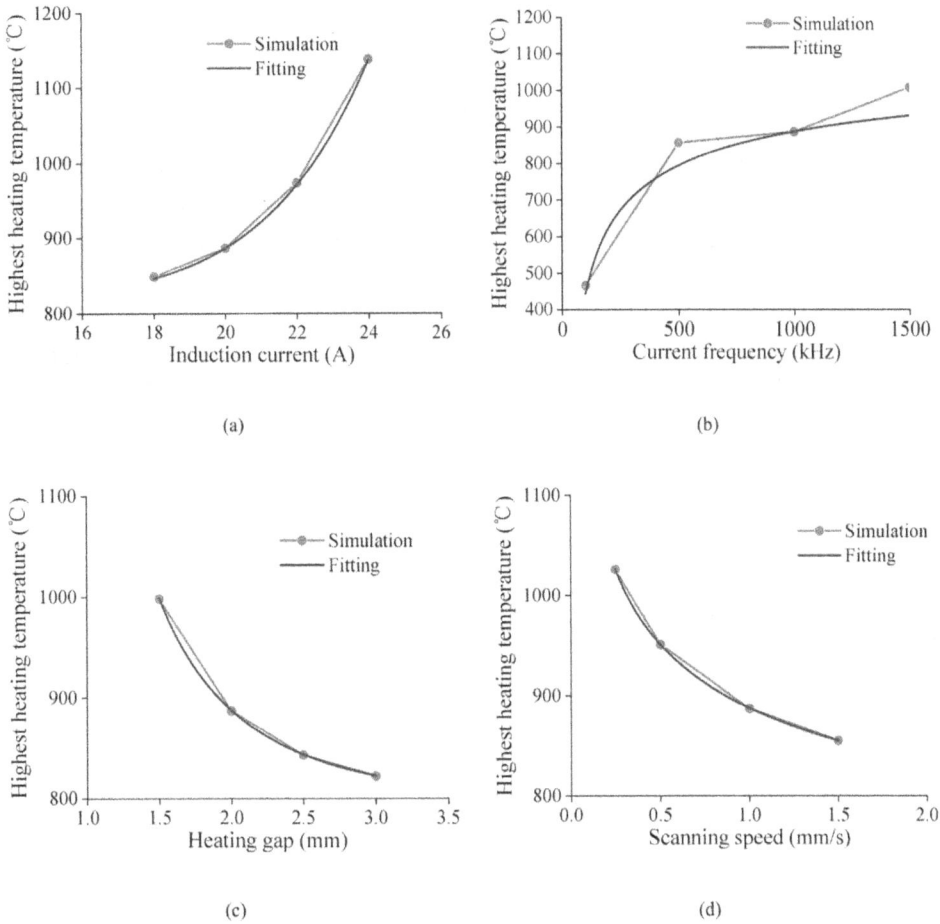

FIGURE 5.14 Simulation and fitting curves varying with different parameters: (a) induction currents, (b) current frequencies, (c) heating gaps, and (d) scanning speeds.

5.3 EXPERIMENTAL VERIFICATION OF HIGH-FREQUENCY BRAZING OF CBN GRAINS

5.3.1 Experimental Verification of the Heating Curve and the Highest Temperature

In the brazing experiments of CBN grains, high-frequency induction heating equipment modelled SPG-06AB III with a frequency of 1 MHz is used. The heating region of the CBN wheel and the inductor are in an argon atmosphere to prevent oxidation. A K-type thermocouple is applied to measure the heating temperature of point A due to its suitable temperature range (0°C–1,300°C), fine linearity, large thermo-electromotive force, high sensitivity, and strong resistance to oxidation.

The wheel substrate rotates and is heated. The distance between the temperature measuring point A and the inductor is 90 mm at the beginning of the heating, and Point A is in the stable heating region based on the above analysis. In the brazing experiments, the heating gap between the inductor and the wheel substrate is adjusted to 2 mm, and the currents with different magnitudes are loaded on both sides of the coil, and then the inductor scans through the heating surface of the metallic wheel substrate at a constant peripheral scanning speed of 0.5 mm/s until the inductor sweeps past point A and the temperature of the point begins to decrease.

The heating curves of Point A obtained in both simulation and experiment are comparatively illustrated in Figure 5.15a. At the initial heating stage (0–140 s), the simulated heating curve is essentially coincident with the experimental one; the temperature increases slowly in the two curves, and the highest temperature variation is 11°C at 140 s. At the rapidly rising temperature stage (140–180 s), the temperature increases rapidly, and the heating rate obtained in the simulation work is always close to the experimental heating rate. The average value of the temperature variation between the simulation and the experiment is 29°C, and the highest temperature variation is 48°C at 162.6 s at this stage. At the cooling stage (after 180 s), the curves obtained in the simulation and the experiment are also nearly coincident.

The highest temperature varying with the current magnitude (18–21 A) in both the simulation and experiment is comparatively illustrated in Figure 5.15b. When the current magnitude is 18 A, the temperatures obtained in the simulation and experiment are 849°C and 866°C, respectively, which are 2% different. When the current magnitude increases to 20 A, the temperatures in the simulation and experiment are 916°C and 918°C, and the difference between them is 0.2%, which is the smallest one compared with those obtained with the other current magnitude. When the current magnitude increases to 21 A, the temperatures in the simulation and the experiment are 975°C and 946°C, respectively, which are 3% different. Based on the above analysis, it is known that the difference in the highest temperature between simulation and experiment is less than 5% with the different current magnitudes ranging from 18 to 21 A. That is to say, the simulation results are generally reasonable and accredited to the current work.

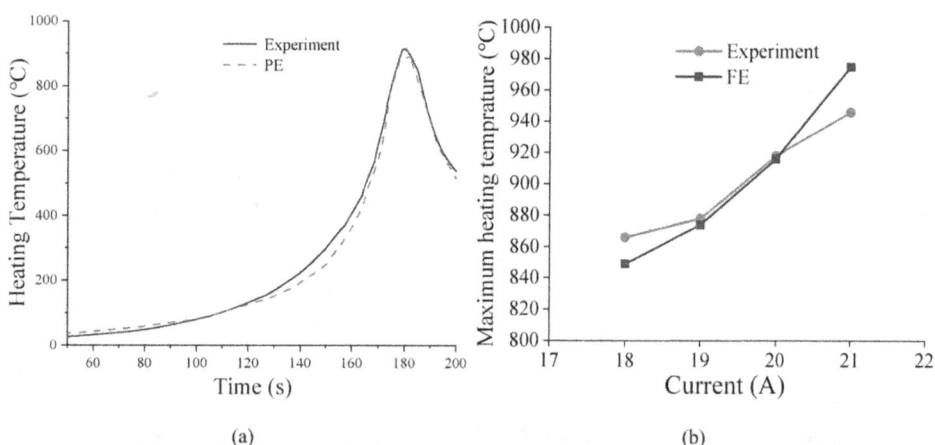

(a)   (b)

FIGURE 5.15 Comparison of the heating temperature curves obtained in simulation and experiment: (a) heating temperature curves of Point A, (b) highest temperatures varying with the current magnitude.

5.3.2 High-frequency Induction Brazing of CBN Grains

Figure 5.16 displays the whole morphology of the brazed CBN grain, which is produced in the case of the optimum parameters as follows: a heating gap of 2 mm, a current frequency of 1 MHz, a current magnitude of 20 A, and a peripheral scanning speed of 0.5 mm/s. It is noted that the polishing of the brazed sample has been done to keep the three different layers (metallic wheel substrate, Ag-Cu-Ti filler alloy, and CBN grain) in a single plane; as such, the joining behaviour of braze partners could be easily observed. In general, a good brazing interface has been formed among wheel substrate/Ag-Cu-Ti filler alloy/CBN grain when applying the above-mentioned optimum parameters.

5.3.3 Optimization of High-Frequency Heating Parameters

It has been pointed out that, in theory, a smaller value of the heating gap between the coil and heated matrix surface helps to concentrate heating power on the brazing interface and improve the heating efficiency. However, in reality, the powder filler layer of the Ag-Cu-Ti alloy applied was nearly 1 mm thick before brazing in the present work. Under such conditions, a heating gap of 2 mm should be chosen to avoid the situation where the inductor is burned out due to the contact between the molten filler and the coil. As for the current frequency, when the frequency applied becomes higher, the skin depth in induction heating gets smaller, the heat generated on the steel wheel matrix is more concentrated on the central heating region, and subsequently, the thermal deformation and residual stresses in the steel wheel matrix become smaller. Moreover, higher heating efficiency can also be obtained with a higher current frequency in the induction heating process. Limited by the experimental apparatus, the ultra-high frequency of 1 MHz is accordingly selected for the optimal parameter in the induction brazing.

The scanning speed is controlled and can be accurately adjusted by the heating device. During the induction heating/brazing process, higher scanning speeds can speed up the rotation of the steel wheel matrix and therefore shorten the brazing time. On the contrary, a higher scanning speed may also cause an insufficient reaction in the brazing interface, which contributes to the undesirable interfacial microstructure with a low bonding strength. Considering the above factors, a range of 0.5–1 mm/s is recommended during

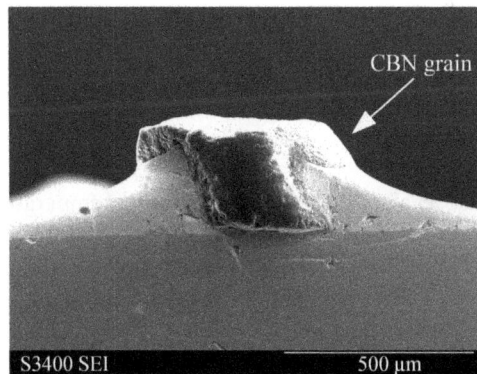

FIGURE 5.16 Morphology of brazed CBN super-abrasive grain.

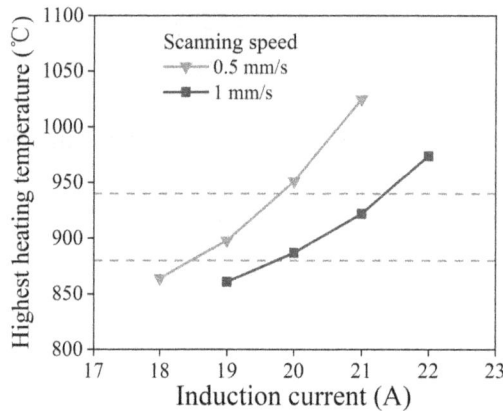

FIGURE 5.17   Heating temperature varies with induction current and scanning speed.

rotating induction heating/brazing, which is the optimum range of the scanning speeds. In the present simulation work, the values of induction currents and scanning speeds are optimized interactively to acquire the ideal brazing temperature range, that is, 880°C–940°C.

As shown in Figure 5.17, the highest temperatures, varying with both induction currents and scanning speeds, are obtained at the stable heating stage in the induction brazing process. Here, the heating gap is 2 mm, and the current frequency is 1 MHz. Apparently, three groups of induction current and scanning speed could be obtained as follows: (i) scanning speed of 0.5 mm/s and induction current of 20 A, (ii) scanning speed of 1 mm/s and induction current of 20 A, and (iii) scanning speed of 1 mm/s and induction current of 21 A. The highest temperatures obtained in the three groups are 898°C, 887°C, and 922°C, respectively.

REFERENCES

Chen, Q. S., P. Gao, and W. R. Hu. 2004. Effects of induction heating on temperature distribution and growth rate in large-size SiC growth system. *Journal of Crystal Growth*, 266: 320–326.

Codrington, J., P. Nguyen, S. Y. Ho, et al. 2009. Induction heating apparatus for high temperature testing of thermo-mechanical properties. *Applied Thermal Engineering*, 29: 2783–2789.

Ding, W. F., J. H. Xu, Z. Z. Chen, et al. 2010. Effects of heating temperature on interfacial microstructure and compressive strength of brazed CBN-AlN composite abrasive grits. *Journal of Wuhan University of Technology-Materials Edition*, 25(6): 952–956.

Eastwood, M. D., and K. R. Haapala. 2015. An induction hardening process model to assist sustainability assessment of a steel bevel gear. *International Journal of Advanced Manufacturing Technology*, 80(5–8): 1–13.

Franco, C., J. Acero, R. Alonso, et al. 2012. Inductive sensor for temperature measurement in induction heating applications. *IEEE Sensors Journal*, 12(5): 996–1003.

Jang, J. Y., and Y. W. Chiu. 2007. Numerical and experimental thermal analysis for a metallic hollow cylinder subjected to step-wise electro-magnetic induction heating. *Applied Thermal Engineering*, 27: 1883–1894.

Li, F., X. K. Li, T. X. Zhu, et al. 2013. Numerical simulation of the moving induction heating process with magnetic flux concentrator. *Advances in Mechanical Engineering*, (3): 907295.

Li, Q. L., J. H. Xu, H. H. Su, et al. 2015. Fabrication and performance of monolayer brazed CBN wheel for high-speed grinding of superalloy. *International Journal of Advanced Manufacturing Technology*, 80: 1173–1180.

Mei, R. B., C. S. Li, X. H. Liu, et al. 2011. Modeling of slab induction heating in hot rolling by FEM. *Engineering*, 2011, 3, 364–370.

Norouzifard, V., and M. Hamedi. 2014. A three-dimensional heat conduction inverse procedure to investigate tool-chip thermal interaction in machining process. *International Journal of Advanced Manufacturing Technology*, 74(9–12): 1637–1648.

Shen, B., B. Song, L. Cheng, et al. 2014. Optimization on the HFCVD setup for the mass-production of diamond-coated micro-tools based on the FVM temperature simulation. *Surface and Coatings Technology*, 253: 123–131.

Sun, R., Y. Qiao, Y. Yang, et al. 2023. Thermal process and flow behavior during laser/ultra-high frequency (UHF) induction hybrid deposition. *International Journal of Heat and Mass Transfer*, 210: 124186.

Sun, R., Y. Shi, Z. Bing, et al. 2019. Metal transfer and thermal characteristics in drop-on-demand deposition using ultra-high frequency induction heating technology. *Applied Thermal Engineering*, 149: 731–744.

Xu W, W. F. Ding, Y. J. Zhu, X. Huang, and Y. C. Fu. 2017. Understanding the temperature distribution and influencing factors during high-frequency induction brazing of CBN super-abrasive grains. *International Journal of Advanced Manufacturing Technology*, 88: 1075–1087.

Zhu, T. X., X. K. Li, F. Li, et al. 2013. The establishment of coupled electromagnetic-thermal analytical model of induction heating system with magnetic flux concentrator and the study on the effect of magnetic permeability to the modeling. *ASME International Manufacturing Science and Engineering Conference*, 6: 10–14.

CHAPTER 6

Rhythmic Grain Distribution on the Wheel Surface

THE ABRASIVE WHEELS WITH uniform grain distribution have more prominent machining performance than those with random grain distribution in several aspects, including grinding forces, grinding temperature, and surface integrity of the ground parts (Tian et al. 2023; Wen et al. 2023; Chen et al. 2023; Yu et al. 2016). Henceforth, how to accomplish the rhythmed grain distribution for single-layer brazed cubic boron nitride (CBN) abrasive wheels would be the notable problem and primary difficulty after the integration between CBN grains and tool substrate has been realized. Figure 6.1 displays the disordered grain distribution of the brazed CBN abrasive wheels when the relevant experimental research in the present investigation is just at an early stage. As a result of the effects of shrinking in volume and the act of flowing in the course of fluidification and solidification of the Ag-Cu-Ti or Cu-Sn-Ti filler powder, many grains have been removed apparently from their

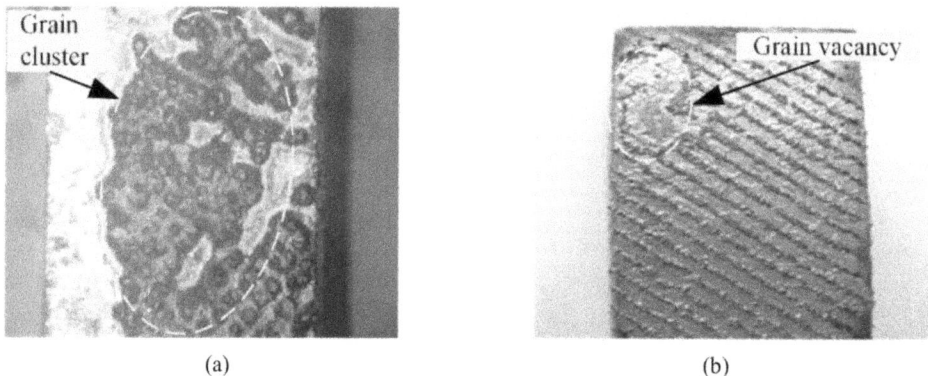

FIGURE 6.1   Surface topographies of brazed CBN specimens without uniform grain distribution: (a) grain cluster, (b) grain vacancy.

DOI: 10.1201/9781032678047-7

original position and gathered into a cluster; thus, in several regional parts of the tool substrate, no grains would take effect during grinding, and consequently, the brazed tool could not exert the expected performance because of the uneven chip space among the grains in the machining process.

To thoroughly solve the particular trouble caused by the filler alloy, a lot of measures have been taken. However, uniform grain distribution is not easy to achieve. Finally, a unique technique was searched, and the brazed CBN specimen with optimum rhythmed grain distribution was authentically accomplished (Figures 6.2 and 6.3). The grains have been arranged evidently with a fixed interval between two adjacent rows, which ensures sufficient space to hold chips. Two typical samples of single-layer brazed CBN abrasive wheels are displayed in Figure 6.4.

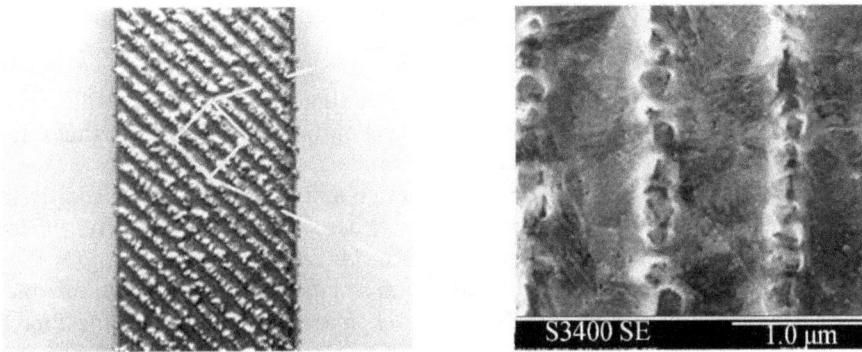

FIGURE 6.2 Surface topographies of brazed CBN specimens with optimum rhythmed grain distribution: (a) whole, (b) region.

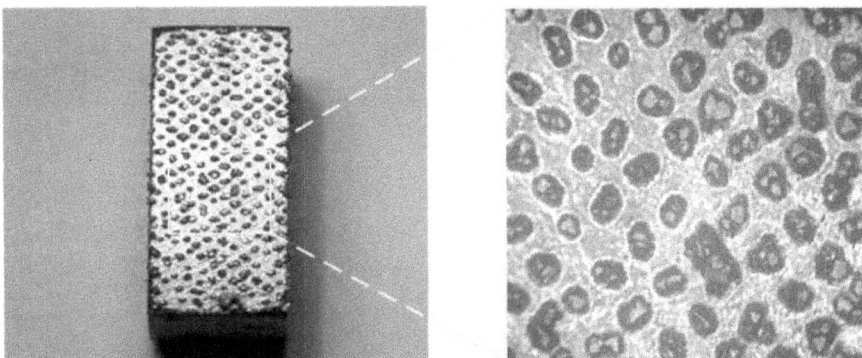

FIGURE 6.3 Uniform grain distribution: (a) whole, (b) region.

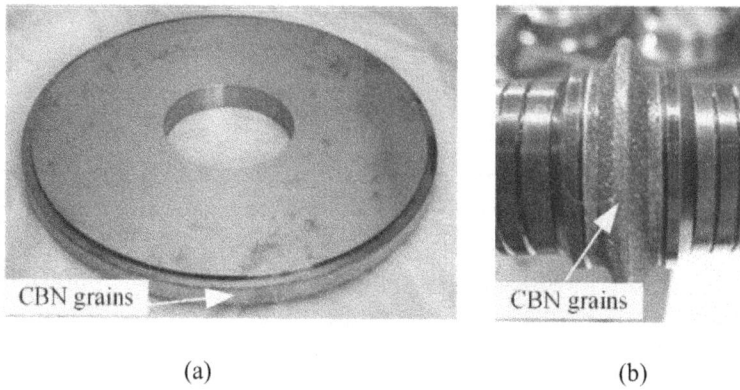

(a)   (b)

FIGURE 6.4   Two typical single-layer brazed CBN abrasive wheels: (a) Wheel I, (b) Wheel I.

REFERENCES

Chen, Z., X. Zhang, D. Wen, et al. 2023. Improved grinding performance of SiC using an innovative bionic vein-like structured grinding wheel optimized by hydrodynamics. *Journal of Manufacturing Processes*, 101: 195–207.

Tian, C., Y. Wan, X. Li, et al. 2023. Permeability design and assessment of the additively manufactured metal-bonded diamond grinding wheel based on TPMS structures. *International Journal of Refractory Metals and Hard Materials*, 114: 106237.

Wen, D., L. Wan, X. Zhang, et al. 2023. Grinding performance evaluation of SiC ceramic by bird feather-like structure diamond grinding wheel. *Journal of Manufacturing Processes*, 95: 382–391.

Yu, H. Y., J. Wang, and Y. S. Lu. 2016. Modeling and analysis of dynamic cutting points density of the grinding wheel with an abrasive phyllotactic pattern. *International Journal of Advanced Manufacturing Technology*, 86: 1933–1943.

PART II

Wear Behaviour and Mechanism

CHAPTER 7

Wear Behaviour and Stresses Effects of Monocrystalline CBN Abrasive Wheels

7.1 INTRODUCTION

7.1.1 Wear Behaviour of Abrasive Wheels

Wear is of particular concern with single-layer brazed cubic boron nitride (CBN) abrasive wheels due to the limited number of available grains and the need for a sufficiently long tool life. The grinding performance of the brazed abrasive wheels would change drastically as the abrasive wheel wears and its topography varies throughout the tool life. The abrasive wheel wear not only controls the ability to hold workpiece size and geometry but also has a remarkable influence on other output parameters such as the ground surface integrity and the very nature of the tool-workpiece interaction. Therefore, the magnitude and mechanism of the gradual wear of the abrasive wheels in the grinding process are very important.

Single-layer brazed CBN abrasive wheels have eliminated premature grain pullout problems, yet they have the advantages of substantially high protrusion of grains and flexible placement of grains across wheel surfaces. However, besides the attritious wear, a single-layer brazed CBN wheel is prone to fracture at the grain-bond junction region (macrofracture) or the grain vertex region (microfracture) (Ding et al. 2015), as displayed in Figure 7.1. Investigations have shown that such macro and microfracture problems are not with CBN grains only but happen to other abrasives such as diamond grains.

In the earlier studies, Zhu et al. (2019) and Lee et al. (2021) pointed out that there were three types of grinding wheel wear: attrition wear, grain fracture, and bond fracture. In general, wheel wear is both physical and chemical, and their influence depends on grinding conditions and the specifics of a wheel-workpiece pair. For example, attrition wear occurs on the grain-workpiece contact interface, which results in the flattening or dulling of abrasive grain. Plastic flow, crumbling, and chemical reactions have a significant effect on attrition wear. Fracture wear is due to the removal of abrasive grains from the wheel,

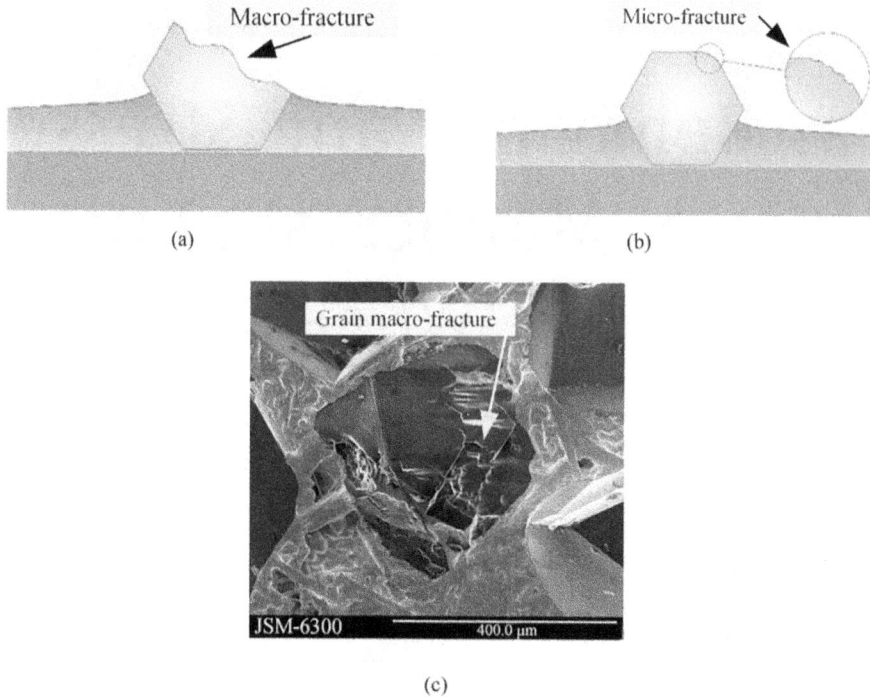

FIGURE 7.1   Fracture wear of brazed CBN grains: (a) schematic of grain microfracture, (b) schematic, and (c) photograph of grain macrofracture.

either by partial grain fracture or by bond fracture. Eiss et al. (1967) investigated some of the basic parameters, such as load, which affect the grain fracture in a grinding process, and provided a correlation to predict grain failure. This explains qualitatively the more significant fracture wear of the brazed CBN grains. Further investigations have explored that grain wear caused by macrofracture at the grain-bond junction reduces the wheel life significantly, which should be minimized, but certain microfracture in the vertex region of grain can enhance the sharpness of a grinding wheel (Qian et al. 2021; Bredthauer et al. 2023; Ding et al. 2014; Jiang et al. 2013).

7.1.2 Stresses Effects on Grain Wear

Stress concentration in grain is the main factor that results in grain fracture during grinding. The stress value in the CBN or diamond grains, which is measured experimentally by the current technique (e.g., Fourier transform infrared spectroscopy, Raman spectroscopy), is only an average over a measured volume and hence does not have sufficient accuracy for the characterization of a small grain whose diameter is usually between 100 and 400 μm. As such, the finite element (FE) method has been widely used. For example, Chen et al. (2009) carried out an FE analysis of the brazing-induced residual stresses in diamond grains and found that the simulation and measurement stresses have similar trends, although the simulated stresses are larger than the experimentally measured stresses. Furthermore, it is reported that the maximum tensile stress, which appeared near the bonding interface, was 349 MPa when the brazing alloy was Ag-Cu-Ti and 1.754 GPa when it was Ni-Cr (Chen et al. 2014). To understand the grain-matrix bonding failure, Zhou et al. (1997) used a two-dimensional FE model to compare the stresses at the grain-matrix interface. Suh et al. (2008) analysed the stress distributions in grains in resin-bonded diamond wheels during the grinding of an optical connector ferrule and reported that the grain breaking could be quantified by the ratio of critical protrusion height to grain diameter. The geometry effect of grains has also been studied, such as spherical and pyramidal grains. In the abovementioned studies, however, the depth of a grain embedded in a bonding layer was beyond 60% of the grain diameter, indicating that the grain protrusion, or the exposed height of the grain above the bonding layer, was low. This means that the investigations were for cases of smaller chip storage spaces, corresponding to a lower material removal rate.

In a brazed CBN grain, the resultant stresses during grinding are the result of those introduced by both brazing and grinding. The former is generated due to the thermal property mismatch between the CBN grain and the bonding material, whereas the latter is caused by the grinding load applied. However, the previous investigations on grain fracture considered only the resultant stresses induced by grinding load, whereas the residual stresses due to wheel fabrication (e.g., sintering, electroplating, or brazing) were overlooked. That is, the coupling effect of the two stress sources on the resultant stress has not been understood. It is essential to clarify the issue with the aid of the FE analysis, involving grain embedding depth, grain wear, bond wear, grain size, and grinding load.

7.2 WEAR PHENOMENON OF MONOCRYSTALLINE CBN ABRASIVE WHEELS IN GRINDING

7.2.1 Experimental Details

This work is undertaken to explore the wear mechanisms and their effect on the grinding behaviour of single-layer brazed CBN abrasive wheels during grinding. Single-layer brazed CBN abrasive wheels manufactured in-house have a diameter of $d_s = 265$ mm and a width of $b = 4.5$ mm. Moreover, replaceable CBN segments are designed for the abrasive wheels to directly observe the changes in the wheel working surface during the grinding process. The CBN abrasive grains with 80/100 US mesh are applied. The grain density on the abrasive layer is 6/mm^2. The brazing process was carried out in a vacuum furnace. Not only the original grain size but also the protrusion height of the brazed CBN grains are measured using KH-7700 optical microscopy.

The ground specimens are commercial K424 cast nickel-based superalloys with the following primary properties: tensile strength 1,010 MPa, yield strength 775 MPa, hardness 37.2 HRC, thermal conductivity 10.85 W/(m·K), density 8.2×10^3 kg/m^3, linear thermal expansion 14.1×10^{-6}/°C. Particularly, K424 superalloy is mostly applied as the material of aero turbine blades owing to its high strength and high heat resistance. It is a typical difficult-to-cut material with the smallest relative cutting coefficient. After mounting, the specimen was ground several times to ensure its flatness and parallel conformity with the wheel feed.

Grinding experiments were conducted in the down-grinding mode on an MMD7125 precision surface grinding machine while applying a commercial 5% dilution of emulsified liquid at 90 L/min. The workpiece is 40 mm long in the grinding direction, 4.5 mm width and 15 mm height. In particular, the workpiece is composed of two blades clamped together with a constantan of 0.02 mm diameter sandwiched between the two blades to form a semi-natural thermocouple junction with the workpiece surface. The temperature signal is captured and recorded directly by the HP3562 dynamic signal analyzer. The thermocouple is calibrated using a high-precision quick calibrator. The result is as follows:

$$T = 30U \ (\text{error} < 5\%) \tag{7.1}$$

where U is the electric potential produced by the thermocouple (mV) and T is the measured temperature (°C).

At the same time, a piezoelectric-type force transducer (Dynamometer 9265B) along with a charge amplifier and a data acquisition system has been used for force measurement in this work.

It is known that the maximum depth at which cutting edges of single-grain penetrate the workpiece is called the uncut chip thickness (or undeformed chip thickness) a_{gmax}, which has an important effect on wheel wear. By approximating the uncut chip cross section as triangular, the maximum uncut chip thickness can be described as

$$a_{gmax} = \left[\frac{3}{C \tan\theta} \left(\frac{v_w}{v_s} \right) \left(\frac{a_p}{d_s} \right)^{1/2} \right]^{1/2} \tag{7.2}$$

where C is the active cutting edge density, θ is the semi-included angle for the uncut chip cross section, v_w is the workpiece velocity, v_s is the wheel speed, a_p is the depth of cut, and d_s is the wheel diameter.

Creep-feed grinding is selected in this experimental work because of its advantages in generating less damage and requiring a lower cost than conventional grinding of a nickel-based superalloy. For realizing the faster wear than the wear behaviour in the conventional grinding process, the grinding experiments were designed and conducted with a combination of wheel speed $v_s = 22.5$ m/s, workpiece infeed speed $v_w = 0.20$ m/min, depth of cut $a_p = 0.20$ mm with the former 200 passes (accumulated removal material volume 7,200 mm³) and another combination of $v_s = 22.5$ m/s, $v_w = 0.40$ m/min, $a_p = 0.20$ mm with the latter 40 passes (accumulated removal material volume 8,640 mm³) until the end of the abrasive wheel life. The former and latter grinding processes corresponded to an uncut chip thickness of 1.08 and 1.53 μm, respectively.

7.2.2 Mechanical Loads of Brazed CBN Abrasive Wheels

The wear behaviour of the CBN abrasive wheels may be taken into account from the viewpoint of the combination of abrasive wheels, workpiece material, grinding parameters, and cooling as well. Especially grinding force and temperature, which correspond to the mechanical and thermal loads, are the major factors in determining and reflecting directly the abrasive wheel wear. These factors are composed of components due to ploughing, shearing, fracturing, wheel loading, friction among the grains, the bonding layer, redeposited debris, and the workpiece material.

The measured normal forces F_n and tangential ones F_t, and the force ratios F_n/F_t versus accumulated removal material volume are given in Figure 7.2. The normal force presses the CBN grains into the workpiece material and penetrates the surface, leading to material removal. In particular, the normal force is also responsible for abrasive wheel wear. Once the thrust on the abrasive wheel exceeds certain limits, the fracture of the abrasive grains and bonding layer may take place. The tangential force determines the power requirement in the grinding process. The intensity of heat generation depends on this force, and it is the most important as far as grinding temperatures.

It is evident from Figure 7.2a that, when the calculated uncut chip thickness is 1.08 μm, the grinding process in this work may be divided into the early unsteady period and the subsequent steady one. In the early period, the normal forces are increased from 42 to 90 N, while the tangential ones are increased from 25 to 40 N. The correspondent force ratios F_n/F_t have increased rapidly from 1.7 to 2.3, as displayed in Figure 7.3b. When the accumulated removal material volume reaches about 252 mm³ (six grinding passes), the grinding process reaches a steady period up to the accumulated removal material volume of 7,200 mm³. At this time, the grinding forces and the force ratios vibrate up and down the average values, such as the normal forces at 50–70 N and the tangential forces at 30–38 N, while the force ratios are varied at the range of 1.7–2.0. However, microscopic observation of the grain wear behaviour reveals dulling of the CBN abrasive grains by attritious wear, thereby causing a slight progressive increase in the forces. Finally, when the uncut chip thickness is increased further to 1.53 μm to obtain accelerating wear, another unsteady-state grinding process is encountered. The final normal force and tangential force are increased to 138 and 62 N, respectively. The force ratio is also increased remarkably to the maximum value of 2.4 until the wheel fails. In general, different from the force ratios F_n/F_t of 4.0–5.0

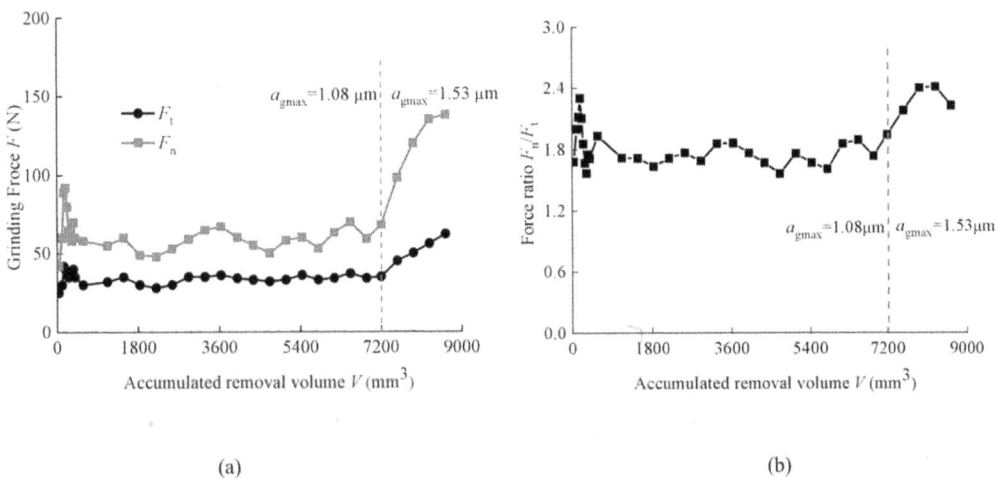

FIGURE 7.2 Grinding forces and force ratios versus accumulated removal material volume: (a) grinding forces, (b) force ratios.

in other investigations (Shaji et al. 2003), the normal force in this experimental work is approximately double the tangential force. This accounts for the grinding process characteristics of single-layer brazed CBN abrasive wheels, i.e., sharp grain cutting edges and sufficient chip accommodation space. All indicate that the single-layer brazed CBN abrasive wheels have good machining performance during grinding cast nickel-based superalloy.

7.2.3 Thermal Loads of Brazed CBN Abrasive Wheels

Figure 7.3 displays the typical temperature curves at the uncut chip thickness of 1.08 and 1.53 μm, respectively. The peaks of the sharp pulses appearing on the signal curves represent the temperature of the individual CBN abrasive grain, while the troughs of the pulses stand for the average temperature distribution of the grinding zone of the machined workpiece surface. The highest temperature of the individual grain is 500°C–600°C, which is 4–5 times higher than the grinding zone temperature range of 100°C to 150°C. In particular, a low grinding zone temperature is a typical feature of the creep-feed grinding process without burning the workpiece. Just as reported in the previous work (Grimmert et al. 2023; Qian et al. 2022), the maximum grinding zone temperature may be less than 120°C during grinding AISI 52100 bearing steel with vitrified CBN abrasive wheels. At this time, the energy partition to the workpiece is only 4.0%–8.5%. This is due to the high thermal conductivity properties of CBN abrasive grains, which enhance heat conduction away from the grinding zone. Furthermore, for the single-layer brazed CBN abrasive wheels, combined with the high values of thermal conductivity of the CBN abrasive grains and Ag-Cu-Ti bonding material, an increased proportion of the grinding energy is taken away by the abrasive wheel. Therefore, a low grinding zone temperature of K424 superalloy may be obtained with brazed CBN wheels.

The effect of the accumulated removal material volume on the grinding zone temperature is demonstrated in Figure 7.4. As a whole, the changes in the grinding force and temperature are similar. In the case of the uncut chip thickness of 1.08 μm, the grinding

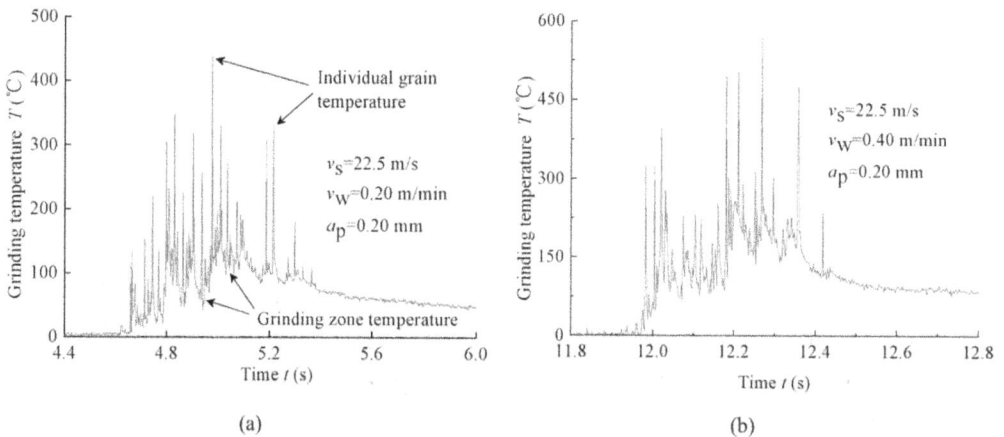

FIGURE 7.3 Typical temperature signals at different grinding parameters: (a) $a_{gmax} = 1.08$ μm, (b) $a_{gmax} = 1.53$ μm.

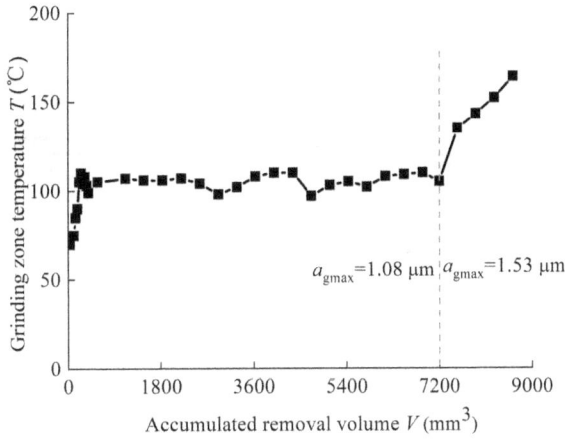

FIGURE 7.4   Grinding zone temperature versus accumulated removal material volume.

temperature tends to increase at a fast speed with increasing accumulated removal material volume in the early unsteady-state grinding period. After several grinding passes, the temperature reaches a stable value of about 100°C. However, when the uncut chip thickness is increased to 1.53 µm, the grinding zone temperature rises to 180°C. The accumulated removal material volume arrives at 8,640 mm³ (250 passes) when the brazed CBN abrasive wheel fails.

7.2.4 Protrusion Height of Brazed CBN Grains

The protrusion height of 200 grains on replaceable segments of the brazed CBN abrasive wheels was characterized using KH-7700 optical microscopy, as illustrated schematically in Figure 7.5. Figures 7.6 and 7.7 demonstrate the statistical results of the size of the original grains and the protrusion height of the brazed grains, respectively.

According to the measured values, it is reasonable to assume that both the original sizes and the protrusion height of CBN abrasive grains follow a normal distribution. The probability density function of the distribution of grain size and protrusion height becomes

$$P(x) = \frac{1}{\sigma\sqrt{2\pi}} e^{-(x-\mu)^2/(2\sigma^2)} \tag{7.3}$$

where μ is the mean grain size or the average protrusion height, and σ is the standard deviation.

The mean grain size and the standard deviation are 171.0 and 16.74 µm, respectively. The protrusion height of the brazed CBN grains and the corresponding standard deviation are 112.9 and 18.20 µm, respectively. Now it is known that the embedding part of the brazed CBN grain almost reaches 70% of the total grain crystal. Moreover, the percentage of brazed grains with a protrusion height ranging from 95 to 130 µm is about 70% of all the grains, while that between 75 and 150 µm is approximately 96% of all the grains. As a consequence, in this work, the change in grain protrusion height versus the accumulated removal material volume was measured and analysed for the particular brazed grains with

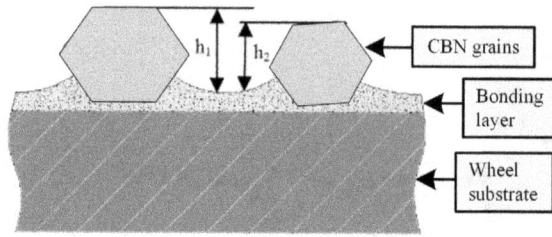

FIGURE 7.5   Schematic illustration of brazed grain protrusion.

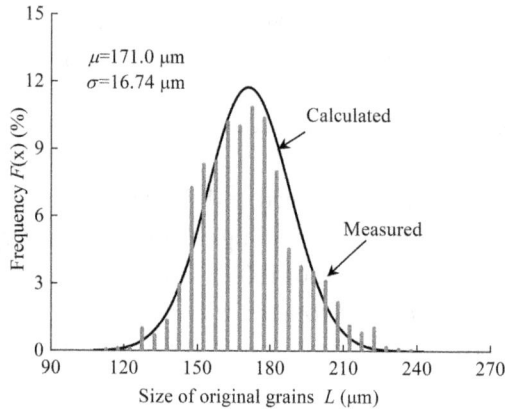

FIGURE 7.6   Size of original CBN grains.

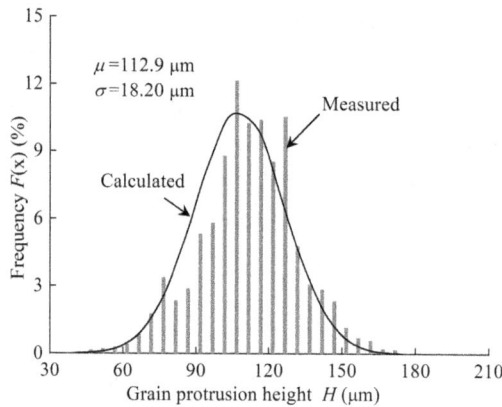

FIGURE 7.7   Protrusion height of brazed CBN grains.

an initial protrusion height of 150, 120, 90, and 75 μm, respectively. The results are shown in Figure 7.8.

As seen in Figure 7.8, a drastic change has been encountered for the grains with an initial protrusion height of 150 and 120 μm at the beginning of the grinding process. The protrusion height has decreased rapidly to about 100 μm due to the grain fracture. The reason is that under such conditions, only small quantities of grains work effectively.

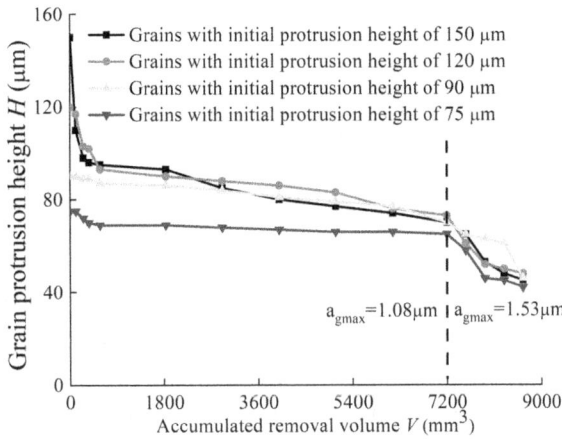

FIGURE 7.8 Grain protrusion height versus accumulated removal material volume.

The mechanical load on the individual grain is so large that the particular grain cracks. For the grains with an initial protrusion height of 90 and 75 µm, the wear curves are generally gentle.

In the steady grinding period, the effects of attritious wear become very obvious for all the grains in the case of the uncut chip thickness of 1.08 µm. Therefore, the grain protrusion height only decreases slowly. However, once the uncut chip thickness is increased to 1.53 µm, all the grains display accelerating wear due to the large mechanical load and the uniform grain protrusion.

7.2.5 General Wear Phenomenon of Brazed CBN Abrasive Wheels

As a whole, grain-cut edges on the wheel surface change their shapes in different patterns with the progress of wheel wear. The overall wheel wear is the culmination of numerous wear events from encounters between individual grains and the ground workpiece. Considering the strong joining between the CBN grains and the bonding layer, the possible wear patterns of a CBN abrasive wheel are illustrated schematically in Figure 7.9. The types of brazed wheel wear include mild wear (i.e., attrition wear, grain microfracture), and severe wear (i.e., grain macrofracture, erosion, or rupture of the bonding material supporting CBN grains).

The morphological changes of 200 grains captured during the grinding process were examined on the replaceable wheel segments. Figure 7.10 demonstrates the corresponding statistical results of the progressive wear patterns of the brazed CBN abrasive grains. Obviously, during grinding K424 cast nickel-based superalloy, there are mainly three patterns of wheel wear, such as attrition wear, grain microfracture, and macrofracture. There is no rupture, but the erosion behaviour of the bonding layer takes place once the grains have been worn entirely. For attrition wear and grain microfracture, their percentages display a fast rise in the early unsteady grinding process and then increase gradually in the subsequent steady period. This is in agreement with the change in grinding force and temperature. In particular, in the final accelerating wear period in the case of an uncut chip

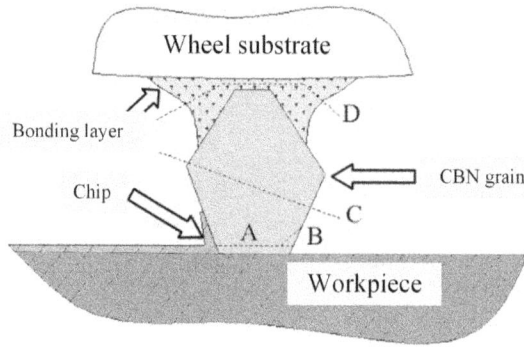

FIGURE 7.9   Schematic illustration of possible wear patterns of brazed CBN abrasive wheels.

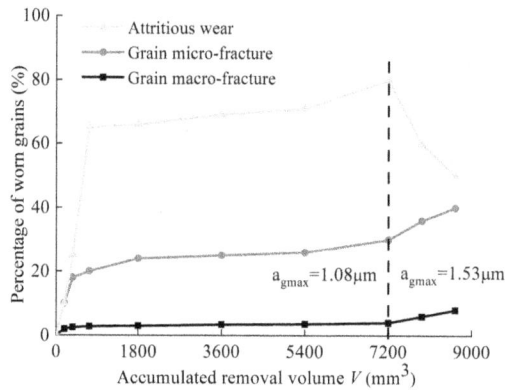

FIGURE 7.10   Grain wear patterns versus accumulated removal material volume.

thickness of $1.53\,\mu m$, the percentage of attritious wear drops rapidly because of the rise of the grain microfracture behaviour. However, it is also observed that the percentage of the macrofractured CBN grains almost stays constant throughout the whole grinding process, and the maximum value arrives at 8% at last.

In the initial wear period, many grains are not involved in cutting because they are blocked by higher grains before them. The grain-workpiece interaction zone is rather small. At this time, the mechanical load of the grains penetrating the workpiece surface is so large that the cutting edges of all the grains with high strength are microfractured or polished to a small wear flat. However, other grains with low strength may crack entirely. For this reason, there appears to be a high slope in the initial period of the progressive wear process, as shown in Figure 7.10. For a new single-layer brazed CBN abrasive wheel, the wear-flat area increases promptly during this period. A few CBN grains protruding the most are knocked away, and more CBN grains become active. Accordingly, the grinding forces increase drastically with the accumulated removal material volume.

In the steady wear period, the wear rate of CBN grains increases slowly. Here, the attritious wear and the grain microfracture have taken the decisive effect, dependent upon the extremely high wear-resistance ability of CBN superabrasive grains. The percentage of grain macrofracture stays at a low level. Therefore, the grinding forces and temperatures almost always remain constant.

In the final accelerating wear period, due to the increased uncut chip thickness and the accumulative wheel attritious wear, the grinding forces are raised and the temperatures are heightened significantly. The grinding process is dominated by the inefficient sliding and ploughing behaviour of grain interactions rather than the efficient cutting behaviour. Additionally, the workpiece material or chips are perhaps adhered to the CBN grains. Moreover, it causes the attritious wear of the bonding layer of the abrasive wheels. In this case, the wear behaviour of the abrasive wheels gets more serious.

7.2.6 Mild Wear Morphology of Brazed CBN Abrasive Wheels

As mentioned above, the mild wear of the brazed CBN abrasive wheel is mainly composed of the attritious wear and the grain microfracture behaviour.

Figure 7.11 shows the attrition wear of brazed CBN grain. In this case, the original sharp CBN grain has turned into a flat and smooth platform. This wear is most probably attributed to the fact that the abrasive grains rub against the workpiece material. The flat grain goes against penetrating the surface of the ground material and hence leads to a large grinding force and a high grinding temperature. Wear-flat zones increase gradually until the sharp CBN grains are flattened entirely. Also, a little workpiece material, which looks more like metal chips, may be found adhered sporadically and physically to CBN grains when the uncut chip thickness is 1.53 μm. Energy dispersive spectrum (EDS) analysis indicates little difference in chemical composition between the adhered material and K424 nickel-based superalloy. A large-scale adhesion phenomenon is not observed on the worn grains in the grinding process.

Though the grain cutting edges become dull because of the ductile attritious wear, the sharpness could be recovered due to the microfracture that occurred on the top surface of

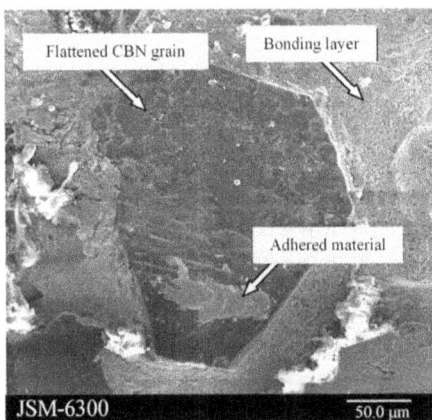
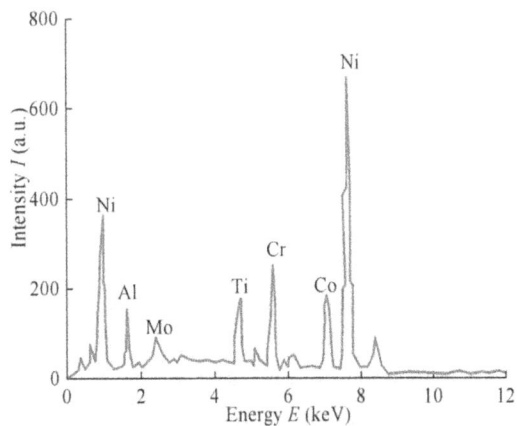

(a)   (b)

FIGURE 7.11 Attritious wear and sporadic adhered material of the brazed CBN grain: (a) Morphology of flattened CBN grain, (b) chemical composition of the adhered material.

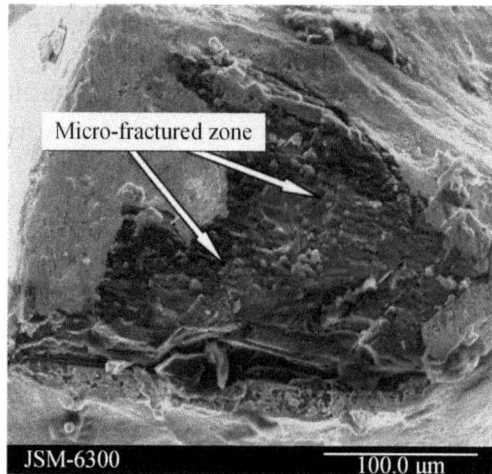

FIGURE 7.12   Microfractured grain.

the grains. For example, the self-sharpening phenomenon owing to the grain microfracture in the abrasive layer has been found using scanning electron microscopy, as displayed in Figure 7.12. Microcracks and fragments are presented on the surface of the CBN grain crystal. Because the individual grain temperature is only 500°C–600°C in the current investigation, thermal fatigue is not the most important reason for the grain fracture. It indeed results from the repeated mechanical impacts on the grains during grinding. Generally, grain microfracture benefits free cutting and is one of the most important factors in controlling the grinding ability of CBN abrasive wheels.

7.2.7 Severe Wear Morphology of Brazed CBN Abrasive Wheels

In addition to the mild wear, some grains endure severe wear under large grinding loads. The severe wear consists of grain macrofracture and erosion or rupture of the bonding material supporting CBN grains.

Figure 7.13 is the SEM image of the macrofractured grain, which is probably the further damage of a previous microfractured grain. Moreover, as shown in Figure 7.14, it is likely that the CBN grain is destroyed using cleavage fracture. This is a common characteristic of the crystalline material; for instance, a blocky-shaped cubo-octahedral single crystal of CBN. The cutting ability of the abrasive wheels is perhaps downgraded or even fallen into failure under extreme conditions. Although macrofracture behaviour of several grains has taken place, the cutting edges of the abrasive grains still work effectively in the current investigation.

At the same time, because of the abrasion effect, the grinding chips always cut back the bonding layer during the grinding operation, especially when the grain protrusion height is small. This may result in the abrasive grains being lost and advancing the wheel failure for the traditional electroplated and vitrified CBN abrasive wheels (Zhang et al. 2022; Macerol et al. 2020). The erosion appearance of the Ag-Cu-Ti bonding layer of a

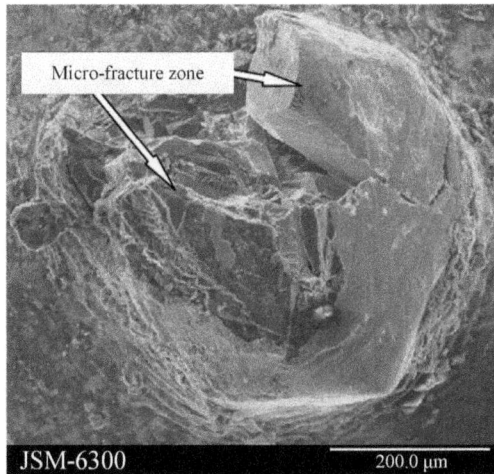

FIGURE 7.13   Macrofractured grain.

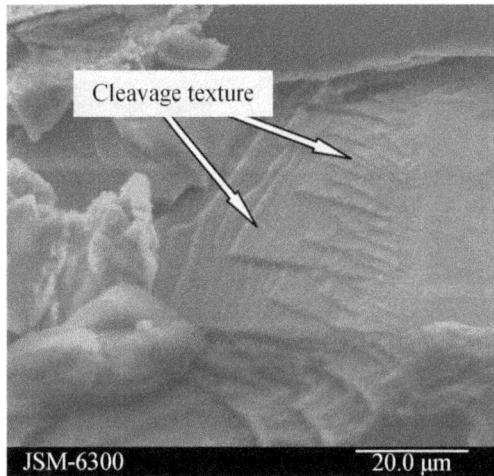

FIGURE 7.14   Cleavage fracture of brazed grains.

single-layer brazed CBN abrasive wheel is displayed in Figure 7.15. The bonding layer of the failed tool has been scratched severely. However, no rupture but weak plastic deformation of the bonding layer is observed. On the other hand, the single-layer brazed CBN wheels ensure that large quantities of abrasive grains cannot be removed from cutting. Thus, the number of active cutting edges is not reduced during grinding. Good machine quality of the workpiece components is obtained. All this contributes to the strong chemical bonding that has been realized between CBN grains and the Ag-Cu-Ti filler alloy in the brazing process. As seen in Figure 7.16, even though the grains have been worn entirely, the root segments of the grains are still retained in the bonding layer. This is very important for the single-layer CBN abrasive wheels to improve the stability of the active grain density and the corresponding uncut chip thickness.

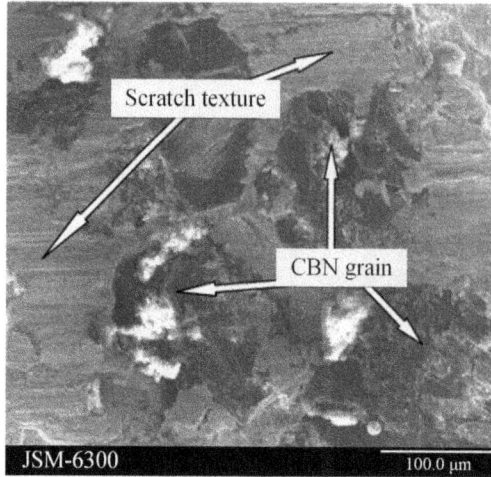

FIGURE 7.15   Erosion appearance of the bonding layer.

FIGURE 7.16   Wear and fracture morphology of the bonding layer.

7.3 STRESSES EFFECTS ON MONOCRYSTALLINE CBN GRAIN WEAR IN GRINDING

7.3.1 FE Modelling

7.3.1.1 Characteristics of a Single-Layer Brazed CBN Abrasive Wheel

A single-layer brazed CBN abrasive wheel of 400 mm in diameter is illustrated schematically in Figure 7.17. The wheel matrix is commercial AISI 1045 steel. The bond material is a Cu-Sn-Ti filler alloy. The CBN grain is assumed to have a regular hexagon shape with an average side length of 200 μm and a maximum grain diameter of 400 μm.

For a brazed CBN grain illustrated in Figure 7.17b, embedding depth and grain wear are key parameters to characterize it. The wear was zero before the grain was damaged. The percentage numbers indicated in the illustration, e.g., 30%, mean that 30% of the initial grain height has been worn out. The grain embedding depth refers to the depth of a grain embedded in the bond, measured from the highest surface point of the grain-bond junction.

Similarly, the embedding depth denoted by, e.g., 40%, indicates that there is 40% of the initial grain height embedded in the bond. In general, the initial embedding depth of grain, e.g., below 50%, is expected to provide sufficient space for storing chips during grinding. When the bond material wears out, the embedding depth of the brazed grain decreases.

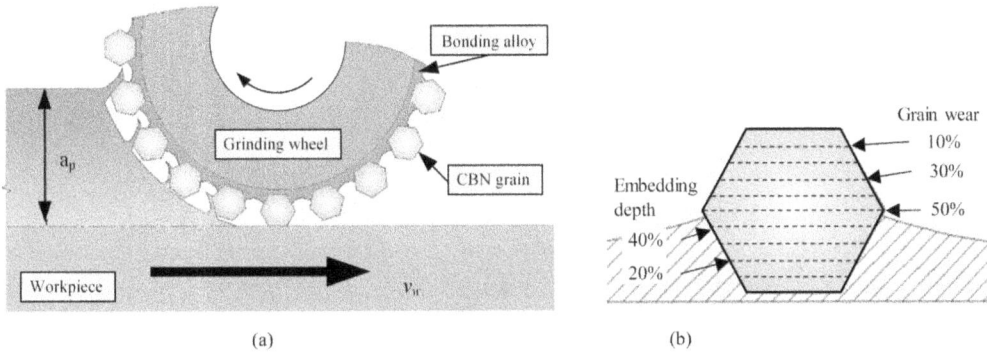

FIGURE 7.17 A schematic diagram of a monolayer-brazed CBN wheel: (a) a monolayer-brazed wheel, (b) a brazed CBN grain.

7.3.1.2 Simulating the Brazing Stresses

The stress analysis with the FE simulation is in two steps: (i) the calculation of brazing stresses in a CBN grain, including the redistribution of the stresses when grain/bond wear occurs, in which the grain/bond wear simulation is by element removal, layer by layer, based on the element death option in ABAQUS; and (ii) the calculation of the stresses in a grain with grinding forces applied, in which the effect of the brazing stresses coupled with the grinding stresses is taken into account.

The basic theory for brazing stress simulation can be found in previous literature (Guo et al. 2022; Zhu et al. 2017) and will not be repeated in this work, as the present focus is on the effect of variation in embedded depth. The grains in wheel fabrication can be uniformly distributed over the surface of a single-layer brazed CBN wheel. The spacing between two adjacent grains is 2 mm, which is far enough such that the brazing stresses around CBN grains do not influence each other. This means that our stress analysis can focus on a single grain and its surrounding bond material.

Figure 7.18a displays the two-dimensional (2D) FE model of a brazed CBN grain with its surrounding bond material. The gap (space) between the top surface of the wheel matrix and the bottom surface of the CBN grain is 10 μm. The size of the wheel matrix applied in the FE simulation is 5 mm in height and 2 mm in width. The area of the matrix region is beyond 200 times larger than that of a CBN grain with a side length of 200 μm. The six vertexes of the grain are rounded with a radius of 2 μm. Figure 7.18b and c display the mesh configurations with grain embedding depths of 30% and 50%, respectively.

In order to simulate the stresses, the boundary conditions to the control volume of the FE model are as follows: the node displacements along X (horizontal) and Y (vertical) directions in the bottom surface are zero; those along X direction in the left and right surfaces are zero too, as shown in Figure 7.18a. The grain-bond-matrix interfaces are assumed to be perfect. The element type used in the brazing residual stress simulation is CPS8R. The mesh contains 14,422 elements with 43,831 nodes when the initial grain embedding depth is 30%.

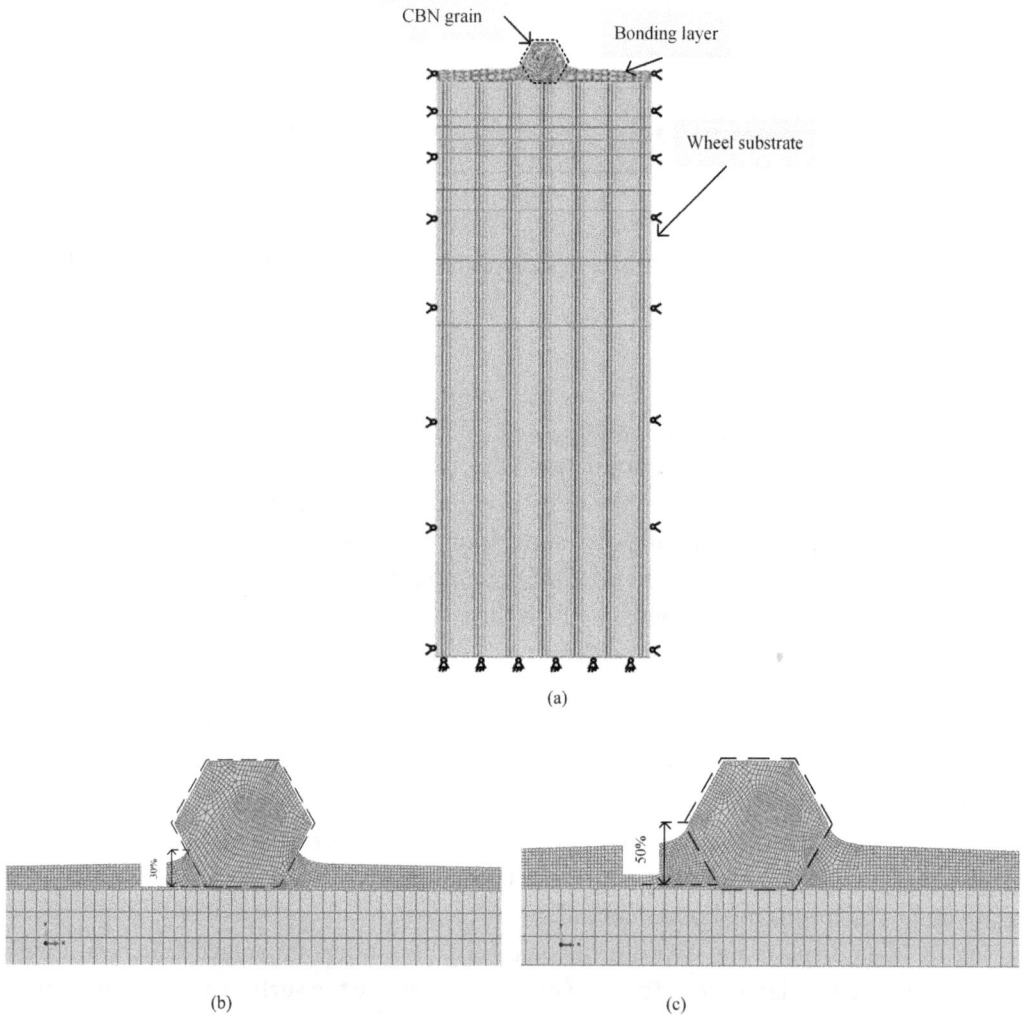

(a)

(b)   (c)

FIGURE 7.18 The FE model for the stresses analysis of a brazed CBN grain: (a) control volume, (b) grain embedding depth of 30%, and (c) grain embedding depth of 50%.

The brazing of CBN grains using Cu-Sn-Ti bonding alloy was usually carried out at 900°C for a holding time of 8 min. Both the heating and cooling rates were 10°C/min in the brazing cycle. Such a brazing cycle was the optimized procedure to acquire good bonding interfaces. In the whole brazing process, the CBN grain is elastic, while the wheel matrix (e.g., AISI 1045 steel) and the bond layer (e.g., Cu-Sn-Ti alloy) are elastic-plastic. During the heating stage of the brazing cycle, both the CBN grain and wheel matrix are not constrained and are thus stress-free. During the cooling stage from the brazing temperature (900°C) to the solidi temperature of the Cu-Sn-Ti alloy (800°C), no residual stress is formed in the brazed joint. Thus, our brazing-induced residual stress calculations will focus on the cooling period from 800°C to room temperature (20°C). The properties of the materials (thermal expansion coefficient, yield strength, Poisson's ratio, and Young's modulus) are taken from the published literature or supplied by the manufacturers, as listed in Table 7.1. In the present simulation, if a property value at a given temperature is not available, the corresponding value at the closest lower temperature is used.

This study also applies the following hypotheses: (i) the physical and mechanical properties of the materials are isotropic; (ii) the temperature distribution in the brazed joint is uniform; (iii) the solutionization and diffusion of the brazed components can be ignored; (iv) the creep behaviour of the materials in the brazing process is negligible; and (v) there are no crystal defects or microcracks in the CBN grains.

TABLE 7.1 Material Properties Used in the FE Calculation of the Brazing-Induced Residual Stresses

Property	Material		
	CBN	Cu-Sn-Ti Alloy	AISI 1045 Steel
Young's modulus E (GPa)	909	113	210 (293 K)
			185 (600 K)
			160 (800 K)
Poisson's ratio μ	0.12	0.29	0.26
Thermal expansion coefficient α (10^{-6} K^{-1})	4.92	18.3	9.1 (293 K)
			13.09 (500 K)
			13.71 (700 K)
			14.18 (800 K)
			14.67 (900 K)
Yield strength σ_s (MPa)		245	360 (293 K)
			263 (600 K)
			179 (800 K)
			78 (900 K)

7.3.1.3 FE Model for Simulating the Resultant Stresses in a Brazed CBN Grain during Grinding

After the analysis of the brazing-induced stresses described above, the grinding forces are applied to the vertex region of the brazed grain. The grinding forces are obtained based on a grinding process simulation using a single CBN grain, as shown in Figure 7.19a, in which a quantitative correlation between the grinding forces and the uncut chip thickness, a_{gmax}, has been given, which has included the effects of abrasive wheel speed, workpiece infeed rate, and depth of cut. For example, when a_{gmax} is 0.6 μm during grinding nickel-based superalloy Inconel718 with a single grain, the simulated normal force F_n is 2.6 N, and the tangential force F_t is 1.4 N. In this work, the above normal and tangential forces are loaded on the grain. The loading zone in the grain vertex region, which is thought to be the contact zone between the ground material and CBN grain, is determined by $a_{gmax} = 0.6$ μm, as illustrated in Figure 7.19b. Accordingly, the resultant stresses coupling the effects of brazing-induced stresses and those of grinding-induced stresses can be analysed.

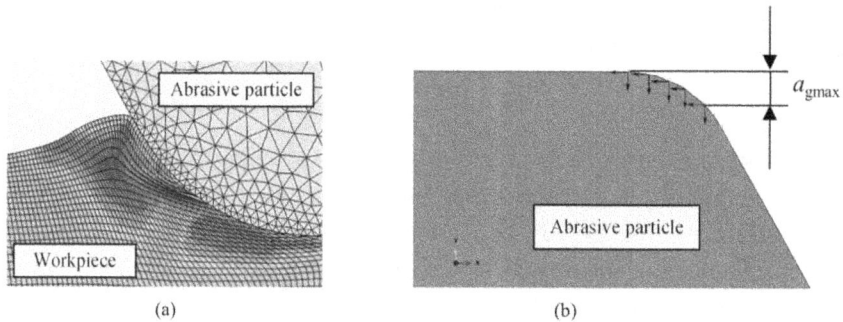

(a)   (b)

FIGURE 7.19   Grinding loads on the grain vertex region: (a) simulation of single-grain grinding, (b) grinding forces loaded in the rounded grain vertex region within the undeformed chip thickness.

7.3.1.4 Region Selection for Stresses Analysis of a Brazed CBN Grain

CBN is a brittle material and has higher strength when compressed. However, a CBN grain can fracture easily under tensile stress. Thus, a tensile failure criterion is suitable for the CBN grain; that is, when the maximum principal stress in the grain reaches a critical value, fracture occurs. For this reason, five reference paths, denoted by I, II, III, IV, and V, respectively, are chosen on the tensile side of a CBN grain for the stress analysis, as illustrated in Figure 7.20, with their positive directions denoted by the arrows. The path I is from the centre point to the side point in the grain bottom surface, which represents the bottom grain-bond interface. Path II is from the grain centre to the side vertex. Path III is from the top surface centre to the top vertex. Path IV is from the bottom surface centre of the grain to its top surface centre, i.e., the central axis of the grain. Path V connects three vertex points, from the bottom to the middle, then to the top, in which the middle vertex is a turning point of the grain geometry. Along Path V, the part of the grain below the junction region of the bond material is embedded in the bond material, representing an actual grain bonding situation in a monolayer-brazed CBN wheel.

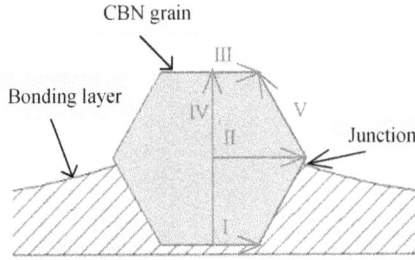

FIGURE 7.20   Five reference paths of a brazed CBN grain.

7.3.2 Effect of Grain Embedding Depth on Brazing Stresses and Resultant Stresses

Figure 7.21 demonstrates the typical contours of the maximum in-plane principal braz-ing stress in the brazed CBN grains with different embedding depths. It is obvious that the stress distributes symmetrically and that significant stress concentration occurs at the grain-bond interface due to the mismatch between the thermal expansion coeffi-cients of CBN and the bond material, Cu-Sn-Ti alloy. Figure 7.22 shows the resultant stress after the application of the grinding forces. Overall, the resultant stresses distribu-tion and the brazing stresses distribution are similar, except in the grain vertex region, where the grinding forces are loaded. Figures 7.23 and 7.24 compare quantitatively the brazing and resultant stresses along Paths I–V, with a variation in the initial embedding depth ranging from 20% to 50%.

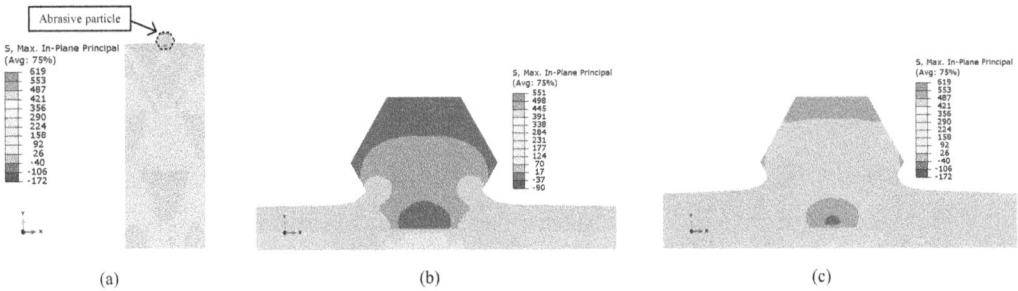

FIGURE 7.21   Variation of the maximum in-plane principal brazing stresses with grain embedding depth: (a) over the whole control volume, (b) grain embedding depth of 20%, and (c) grain embed-ding depth of 40%.

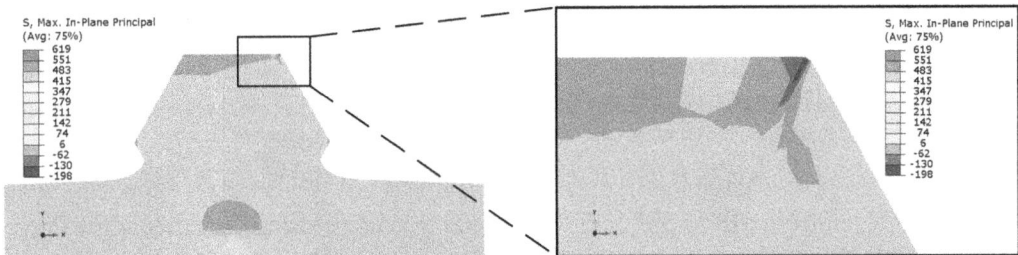

FIGURE 7.22   Resultant stresses distribution in the CBN grain (initial embedding depth of 40%): (a) stresses in the grain-bond joint, (b) stresses distribution around the grain vertex.

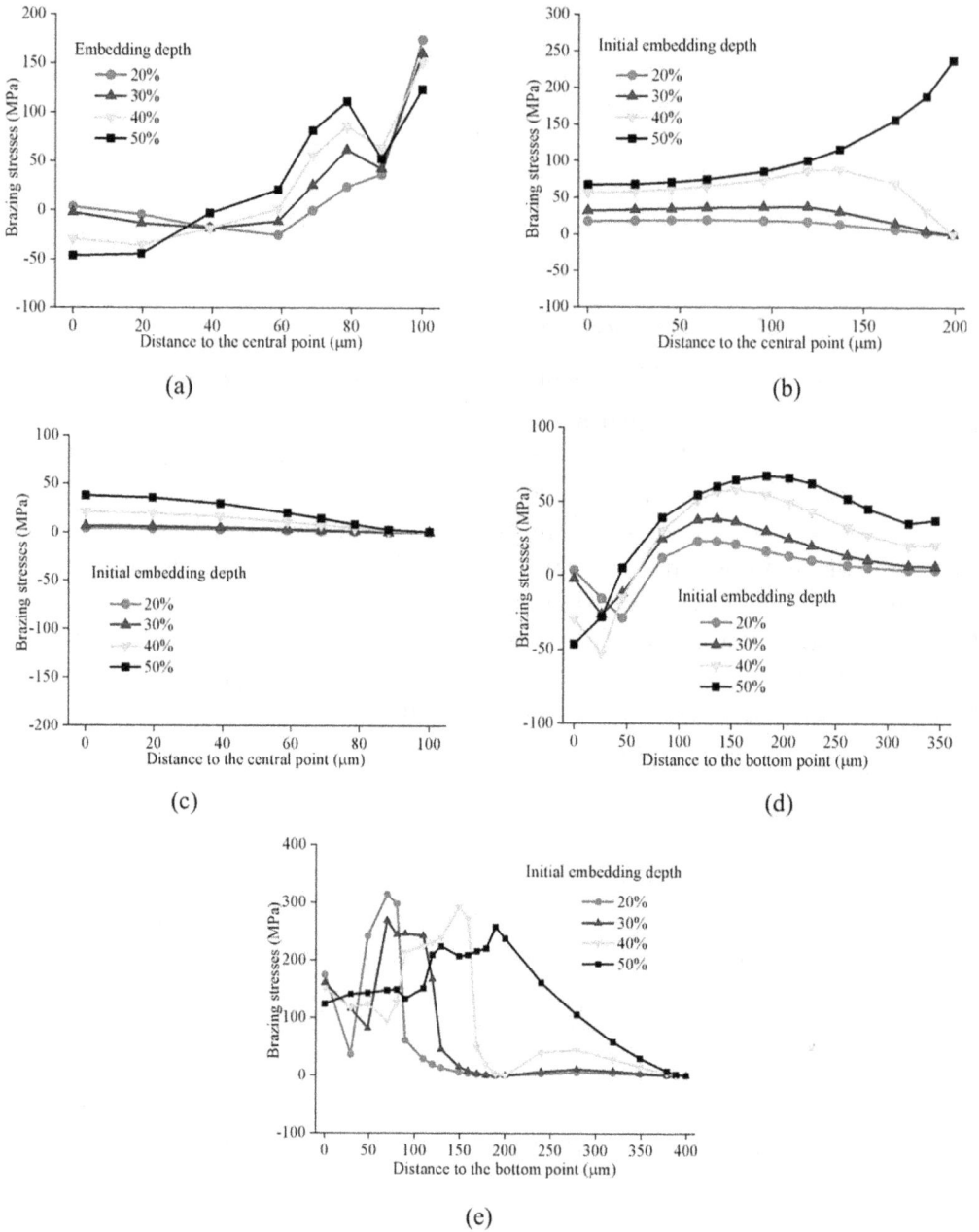

FIGURE 7.23 Variation of brazing stresses with the initial embedding depth: (a) Brazing stresses along Line I, (b) Brazing stresses along Line II, (c) Brazing stresses along Line III, (d) Brazing stresses along Line IV, and (e) Brazing stresses along Line V.

Figure 7.23a shows that along Path I, the brazing stresses are generally compressive in the central region from 0 to 40 μm in the horizontal axis when the initial embedding depth is 40% or 50%. However, when the initial embedding depth is 20% or 30%, the stresses are close to zero in the region from 0 to 60 μm. Beyond this region, the stresses become tensile.

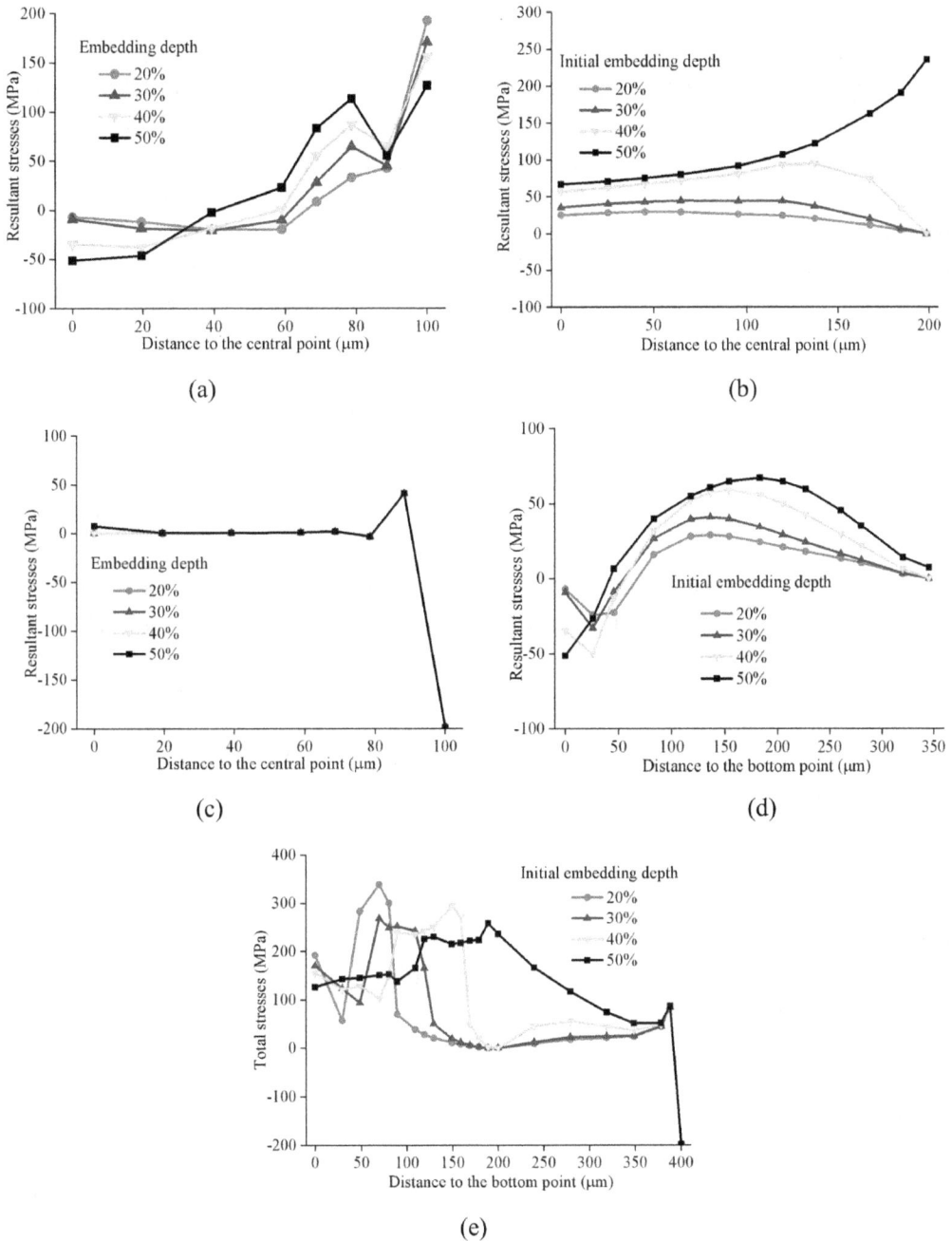

FIGURE 7.24 Variation of resultant stresses with the initial embedding depth: (a) Resultant stresses along Line I, (b) Resultant stresses along Line II, (c) Resultant stresses along Line III, (d) Resultant stresses along Line IV, and (e) Resultant stresses along Line V.

Around the end of Path I, the largest brazing-induced stress decreases with increasing the embedding depth. Comparing with Figure 7.24a, it can be seen that the grinding forces have a very small influence on the resultant stresses along with the bottom grain-bond interface.

Along Path II, the brazing stresses are always below 75 MPa when the initial embedding depth is 20%–40%, as shown in Figure 7.23b, and almost vanish in the neighbourhood of the path end around the vertex of the grain. The stresses are so small that they do not contribute to the grain fracture. When the grain embedding depth increases to 50%, however, tensile stresses emerge, which increase from the grain centre to its surface along Path II, reaching 240 MPa. This is mainly caused by the higher thermal shrinkage of the bond material (e.g., Cu-Sn-Ti alloy). Figure 7.24b shows that grinding forces do not add too much to the resultant stresses, which is similar to the Path I situation observed above.

Figure 7.23c demonstrates the variation of brazing stresses along Path III when the grain embedding depth increases, which shows that the stresses are tensile, though the magnitudes are not so large. This is understandable because the thermal shrinkage of the bond material applied makes the grain vertex region expand slightly, and it is reasonable to expect that the peak tensile stresses always appear at the centre of the grain (beginning of Path III). Then they gradually decrease with the increasing distance to the original point along Path III. The application of the grinding forces will certainly change the resultant stresses at the vertex of Path III, as shown in Figure 7.24c.

Along Path IV, the magnitudes of the brazing stresses are generally small, which vary from −50 to 70 MPa with the different embedding depths, as shown in Fig. 7.23d. When compared Figure 7.23d with Figure 7.24d, it can be seen that the application of the grinding forces has very little influence on the brazing stresses on Path IV, indicating that brazing stresses dominate this region.

On Path V, the brazing stresses are always tensile and reach their peaks at the grain-bond interface in the junction region, as shown in Figure 7.23e. It can be seen that the magnitudes of the brazing tensile stresses along Path V are generally larger than the corresponding stresses on Paths I–IV. Again, the brazing stresses on Path V are due to the mismatch of thermal properties of the grain material (CBN), the bond materials (Cu-Sn-Ti) and the wheel matrix (AISI 1045 steel). Such tensile stresses increase the possibility of the brittle CBN grain fracturing during grinding, because Path V is actually on the rake surface of the grain (cutting edge). When grinding forces are applied, as shown in Figure 7.24e, the peak resultant tensile stresses at the grain-bond junction increase, though not significantly, and tension takes place in the region close to the grain vertex where grinding forces are applied. The increase in tensile stress is mainly because of the applied tangential force F_t during grinding.

The above discussion shows that brazing stresses dominate the deformation state of the grain. The rake surface of the grain (Path V) experiences tensile stresses, which makes the grain easier to fracture and wear. As such, only the stress distribution along Path V will be considered in the wear analysis of the grain in the following section.

7.3.3 Effect of Grain Wear on Redistributed Brazing Stresses and Resultant Stresses

When a grain wears out, the stresses distribution in the grain will change. In the simulations, the grain wear is realized by removing the elements layer by layer. Figure 7.25a shows that the peak values of the redistributed brazing stress in the grain-bond junction region decrease when the grain wears more, from, e.g., 280 MPa at the grain wear of 10% to 210 MPa at the grain wear of 50%. When the grinding forces are applied, the resultant peak

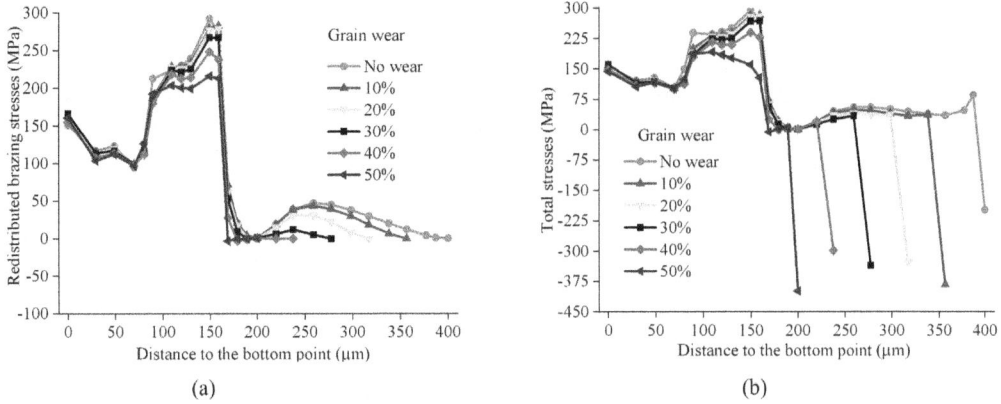

FIGURE 7.25   Effect of grain wear on the stresses distributions along Path V (initial embedding depth of 40%): (a) redistributed brazing stresses, (b) resultant stresses during grinding.

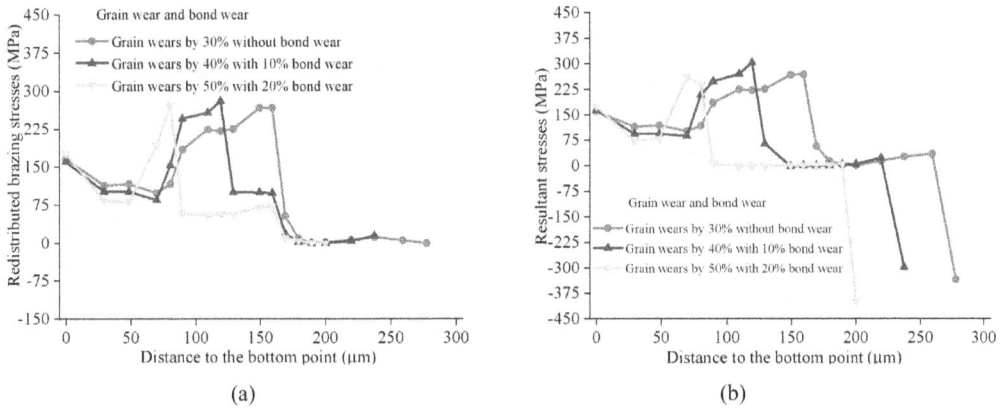

FIGURE 7.26   Effect of simultaneous grain and bond wear on the stresses distribution along Path V (initial embedding depth of 40%): (a) redistributed brazing stresses, (b) resultant stresses during grinding.

stresses at the junction will have noticeable changes when the grain wear reaches 50%, as shown in Figure 7.25b. The large compressive stresses near the grain vertex region are due to the grinding forces applied.

In the above, only the stresses redistributions by grain wear are examined. In an actual grinding process, grain wear and bond wear happen simultaneously. Additionally, in-situ electrochemical wheel dressing may be applied to remove the bond material to obtain an essential grain protrusion height. It is, therefore, necessary to understand the effect of simultaneous grain wear and bond wear on the stress distributions.

Figure 7.26 compares quantitatively the stresses distributions along Path V in the CBN grains under three wear conditions: (i) the grain wears by 30% without bond wear; (ii) the grain wears by 40% with 10% bond wear, and (iii) the grain wears by 50% with 20% bond wear. It can be seen that simultaneous grain and bond wear brings about a significant redistribution of both brazing and resultant stresses, although the changes in their peak values are minimal.

7.3.4 Effect of Grain Size on Brazing Stresses and Resultant Stresses

To meet different grinding requirements in terms of quality and efficiency, different brazed CBN wheels may contain grains of different sizes. To understand the grain size effect, let us consider a range of CBN grains with side lengths ranging from 50 to 200 μm, which corresponds to the most common grain diameters ranging from 100 to 400 μm. The embedding depth is fixed at 40%.

Figure 7.27 compares the brazing and resultant stresses along Path V in the grains of different sizes. To easily discuss the stresses of different grain sizes, the horizontal axis of Figure 7.27 is normalized by the total length of Path V. It can be seen that the peak values of the brazing-induced tensile stresses always occur in the grain-bond junction region, as shown in Figure 7.27a. A larger grain experiences a smaller peak of the brazing stresses.

Figure 7.27b shows that grain size has a significant effect on the resultant stresses. With the grain of 50 μm in side length, both the compressive and tensile stresses are the highest, reaching −1,012 and 630 MPa, respectively, in comparison with those of −198 MPa and 70 MPa in the grain of 200 μm in side length. Also, the resultant stresses in the grain-bond junction region during grinding have also shown the grain size effect, whose peak value increases from 420 to 550 MPa.

7.3.5 Effect of Grinding Load on Resultant Stresses

Four groups of normal F_n and tangential F_t grinding forces are examined, as listed in Table 7.2, with an initial grain embedding depth of 40% and a grain side length of 200 μm. Here, the grinding forces are obtained at the given uncut chip thickness by the grinding simulation. Furthermore, when the uncut chip thickness increases, the loading zone in the grain vertex region also enlarges.

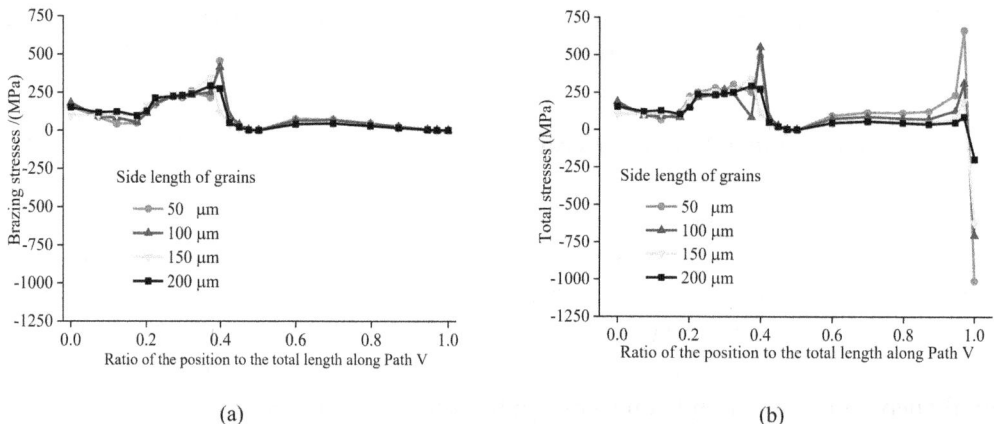

(a)   (b)

FIGURE 7.27 Variation of resultant stresses along Path V with grain size (initial embedding depth of 40%): (a) brazing stresses, (b) resultant stresses.

TABLE 7.2 Grinding Forces versus Uncut Chip Thickness of a Grain

Uncut Chip Thickness a_{gmax} (μm)	Normal Grinding Forces F_n (N)	Tangential Grinding Forces F_t (N)
0.6	2.6	1.4
2	3.5	1.5
4	4.6	2.2
8	9.4	4.9

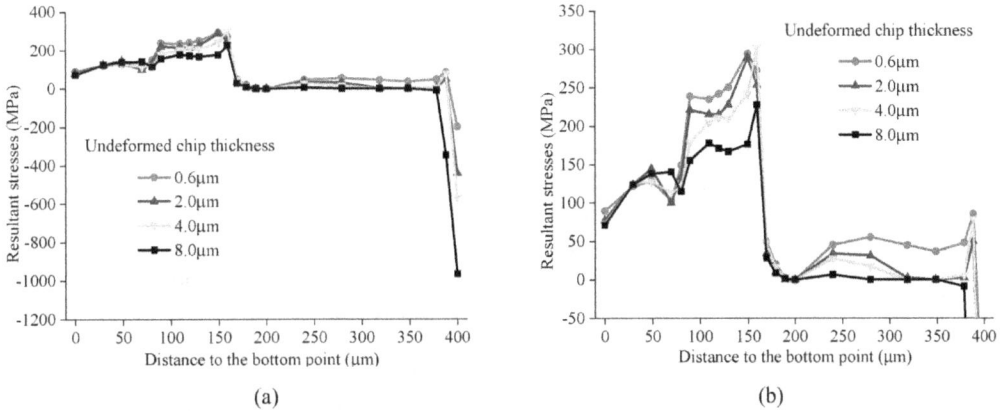

FIGURE 7.28 Variation of resultant stresses along Path V with grinding loads (initial embedding depth of 40%): (a) full stresses range, (b) narrowed stresses range.

Figure 7.28 shows the effect of grinding forces on the resultant stresses along Path V of the grain, where Figure 7.28b includes a narrowed range of the stresses range to try to display the effect on stresses peak values. It can be seen that both the tensile stresses in the junction region and the peak compressive stresses in the grain vertex region are influenced. When the undeformed chip thickness increases from 0.6 to 8.0 μm, the resultant compressive stresses in the grain vertex region vary significantly from −198 to −1,012 MPa, but the peak tensile stresses in the junction region decrease from 290 to 220 MPa. The decrease in peak tensile stress mentioned above is mainly attributed to the increasing compressive stress from the normal force F_n, the effect of which is larger than that of the tangential force F_t during grinding.

Moreover, in the zone from the junction region (e.g., 150 μm at the horizontal axis) to the grain vertex region, tensile stresses are always below 50 MPa. These indicate that grain fracture wear should generally take place in the junction region (macrofracture) and the vertex region (microfracture) of a grain. On the other hand, because the resultant tensile stresses in the junction regions decrease with increasing simultaneously the grinding force and loading zone, the possibility of grain macrofracture drops. Thus, to effectively control the macrofracture wear of a brazed grain, a larger uncut chip thickness is more favourable.

REFERENCES

Bredthauer M., P. Snellings, P. Mattfeld, et al. 2023. Wear-related topography changes for electroplated cBN grinding wheels and their effect on thermo-mechanical load. *Wear*, 512–513: 204543.

Chen, Y., Y. C. Fu, H. H. Su, et al. 2014. The effects of solder alloys on the morphologies and mechanical properties of brazed diamond grits. *International Journal of Refractory Metals and Hard Materials*, 42: 23–29.

Chen, Y., J. H. Xu, Y. C. Fu, et al. 2009. Finite element analysis of residual stress in diamond/steel brazed joint. *Material Science Forum*, 626–627: 195–200.

Ding, K., Y. C. Fu, H. H. Su, et al. 2014. Wear of diamond grinding wheel in ultrasonic vibration-assisted grinding of silicon carbide. *International Journal of Advanced Manufacturing Technology*, 71: 1929–1938.

Ding, W. F., Y. J. Zhu, L. C. Zhang, et al. 2015. Stress characteristics and fracture wear of brazed CBN grains in monolayer grinding wheels. *Wear*, 2015, 332–333: 800–809.

Eiss, N. S. 1967. Fracture of abrasive grain in grinding. *Transaction of the ASME-Journal of Engineering for Industry*, 89: 463–469.

Grimmert, A., F. Pachnek, and P. Wiederkehr. 2023. Temperature modeling of creep-feed grinding processes for nickel-based superalloys with variable heat flux distribution. *CIRP Journal of Manufacturing Science and Technology*, 41: 477–489.

Guo, S., L. Sun, J. Fang, et al. 2022. Residual stress, microstructure and corrosion behavior in the 316L/Si3N4 joint by multi-layered braze structure-experiments and simulation. *Ceramics International*, 48(22): 32894–32907.

Jiang, J. L., P. Q. Ge, W. B. Bi, et al. 2013. 2D/3D ground surface topography modeling considering dressing and wear effects in grinding process. *International Journal of Machine Tools and Manufacture*, 74: 29–40.

Lee, E. T., Z. Fan, and B. Sencer. 2021. Estimation of cBN grinding wheel condition using image sensor. *Procedia Manufacturing*, 53: 286–292.

Macerol, N., L. Franca, and P. Krajnik. 2020. Effect of the grit shape on the performance of vitrified-bonded CBN grinding wheel. *Journal of Materials Processing Technology*, 277: 116453.

Qian, N., Y. Fu, A. Khan, et al. 2022. Holistic sustainability assessment of novel oscillating-heat-pipe grinding-wheel in Earth-friendly abrasive machining. *Journal of Cleaner Production*, 352: 131486.

Qian, N., Y. Fu, F. Jiang, et al. 2021. CBN grain wear during eco-benign grinding of nickel-based superalloy with oscillating heat pipe abrasive wheel. *Ceramics International*, 48(7): 9692–9701.

Shaji, S., and V. Radhakrishnan. 2003. An investigation on solid lubricant moulded grinding wheels. *International Journal of Machine Tools and Manufacture*, 2003, 43: 965–972.

Suh, C. M., K. S. Bae, and M. S. Suh. 2008. Wear behavior of diamond wheel for grinding optical connector ferrule-FEA and wear test. *Journal of Mechanical Science and Technology*, 22: 2009–2015.

Zhang, C., S. Qu, W. Xi, et al. 2022. Preparation of a novel vitrified bond CBN grinding wheel and study on the grinding performance. *Ceramics International*, 48(11): 15565–15575.

Zhou, Y., P. D. Funkenbusch, and D. J. Quesnel. 1997. Stress distributions at the abrasive-matrix interface during tool wear in bound abrasive grinding-a finite element analysis. *Wear*, 209: 247–254.

Zhu, Y., W. Ding, Z. Rao, et al. 2019. Self-sharpening ability of monolayer brazed polycrystalline CBN grinding wheel during high-speed grinding. *Ceramics International*, 45(18A): 24078–24089.

Zhu, Y., W. Ding, T. Yu, et al. 2017. Investigation on stress distribution and wear behavior of brazed polycrystalline cubic boron nitride superabrasive grains: Numerical simulation and experimental study. *Wear*, 376–377(B): 1234–1244.

CHAPTER 8

Wear Behaviour of Polycrystalline CBN Abrasive Wheels

8.1 INTRODUCTION

Compared with the monocrystalline cubic boron nitride (CBN) grains, the unique microstructure of polycrystalline CBN grains overcomes the drawback of anisotropy, which tends to cause cleavage in grinding due to the limited slipping planes. The polycrystalline CBN abrasive tools mainly work in extreme conditions such as high temperatures, friction, and impact load environments, which could lead to brittle fracture of the abrasive grains, resulting in a reduction in the number of effective abrasive grains to shorten the tool life. Therefore, it is necessary to investigate the crack propagation and fracture toughness of polycrystalline CBN abrasive materials as well as the influencing factors.

The mechanical properties of polycrystalline CBN abrasive grains mainly depend on the internal microstructure, which consists of grain boundaries, grains with different sizes, shapes, and constituents, inclusions, and porosity. Concerning the influence of the internal microstructures of polycrystalline CBN on its mechanical properties, several experimental studies have been conducted. For example, Liu et al. (2016) studied the effect of grain size on the mechanical properties of CBN-Si composite, in which it was found that the hardness and thermal stability increased with the increase of grain size and the fracture toughness decreased with the increase of grain size. McKie et al. (2011) investigated the influence of grain size and Al content on the hardness, fracture toughness, and flexural strength of the CBN-Al composite, based on which they pointed out that the hardness of the CBN-Al composite decreased with the increase in CBN grain size and Al content, fracture toughness depended on the combined relationship between the grain size and Al content, and there was no significant relationship between flexural strength, grain size, and Al content. Yue et al. (2016) studied the influence of the mass ratio of Ti_3SiC_2 on the flexural strength, compressive strength, and Vickers hardness of

DOI: 10.1201/9781032678047-10

the CBN-Ti$_3$SiC$_2$ composite. It was found that the flexural strength decreased with the increase of the Ti$_3$SiC$_2$ mass ratio, and the compressive strength and hardness increased with the increase of the Ti$_3$SiC$_2$ mass ratio.

The experimental studies mentioned above reveal the influence of the various microstructural characteristics of polycrystalline CBN on its mechanical properties. However, it is usually time-consuming and cost-expensive to sinter polycrystalline CBN and measure the various mechanical parameters of polycrystalline CBN. Meanwhile, the control of the input parameters is limited in an experimental operation, because it is difficult to change one single parameter while keeping other parameters unaffected. However, by resorting to the finite element method, these problems would not exist. Using the finite element method, various material parameters and load conditions can be set to simulate the mechanical behaviour of the material under various working conditions, and the simulation results could be conveniently obtained in the post-processor. The cohesive model is a very effective method in simulation research concerning crack propagation and fracture toughness of polycrystalline materials. After the successful implementation of the cohesive model into the finite element method, it has been widely used in the crack propagation simulation of brittle polycrystalline materials. Sfantos and Aliabadi (2007) developed a cohesive grain boundary element formulation to model intergranular microcracking evolution in polycrystalline brittle materials. The stochastic effects of each grain morphology-orientation, internal friction, and randomly distributed pre-existing flaws on the overall behaviour and microcracking evolution of a polycrystalline brittle material were studied, respectively. Yin et al. (2012) inserted cohesive elements with the bilinear softening law into both the mastic and the interfaces between the mastic and aggregates to simulate crack initiation and propagation of asphalt mixture under uniaxial tensile loading. The influence of aggregate distribution and parameters of the cohesive model on the performance of the asphalt mixture were evaluated. Zhou et al. (2012) utilized a cohesive element method to simulate the phenomenon of crack propagation in the microstructures of ceramic tool materials. The influence of grain size, grain boundary strength, and microcracks on cracking patterns and fracture toughness were analysed, respectively. Kraft and Molinari (2008) developed a two-dimensional finite element model using a cohesive interface approach to investigate transgranular fracture in polycrystalline alumina under tensile loading. The effects of grain boundary distributions on mesoscopic failure strength and fracture energy and the resulting percentages of transgranular fracture were examined. These studies revealed that the cohesive model could be used for a variety of materials to simulate a variety of fracture mechanisms.

This work mainly deals with the effects of grain size, the bonding strength of the grain boundary, and grain boundary stiffness on the fracture toughness of polycrystalline CBN abrasive materials under tensile loading. The potential crack paths of polycrystalline CBN are represented through embedding cohesive elements based on the traction-separation law both in grains and along grain boundaries. Simulation calculations are performed through the general finite element analysis software Abaqus. In this research work, the energy dissipations due to fractures of various microstructures as well as the length of cracks and the percentage of transgranular fractures are calculated. The relationship between fracture toughness and crack characteristics is analysed in detail.

8.2 WEAR PHENOMENON OF POLYCRYSTALLINE CBN ABRASIVE WHEELS IN GRINDING

8.2.1 Experimental Details

The experimental work is first carried out to detect the grain wear of the brazed polycrystalline CBN abrasive wheel heads by comparison with their brazed monocrystalline CBN counterparts during grinding (Ding et al. 2011). The size of the polycrystalline CBN abrasive grains used in this work was 300–400 μm, which was equal to that of the monocrystalline CBN grain of 40/50 US mesh. The binder material in the polycrystalline CBN grains was AlN. The size of the microcrystalline CBN particles in polycrystalline CBN grains was about 5–10 μm. Figure 8.1a shows the morphology of a brazed polycrystalline CBN abrasive wheel head, which has a wide application prospect in polishing the tool marks of the turning and milling surfaces of the machined components. As seen in Figure 8.1b, good joining has been realized between the polycrystalline CBN grains and Ag-Cu-Ti filler layer.

Figure 8.2 demonstrates the experimental setup. The surface grinding machine, model HZ-Y150 was modified to realize a constant-force grinding technique.

(a)   (b)

FIGURE 8.1   Brazed polycrystalline CBN abrasive wheel head: (a) brazed tool, (b) brazed grain.

FIGURE 8.2   Experimental setup.

The spindle model GD0800C was applied. The length and width of the machined components of the titanium alloy Ti-6Al-4V were 110 and 8 mm, respectively. The diameter of the brazed wheel heads was 12 mm. The constant normal grinding force was provided through the gravity of the machined component and the additive. In particular, the workpiece and the additive may be moved together up and down along the guide rail depending on the opposite force of the abrasive tools on the machined components. Under such conditions, the grain penetration depth and corresponding stock removal of the ground material were varied gradually with the progressive wear of the abrasive wheel heads.

Grinding tests were carried out in the down mode at a fixed wheel velocity of approximately 10 m/s (17,000 rpm), a workpiece speed of 4 m/min, and a constant normal force of 3 N. A commercial 5% dilution of emulsified liquid was provided for cooling at 5 L/min. Five brazed wheel heads with CBN abrasive grains of 40/50 US mesh were also utilized for comparison with the five counterparts with polycrystalline CBN grains in the present experiments.

The measurements were taken every 20 grinding passes until the surface roughness of the ground components was below Ra 1.6 μm. The measured experimental data were mainly composed of the stock removal and the progressive weight loss of the worn wheel heads. Especially, the loss weight of the abrasive tools was measured using an electrical balance (Adventurer AR1140) with a readability of 0.0001 g. The average results of the brazed abrasive tools were obtained. The wear morphology of the abrasive grains was examined using KH-7700 optical microscopy (OM) and QUANTA-200 scanning electron microscopy (SEM). The chemical composition of the adhered material was detected by EDAX energy-dispersive spectroscopy (EDS). The surface roughness of the samples was measured perpendicular to the grinding direction using a MAHR perthometer M2 surface roughness tester.

8.2.2 General Wear Phenomenon of Polycrystalline CBN Abrasive Wheels

Figure 8.3 shows the stock removal and the accumulated loss weight of the polycrystalline CBN and monocrystalline CBN abrasive wheel heads versus the grinding passes, respectively. For brazed monocrystalline CBN abrasive tools, the stock removal reaches about 2,800 mm³ with 120 grinding passes. Under the same condition, the stock removal of the brazed polycrystalline CBN tools arrives at 3,000 mm³. Moreover, for the brazed polycrystalline CBN ones, the stock removal finally reaches 3,500 mm³ and increases by 25% with a total of 160 grinding passes. On the other hand, because the material removal rates may be reflected by the slopes of the curves of stock removal, it is found that the material removal rates of polycrystalline CBN tools are higher than those of monocrystalline CBN tools. On the other hand, according to Figure 8.3b, the wear behaviour of both the polycrystalline CBN and monocrystalline CBN abrasive tools is in three stages: the initial stage with rapid wear, followed by the steady stage with a more or less constant wear rate, and finally the abrupt stage with accelerating wear. In general, the wear behaviour of brazed monocrystalline CBN abrasive tools is severer than that of their brazed polycrystalline CBN counterparts. The brazed monocrystalline CBN tools end their service life within about 120 grinding passes, while the polycrystalline CBN tools end within nearly 160 passes. In other words, the service life of the abrasive tools with

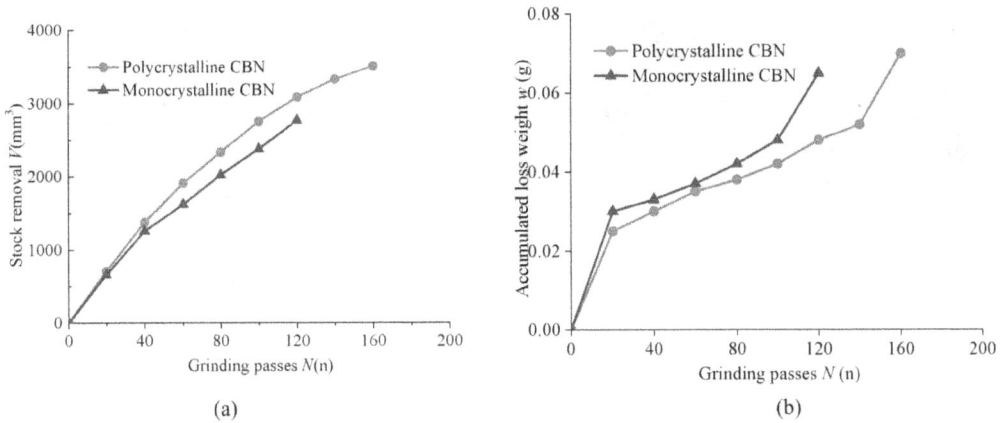

FIGURE 8.3 Comparison of grinding performance of brazed PCBN and CBN abrasive wheel heads: (a) stock removal, (b) tool wear.

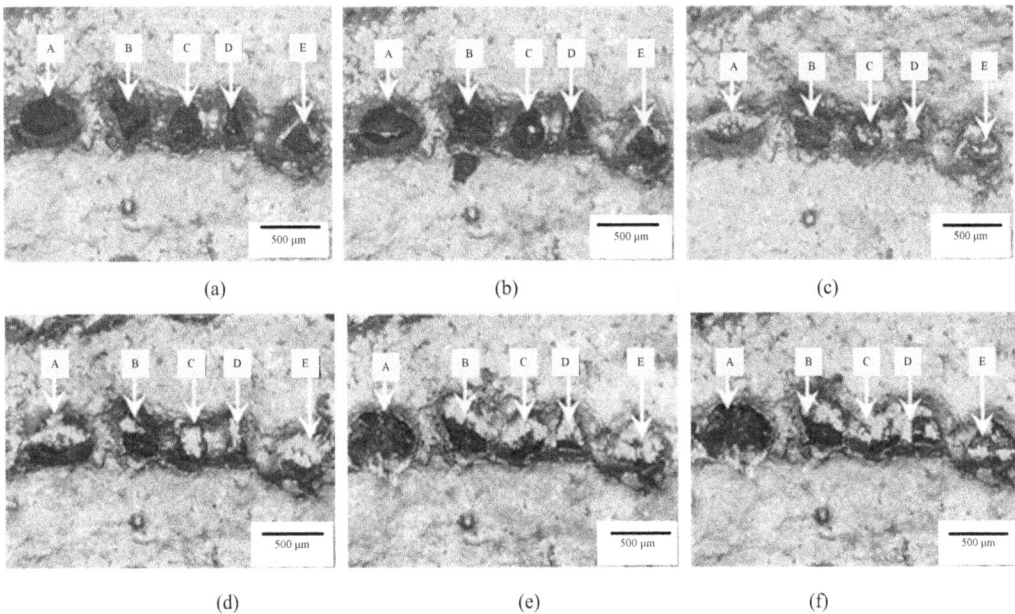

FIGURE 8.4 Morphological evolution of the brazed polycrystalline CBN abrasive tool: (a) initial, (b) 20 passes, (c) 40 passes, (d) 80 passes, (e) 120 passes, (f) 160 passes.

polycrystalline CBN grains is prolonged by about 30% compared to that of the tools with monocrystalline CBN grains. Now it is obvious that, in comparison with brazed monocrystalline CBN abrasive tools, polycrystalline CBN counterparts have a higher material removal rate and weaker tool wear. This indicate the better grinding ability of polycrystalline CBN abrasive grains.

The surface topography of an abrasive tool is influenced by the applied grinding condition and the corresponding tool wear. Accordingly, the typical wearing morphology of five polycrystalline CBN grains was observed on the working surface of a brazed wheel head, as shown in Figure 8.4. For the sake of easier discussion, each grain is numbered,

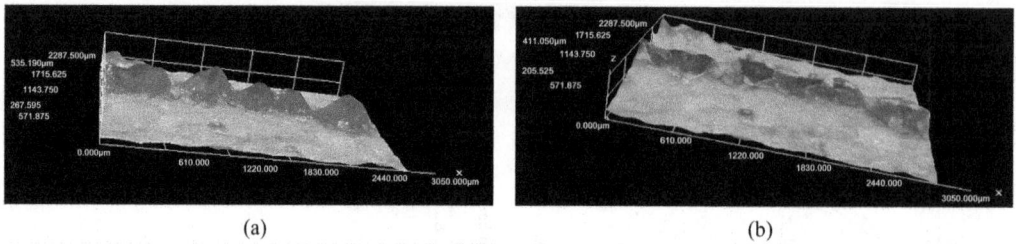

(a)   (b)

FIGURE 8.5   3D contour images of surface topography of brazed PCBN abrasive wheel heads: (a) initial, (b) 160 passes.

i.e., #A, #B, #C, #D, and #E. Figure 8.5 demonstrates the 3D contour images of the initial and final working surface topography of the brazed polycrystalline CBN abrasive tools, which could provide an additional remark for Figure 8.4a and f.

As seen in Figures 8.4a and 8.5a, all the polycrystalline CBN grains are protruding over the filler layer at the initial stage. After 20 grinding passes, one grain (#B) has experienced macrofracture. The fragment is still retained besides the grain, as demonstrated in Figure 8.4b. Other grains, i.e., #A, #C, #D, and #E, do not change their morphology throughout this period. This is accounting for the fact that the four grains protrude less than the grain #B that has been morphologically changed in their surroundings.

In the steady grinding stage, all the polycrystalline CBN grains have been engaged in grinding behaviour after the surrounding grains have worn down to the average protrusion level of those grains. Figure 8.4c–e shows the pictures captured to compare the morphological change of the working surface of the abrasive tools at this stage. No grain pullout phenomenon is observed throughout the tracing period. All five grains in the special zone display various amounts of microfracture and wear-flats in the case of 120 grinding passes. Grain #A endured the attrition wear before the 80 grinding passes, then macrofracture happened suddenly. Grain #C, #D, and #E have been flattened continuously.

Figures 8.4f and 8.5b are the final surface topographies of the worn polycrystalline CBN wheel heads. A little change in the wear morphology of grain #A is found, which implies that this grain has not worked much to grind material in the final period. However, the remarkable morphological evolution of other grains has been observed. Moreover, some ground material has adhered to the surface of the polycrystalline CBN grains.

8.2.3 Wear Morphology of Brazed Polycrystalline CBN Grains

Figure 8.6a displays the attrition wear of a brazed polycrystalline CBN grain in the steady grinding stage, which provides direct evidence to testify to the existence of macroscopical grain flattening. A higher magnification is required for a more in-depth assessment of the microcrystalline CBN particles on the flattened surface of the polycrystalline CBN grains. Therefore, the regional morphology of the fracture surface of the polycrystalline CBN grain is examined, as shown in Figure 8.6b.

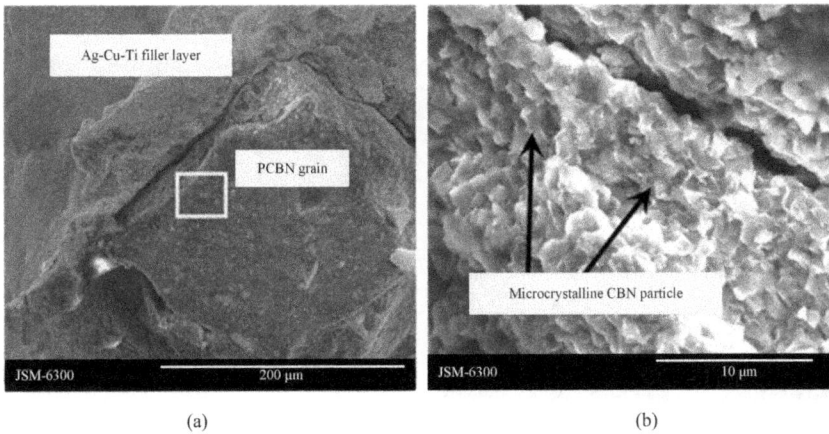

FIGURE 8.6   Attrition wear of brazed PCBN abrasive grains: (a) whole, (b) regional.

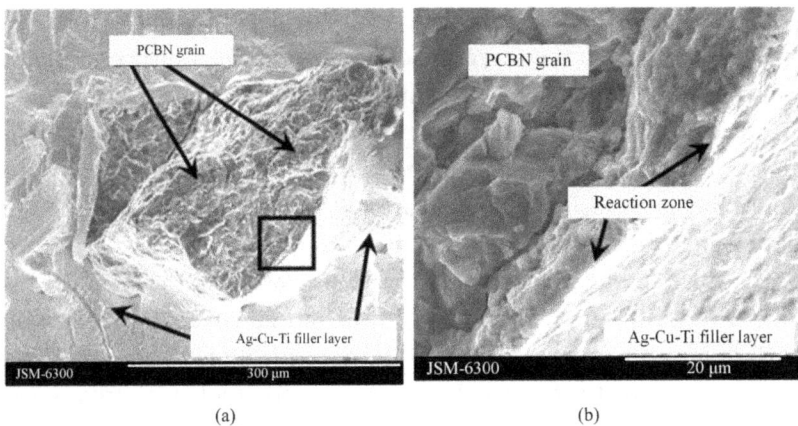

FIGURE 8.7   Joining interface and reaction zone between PCBN abrasive grain and Ag-Cu-Ti filler layer: (a) joining interface, (b) reaction zone.

Many microcrystalline CBN particles have protruded on the fracture surface. Consequently, brazed abrasive tools with PCBN grains are always conducive to the maintenance of sharp cutting edges and highly resistant to attrition wear over long periods of grinding.

In the process of grinding, individual grains on the working surface of the abrasive tools serve as the grinding elements for chip formation and material removal. To realize the grinding behaviour using the fracture of the microcrystalline CBN particles in the polycrystalline CBN abrasive grains, the grains should be embedded firmly in the working layer. That is, any premature pullout of abrasive grains is expected to be arrested. Ag-Cu-Ti filler alloy may form chemical bondage between the polycrystalline CBN grains and the abrasive tool hub. At the same time, appropriate residual thermal stresses also enhance the mechanical properties of the brazed joints. Figure 8.7 shows the joining interface and

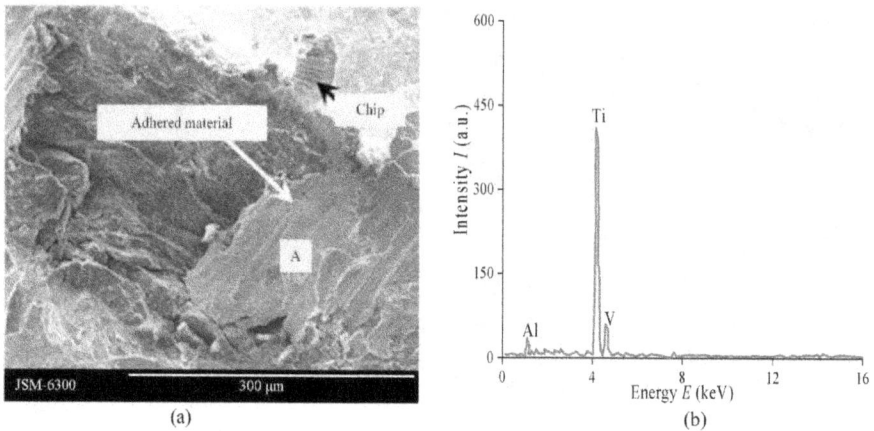

FIGURE 8.8   Adhered material and chemical composition of brazed polycrystalline CBN abrasive grains: (a) adhered material.

the correspondent chemical reaction zone between the polycrystalline CBN grain and the Ag-Cu-Ti filler layer. Though the severe macrofracture of the polycrystalline CBN grain (i.e., grain #A in Figure 8.4) has happened at bond-level, grain pullout from the working surface of the brazed tool has not been observed yet.

In the final accelerating wear stage, the polycrystalline CBN abrasive tools are loaded easily with the adhered deposit material of various amounts. Generally, the adhered material is small in some zones and covers a small portion of the abrasive grain, as shown in Figure 8.8a. The adhered material contains striations parallel to the grinding direction on its top surface. According to the literature (Souza and de Silva, 2019), chips adhering to the grain surface may not be removed from the abrasive tools by the cooling fluid. If an adhered chip is retained on the first rotation of the abrasive tool, it is perhaps mechanically pressed onto the grinding surface and clearance face of the grains by the working blocks on subsequent rotations. Thus, successive rotations of the abrasive tool cause another chip to be easily pressed onto the adhered chips. The nature of the layering in the deposit bulk is remarkable in Figure 8.8a. The continued growth of the adhered material is therefore realized through chip-to-chip adhesion. The energy dispersive spectrum (EDS) is utilized to characterize the deposit in zone A of Figure 8.8a. According to the obtained result displayed in Figure 8.8b, the chemical composition of the adhered material mainly consists of Ti, Al, and V, which is quite similar to that of the gound material of the Ti-6Al-4V alloy. In addition, a piece of chip, which has a typical shear-localized morphology, is also observed in the grain margin of the post-grinding polycrystalline CBN abrasive tools (Figure 8.8a).

The difference in grain wear morphology is noticeable between the brazed polycrystalline CBN grains and the brazed monocrystalline CBN ones. For the brazed monocrystalline CBN abrasive tools, it is evidenced by attrition wear (Figure 8.9a) with a small load on an individual grain and fracture wear (Figure 8.9b) with an increased load on an individual grain. As seen in Figure 8.9a, a flat surface has been produced on the cutting edges of the CBN grain. And according to Figure 8.9b, though newly sharp cutting edges are

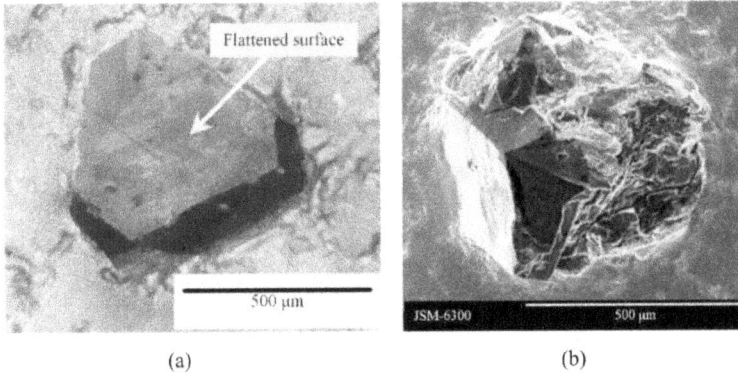

FIGURE 8.9 Wear morphology of brazed monocrystalline CBN abrasive grains: (a) attrition wear, (b) fracture wear.

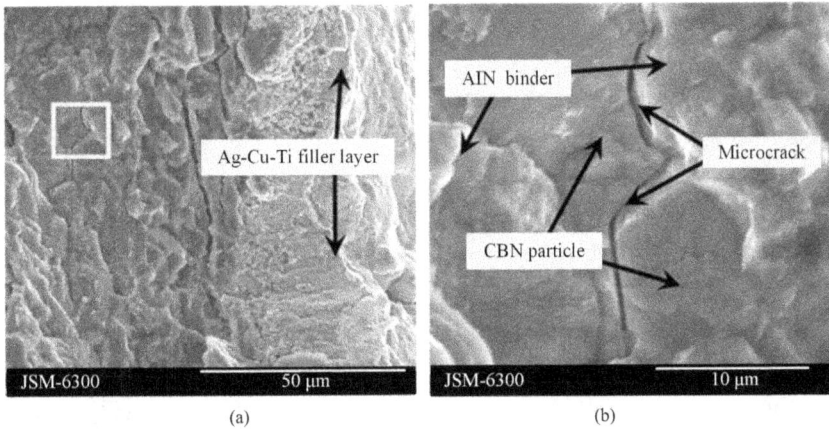

FIGURE 8.10 Microfracture morphology and microcrack path of the PCBN abrasive grains: (a) microfracture morphology, (b) microcrack path.

perhaps created through the grain fracture, almost the entire grain is put away. Under such circumstances, the good grinding ability and accuracy of the CBN abrasive tools are no longer provided.

8.2.4 Fracture Mechanism of Brazed Polycrystalline CBN Grains

The chip formation during grinding has three periods: sliding, ploughing, and cutting. Only the last one removes material from the machined workpiece. When the abrasive grains lose their sharpness, more force is required in the first two periods, and therefore the whole cutting force increases. The abrasive grains may endure fracture.

The self-sharpening effect of polycrystalline CBN abrasive grains using microfracture behaviour is mainly dependent upon the level of the thermal-mechanical loads during grinding. According to the crack mechanism of the particle-reinforced composites, the fracture patterns of the polycrystalline CBN grains include two possible modes: intergranular and transgranular (Zhao and Wang, 2009). Figure 8.10 displays the representative curving path of the microcracks in the abrasive grain. The fracture behaviour of polycrystalline CBN grains is dominated by the joining effect of the AlN binder material among the adjacent microcrystalline CBN grains.

The application of AlN binder material as the bonding system in the polycrystalline CBN abrasive grains makes it possible to create a series of microscopic deviations of the microcrack from the actual path of its propagation. This is accomplished through the formation of bridging sites and certain residual thermal stresses. Therefore, the propagation of the microcrack along a straight line is locally blocked. The frontline of a growing microcrack with a width of 0.5 μm cuts the encountered microcrystalline CBN particles or passes along their boundaries, as shown in Figure 8.10. Under such conditions, additional energy has to be provided for the further propagation of the microcrack. The fracture toughness of PCBN abrasive grains is about 6.8 MPa·m$^{1/2}$, while that of the monocrystalline CBN grains is only 2.5 MPa·m$^{1/2}$. As a result, in the case of strong fixing of the polycrystalline CBN grains in the Ag-Cu-Ti filler layer, the formation and propagation of the microcracks in the polycrystalline CBN abrasive grains are controlled effectively. This is beneficial to decrease the macrofracture behaviour of the monocrystalline CBN abrasive grains. Therefore, the service life of the polycrystalline CBN abrasive tools is prolonged remarkably.

8.3 STRESSES EFFECTS ON POLYCRYSTALLINE CBN GRAIN WEAR

8.3.1 Numerical Microstructure Construction of Polycrystalline Grains

Because the polycrystalline CBN grains are comprised of microcrystalline CBN particles and AlN ceramic binder, the numerical microstructure should be investigated first to simulate the stress distribution in the finite element model (Zhu et al. 2017). The Voronoi tessellation method has been applied in several studies to simulate numerical microstructures in finite element analysis (Alveen et al. 2013). For example, Warner and Molinari (2006) and Zhou et al. (2012) have used finite element analysis based on the Voronoi tessellations method to investigate the failure behaviour of brittle materials. Also, Li et al. (2006) combined finite element analysis and the Voronoi tessellation technique to investigate the mechanical properties of the tungsten-based bulk metallic glass matrix composites. In general, the simulation results agreed well with those tested in the experiment. Accordingly, the numerical microstructures of polycrystalline CBN grains can be created effectively using the Voronoi tessellation method.

In this work, the finite element model combined with Voronoi tessellation techniques was established for analysing stress distribution in polycrystalline CBN grains, which contain the residual stress produced during the brazing process and the resultant stress generated in grinding. Additionally, the influence of embedding depth, volume fraction, and grinding loads on the stress distribution and grain fracture wear were discussed. Finally, the grinding experiment was conducted to verify the stress simulation results by characterizing the wear topography evolution of brazed polycrystalline CBN grains.

To investigate stress distribution for polycrystalline CBN grains, a random microstructure model has been constructed using the Voronoi tessellation method in this work. The

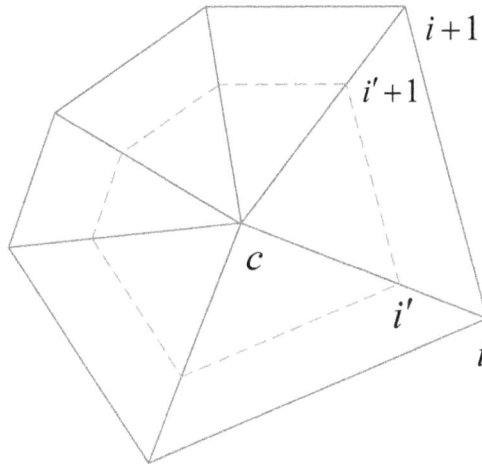

FIGURE 8.11   Schematic of the Voronoi cell.

typical Voronoi tessellation method could be generally described as follows: (i) a set of seeds are randomly spread in a 2D space; (ii) space is partitioned into polygonal Voronoi cells with the partitioning line of the perpendicular bisector of the segment that joins two seeds (Li et al. 2006).

At the same time, to build the inner microcrystalline CBN particles and AlN binder for polycrystalline CBN grains according to the Voronoi tessellation method, a new parameter V_f, which describes volume fraction, is introduced. Figure 8.11 schematically displays a convex polygon of the original particle with a solid line, which is similar to the shape utilized in the Voronoi cell. The coordinate centroid, c, can be calculated as follows:

$$\left\{ \begin{array}{l} x_c = \dfrac{1}{A}\sum_{i=1}^{n}(x_i + x_{i+1})(x_i y_{i+1} - x_{i+1} y_i) \\[4mm] y_c = \dfrac{1}{A}\sum_{i=1}^{n}(y_i + y_{i+1})(x_i y_{i+1} - x_{i+1} y_i) \end{array} \right. \tag{8.1}$$

where x_i and y_i are the coordinates of vertex i, n is the sum of vertices number in the original particle; A refers to the area of the original particle, the value of which can be calculated as:

$$A = \frac{1}{2}\sum_{i=1}^{n}(x_i y_{i+1} - x_{i+1} y_i) \tag{8.2}$$

Afterwards, the segments that connect the centroid with each vertex are drawn. N triangles have been partitioned from the outer particle. The dashed lines, as displayed in Figure 8.11, describe the newly-born particle. Assume that each of the vertices in the newly-born particle is located in the dashed line, and its edge is parallel to the corresponding one of the original particles. The shape of the newly born particle will be similar to that of the original particle, and the volume fraction V_f can be determined as follows:

$$V_f = \frac{A_{i',i'+1,c}}{A_{i,i+1,c}} \tag{8.3}$$

where $A_{i',i'+1,c}$ is the area of the newly born particle and $A_{i,i+1,c}$ is the area of the original particle. Due to the similarity of the newly born particle and the original particle, a relationship between the coordinates of the newly born particle and the original particle exists as follows:

$$\frac{A_{i',i'+1,c}}{A_{i,i+1,c}} = \frac{(x_c - x_{i'})^2}{(x_c - x_i)^2} = \frac{(y_c - y_{i'})^2}{(y_c - y_i)^2} \tag{8.4}$$

The coordinates of vertes i' can be obtained by combining Eqs. (8.3) and (8.4):

$$\begin{cases} x_{i'} = x_c - \sqrt{V_f}(x_c - x_i) \\ y_{i'} = y_c - \sqrt{V_f}(y_c - y_i) \end{cases} \tag{8.5}$$

Accordingly, the size and shape of a newly born particle can be determined by utilizing Eqs. (8.1) and (8.5).

In this work, the Voronoi tessellation process was coded with the MATLAB 2010b software. In the FEA model, the maximum size of the microcrystalline CBN particles within a polycrystalline CBN grain is chosen as 20 μm. As a result, several 625 randomly distributed seeds can be created in an area of 0.5×0.5 mm, as displayed in Figure 8.12a; meanwhile, a 2D Voronoi tessellation process has been performed with these seeds (Figure 8.12b). Then, the volume fraction V_f is taken into consideration based on Eqs. (8.1) and (8.5), and the results are shown in Figure 8.12c. Finally, the 2D Voronoi diagram has been transferred to the finite element software ABAQUS.

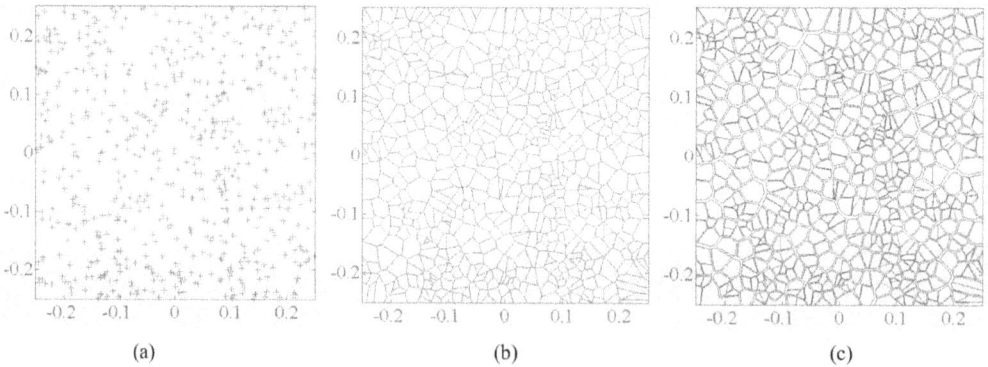

FIGURE 8.12 Evolution of the Voronoi tessellation diagram: (a) seeds spreading, (b) the Voronoi diagram, (c) generation of the microcrystalline CBN particles within a PCBN grain.

8.3.2 Finite Element Model of the Brazing Stress within Polycrystalline Grains

To be consistent with the polycrystalline CBN grains used in the grinding experiment, the 2D Voronoi diagram has been trimmed off, and the shape of a regular hexagon is constructed with a grain side length of 200 μm (maximum diameter of 400 μm). A square of the AISI 1045 steel substrate with a height of 5 mm and a width of 2 mm is built; as such, the area of the metallic substrate is 10 mm², which is about 200 times larger than that of the polycrystalline CBN grain. Under such conditions, the size of the metallic substrate used will not affect the stress distribution in the polycrystalline CBN grain. Particularly, the concept of embedding depth refers to the depth of a grain embedded in the bonding alloy. For instance, the 40% embedding depth means that the distance from the highest interface between the grain and the Ag-Cu-Ti bonding alloy to the bottom of the grain is about 40% of the diameter of the grain, as displayed in Figure 8.13.

The element type used is CPS8R. Figure 8.14 shows the meshed finite element model with the embedding depth of 40%, the microcrystalline CBN particle size of 20 μm,

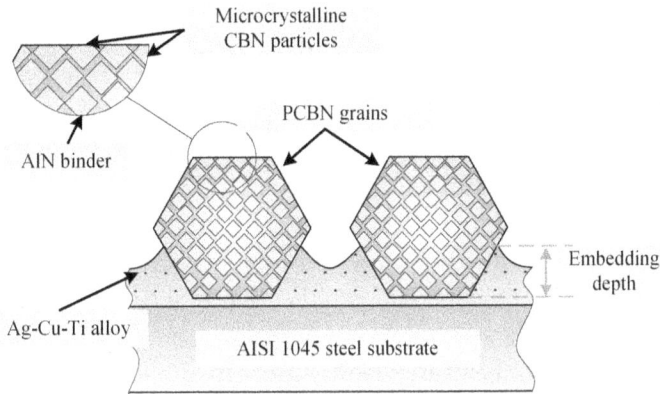

FIGURE 8.13   Schematic diagram of brazed polycrystalline CBN grains.

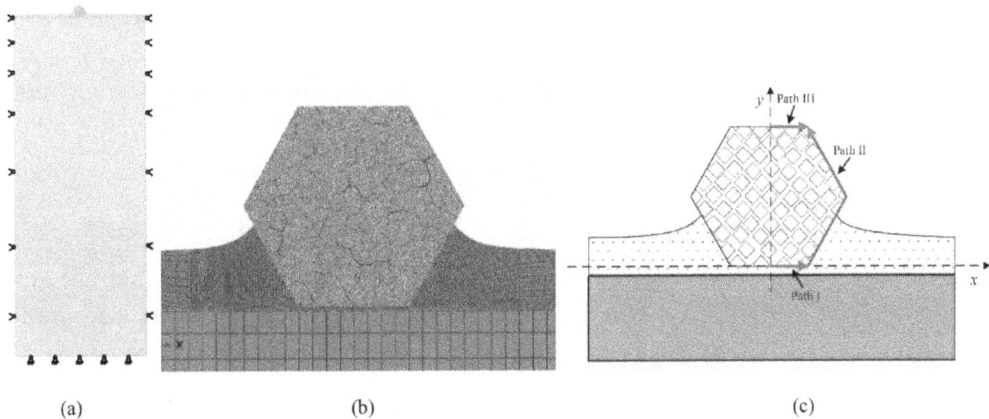

FIGURE 8.14   Finite element model for the brazing-induced stress distribution in a PCBN grain: (a) boundary conditions, (b) meshing diagram, (c) three paths selected for stress distribution analysis.

and the grain side length of 200 µm, which contains 16,382 elements and 59,054 nodes. Moreover, as can be seen from Figure 8.14a, the node displacement on the bottom surface and that on the right and left surfaces along the x (horizontal) axis are all defined as zero. The node displacement along the y (vertical) axis on the bottom surface is also defined as zero.

Also, three reference paths are selected to characterize the brazing-induced residual stress in the polycrystalline CBN grain, as illustrated in Figure 8.14c. They are represented as Paths I, II, and III, respectively, with directions denoted by arrows. For example, Path I is along the bottom of the grain, which starts from the centre of the grain edge to the vertex of the hexagon. Path II is the most complex one, which starts from the grain bottom vertex to the grain middle vertex, then to the top vertex; particularly, Path II includes the embedding area, bonding interface, and grinding force area, and three vertex points are also connected. Path III is similar to Path I, and it is along the grain top boundary, which is from the grain edge centre to the vertex.

8.3.3 Materials Properties and Grinding Loads Utilized in the Finite Element Model

$(Ag_{72}Cu_{28})_{95}Ti_5$ (wt.%) alloys are utilized as the brazing filler material to join polycrystalline CBN grains and AISI 1045 steel substrate in the vacuum furnace, a brazing temperature of 920°C with a hold time of 5 min is used as the optimal operation parameters. To ensure a good joining at the interface of polycrystalline CBN grain-bonding alloy and bonding alloy-metallic substrate, the heating and cooling rate in the brazing cycle is set at 10°C/min. Moreover, the differential thermal analysis (DTA) has indicated that the solidification temperature of the Ag-Cu-Ti filler alloy is around 800°C. Meanwhile, the brazing-induced residual stress would not be produced in both the heating stage and the cooling stage from the brazing temperature to the solidification temperature. As a consequence, a cooling condition from solidification temperature (800°C) to room temperature (20°C) is used in the present simulation.

As listed in Table 8.1, the material properties (e.g., Young's modulus, Poisson's ratio, thermal expansion coefficient, and yield strength) are provided either from the manufacturers or from the literature. Particularly, in order to save the computation load, some assumptions are made. For example, all materials used in the model are regarded as isotropic and rigid plastic. Meanwhile, the polycrystalline CBN grain microcracks are ignored, and it is considered a continuum body. In addition, the material creep behaviour, solubilization behaviour, and diffusion behaviour during the brazing process are not taken into account due to their limited influence.

After the brazing-induced stress distribution has been analysed, the grinding forces can be loaded on the vertex region on the grain top surface. The cutting depth of a single grain in the polycrystalline CBN abrasive wheel and the corresponding grinding load are the same as those described in Sections 7.3.1 and 7.3.5. Therefore, no more information was provided here.

TABLE 8.1 Materials Properties Utilized in the Present Finite Element Simulation

Properties	Microcrystalline CBN Particle	AlN Binder	Ag-Cu-Ti Filler Alloy	AISI 1045 Steel
		Materials		
Young's modulus E (GPa)	909	320	100 (293 K) 90 (500 K) 80 (700 K) 70 (800 K) 67 (900 K)	210 (293 K) 185 (600 K) 160 (800 K)
Poisson's ratio μ	0.12	0.23	0.35	0.26
Thermal expansion coefficient α (10^{-6} K^{-1})	4.92	4.5	16.7 (293 K) 18.0 (500 K) 19.0 (700 K) 20.0 (800 K) 20.5 (900 K)	9.1 (293 K) 13.09 (500 K) 13.71 (700 K) 14.18 (800 K) 14.6 7(900 K)
Yield strength σ_s (MPa)	-	-	230 (293 K) 170 (473 K) 100 (700 K) 70 (800 K) 20 (900 K)	360 (293 K) 263 (600 K) 179 (800 K) 78 (900 K)

8.3.4 Characterization of Stress Distribution in Polycrystalline CBN Grains

As a typical brittle material, the polycrystalline CBN grains can bear high compressive stress but fracture easily under large tensile stress (Zheng et al. 2012; Zhang et al. 2016). Therefore, the values of the residual stress in brazed polycrystalline CBN grains could be evaluated by the maximum in-plane principal stress. Figures 8.15 and 8.16 show the contours of the maximum in-plane principal stress simulated during the brazing process and the grinding process, respectively. Here the embedding depth is 40%, and the volume fraction is 80% with the uncut chip thickness of 0.6 μm. Tensile stress has been observed in the microcrystalline CBN particles and compressive stress in the AlN binders. Moreover, the brazing-induced tensile stress in PCBN grains greatly concentrates in the interfaces among the microcrystalline CBN particles, AlN ceramic binder, and Ag-Cu-Ti filler alloy (Figure 8.15b), which is caused by the thermal expansion coefficient mismatch. A large compressive stress occurs at the vertex of the PCBN grain, where the grinding force is loaded, as displayed in Figure 8.16b.

Furthermore, the microcrystalline structure of the polycrystalline CBN grains may affect their fracture behaviour. However, it is out of the scope of the current study, which is investigating the effect of the embedding depth, volume fraction, and grinding loads as brazing and grinding parameters on the stress distribution. Hence, the effect of the microcrystalline structure will be investigated in further research.

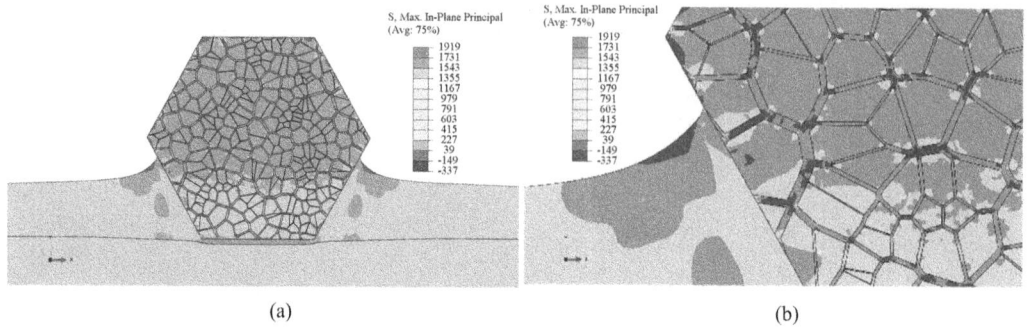

FIGURE 8.15 Contour maps of residual stress distribution from the brazing process: (a) single-grain bond, (b) grain bond interface.

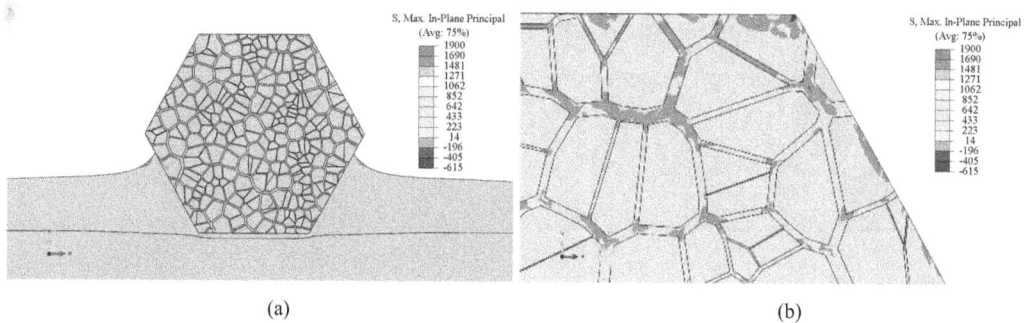

FIGURE 8.16 Contour maps of the resultant stress distribution in the grinding wheel: (a) single-grain bond, (b) stress concentration from the polycrystalline structure.

8.3.5 Influence of Embedding Depth on Stress Distribution in PCBN Grains

Figures 8.17–8.19 quantitatively compare the maximum in-plane principal stress distribution along Paths I, II, and III, respectively, with the grain embedding depth varying from 20% to 50%. Here, the volume fraction is 80%, and the a_{gmax} is 0.6 μm.

It can be observed from Figure 8.17a and b that along Path I, not only the brazing-induced residual stress but also the resultant stress in grinding just slightly changed. However, because grain fracture wear probably happens when the tensile stress in the grain is greater than its fracture strength; as such, the maximum tensile stress should be discussed. Two points along Path I (denoted as Point A and Point B in Figure 8.17a) where the local maximum tensile stress occurs are selected, as displayed in Figure 8.17c. It should be noted that the resultant stress at Points A and B changes less compared with the brazing-induced residual stress. Therefore, only the brazing-induced stress is analysed here. With an embedding depth of about 20%, the tensile stress at point A is 1,532 MPa and that at point B is 1,073 MPa, respectively. The tensile stress at point A increases by 6.9% to 1,638 MPa when the embedding depth increases to 40%. When the embedding depth further increases to 50%, the stress at point A decreases by 3.7% to 1,578 MPa. However, the stress at Point B stays almost constant.

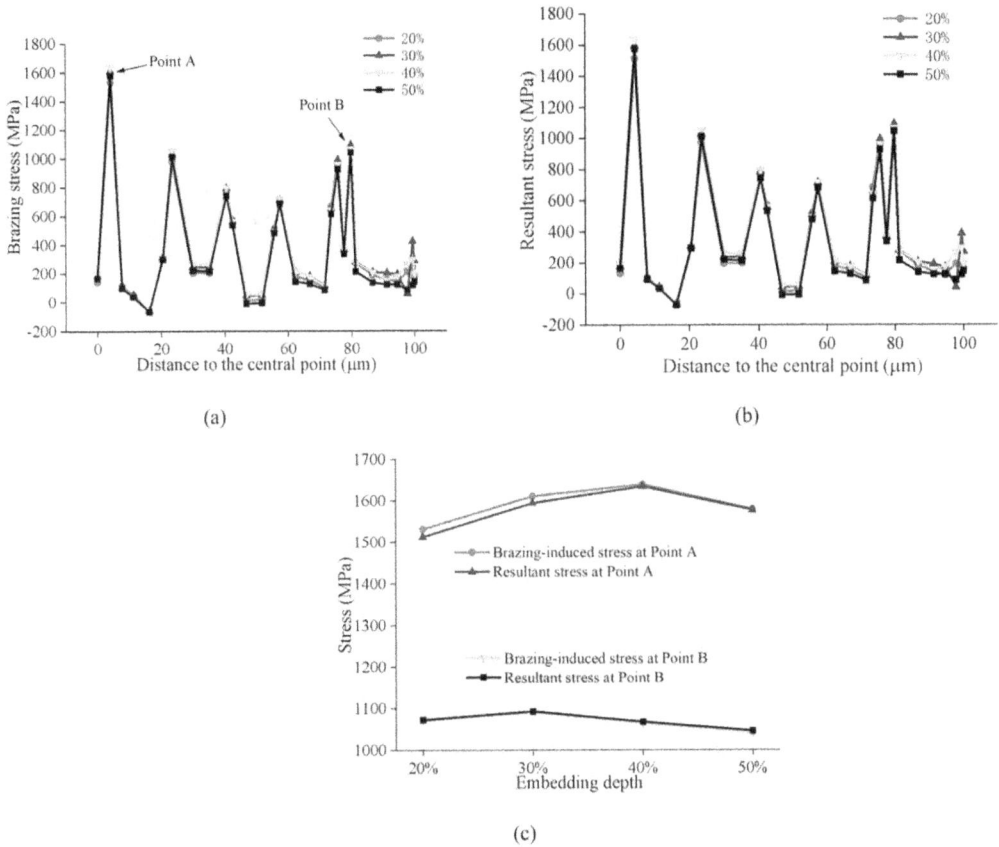

FIGURE 8.17 Maximum in-plane principal stress distribution along Path I in PCBN grains with different embedding depth: (a) brazing-induced stress, (b) resultant stress in grinding, (c) stress at points A and B.

The stress distribution along Path II in polycrystalline CBN grains with different embedding depths is displayed in Figure 8.18. The brazing-induced stress along Path II is tensile, and the peak values are over 1,000 MPa, which occurs at the interface of the microcrystalline CBN particles/AlN binders. When the grinding forces are applied, large tensile and compressive stress occurs at the grain vertex, as illustrated in Figure 8.18b. The stress near the interface is affected by the embedding depth. Hence, the maximum stress in Path II (donated as point C), the stress in the vertex, and the maximum stress near the interface are selected to investigate the stress distribution in the grain, as illustrated in Figure 8.18c. The tensile stress is 1,315 MPa at point C, 35 MPa at the vertex, and 821 MPa at the interface, respectively. Obviously, with the change in embedding depth, the brazing-induced stress at point C and the vertex almost keep constant. On the other hand, when the embedding depth increases to 30%, the tensile stress at the interface increases by 8.9% to 894 MPa.

In contrast, the stress decreases by 28% to 643 MPa when the embedding depth increases to 50%. In addition, the resultant stress at point C and the interface change slightly after the

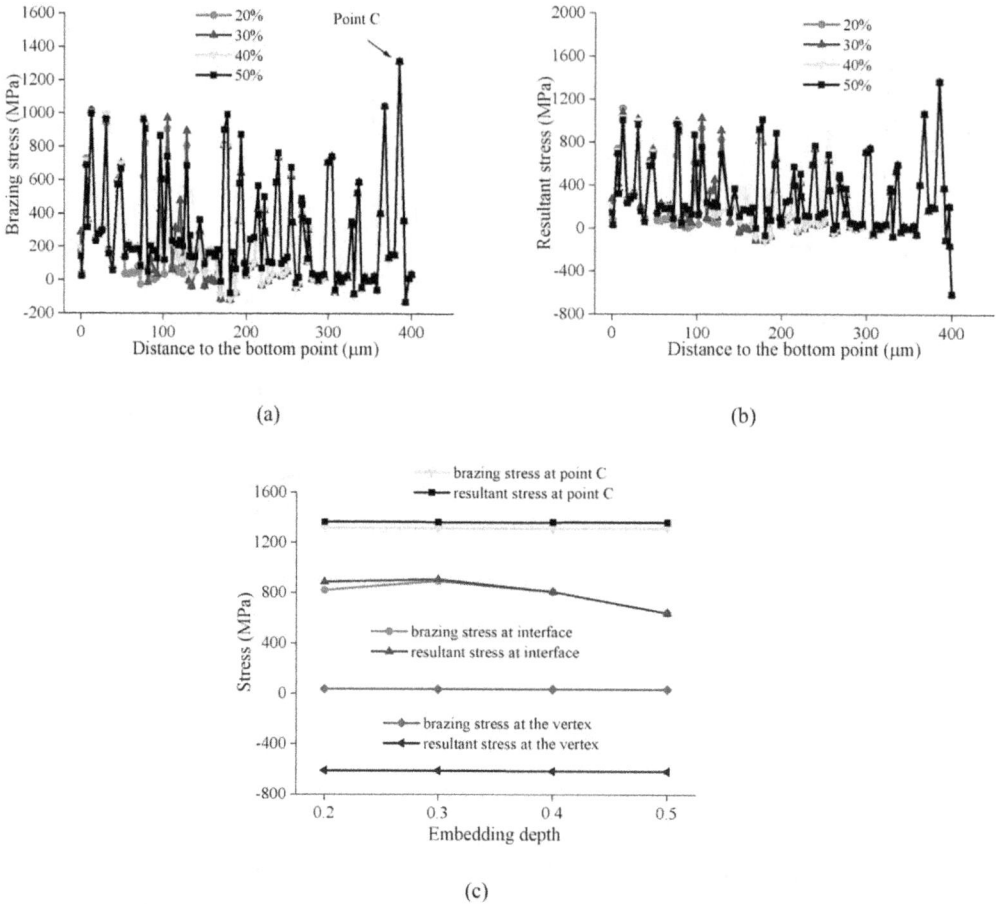

(a)

(b)

(c)

FIGURE 8.18 Maximum in-plane principal stress distribution along Path II in PCBN grain with different embedding depth: (a) brazing-induced stress, (b) resultant stress in grinding, (c) stress at point C and the interface.

grinding forces are applied. On the other hand, the resultant stress at the vertex changes from tensile stress (35 MPa) to compressive stress (−610 MPa). It's noted that point C is close to the vertex. The tensile stress at point C is over 1,300 MPa, which may cause the polycrystalline CBN grain microfracture wear.

Figure 8.19 shows the stress distribution along Path III in polycrystalline CBN grains with different embedding depths. Along Path III, the influence of the embedding depth on the stress distribution is generally small. The maximum tensile stress is above 1,200 MPa, and the compressive stress is around −610 MPa.

As discussed above, the large stress usually exists in three paths in the polycrystalline CBN grain. The large stress along Path I and in the interface along Path II could result in grain macrofracture wear, and the stress near the vertex and along Path III will cause microfracture wear. When the embedding depth is 50%, the stress along Path I and in the interface along Path II is lower than the other comprehensively.

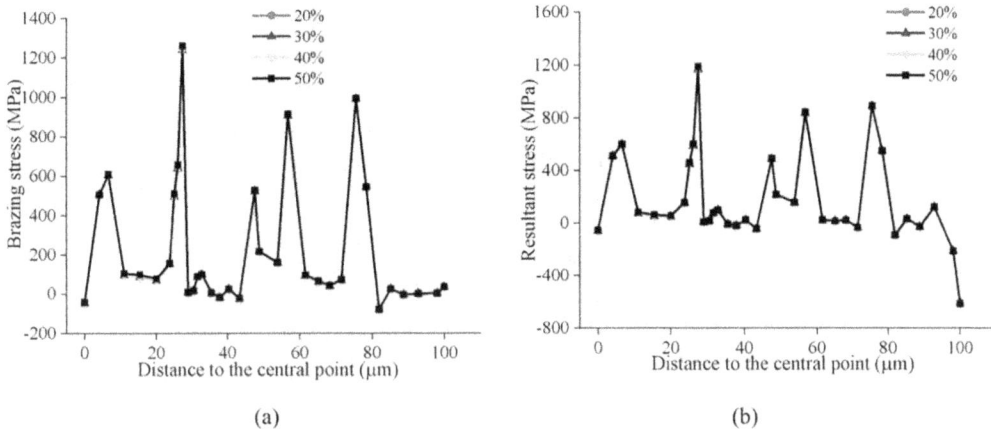

FIGURE 8.19   Maximum in-plane principal stress distribution along Path III in PCBN grains with different embedding depth: (a) brazing-induced stress, (b) resultant stress in grinding.

As such, the 50% embedding depth is selected. Meanwhile, the stress distribution along Path II is complex, which will affect the fracture wear of the PCBN grain. Therefore, the stress distribution along Path II will be discussed in detail.

8.3.6 Influence of Volume Fraction on Stress Distribution within PCBN Grains

The polycrystalline CBN grains are combined with microcrystalline CBN particles and AlN binder, and the ratio of the materials in the polycrystalline CBN grains may affect the mechanical properties of the grains. Figure 8.20 displays the polycrystalline CBN grain with a volume fraction range of 60% to 90%.

Figure 8.21 shows the effects of volume fraction on the stress distribution along Path II after the brazing process and in the grinding process, respectively. Here, the a_{gmax} is 0.6 μm, and the embedding depth is 40%. To avoid grain macrofracture wear and improve the life of grinding wheels, the stress along Path II in the range from 0 to 320 μm should be kept. Figure 8.21b shows the largest tensile stress in the range from 0 to 320 μm, the range from 320 to 400 μm along Path II, and the stress at the vertex, respectively. After the brazing process, the maximum tensile stress along Path II in the range from 0 to 320 μm is 1,590 MPa when the volume fraction is 60%. The maximum tensile stress in the range from 320 to 400 μm is 1,391 MPa. When the volume fraction increases, the largest tensile stress in the range from 0 to 320 μm greatly decreases by 50.7% to 783 MPa. However, the stress in the range from 320 to 400 μm increases by 18.9% to 1,654 MPa first, then decreases by 50.5% to 818 MPa. The stress at the vertex keeps in the range of 50 to 200 MPa, which are tensile ones.

When the grinding forces are applied, the largest tensile stress in the range from 0 to 320 μm and the range from 320 to 400 μm almost keep constant. However, the stress at the vertex changes from tensile stress to compressive stress. The largest stress changes are around −600 MPa.

It is obvious that, when the volume fraction is 80%, the resultant stress in grinding is low in the range of 0 to 320 μm and large in the range of 320 to 400 μm, which indicates that grain microfracture wear may occur.

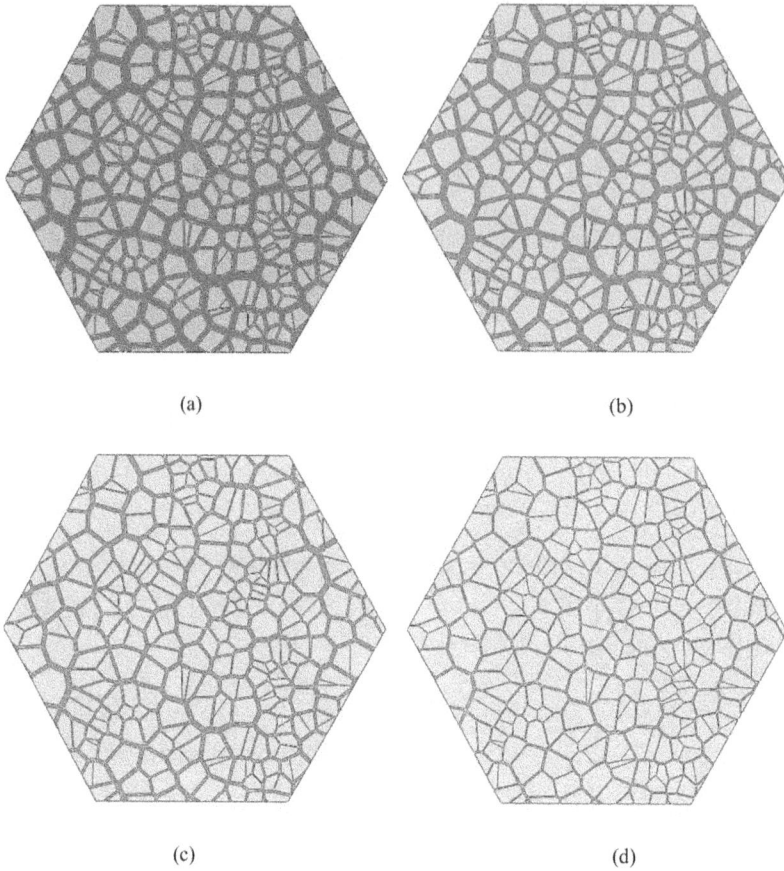

(a)   (b)

(c)   (d)

FIGURE 8.20 The component of PCBN grains with different volume fractions: (a) V_f=60%, (b) V_f =70%, (c) V_f =80%, (d) V_f =90%.

8.3.7 Influence of Grinding Loads on Resultant Stress in Polycrystalline CBN Grains

During the grinding process, different grinding parameters may be utilized to meet various machining requirements. In this case, the effect of grinding loads on the resultant stress is investigated. In this section, four groups of grinding forces (e.g., normal F_n and tangential F_t) are applied to the grain vertex region after the brazing process. The volume fraction is fixed at 80%, and the grain embedding depth is selected at 40%. Moreover, the area of the loading zone will increase with the increase in a_{gmax}.

Figure 8.22 shows the resultant stress distribution along Path II. The stress distribution on the grain vertex is significantly affected by the grinding force. As illustrated in Figure 8.22b, the resultant stress in the area (x=385 μm) is 1,361 MPa when a_{gmax} is 0.6 μm. With a_{gmax} increasing to 8 μm, the stress can decrease by 75.7% to 330 MPa. On the other hand, the maximum compressive stress changes from −614 MPa to −1,220 MPa, with an a_{gmax} increase from 0.6 to 8 μm. Therefore, to reduce the grain macrofracture wear, the a_{gmax} should be selected around 0.6 μm.

The influence of the embedding depth, volume fraction, and uncut chip thickness on the stress distribution in the polycrystalline CBN grain has been discussed. To increase the

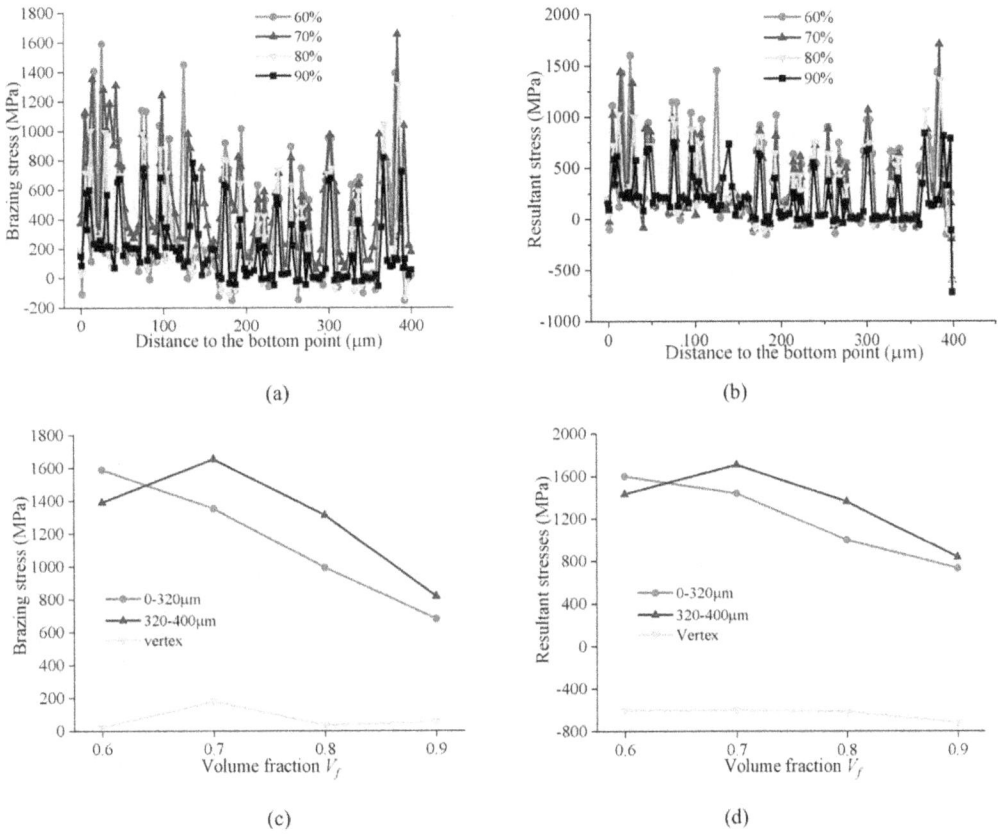

(a)

(b)

(c)

(d)

FIGURE 8.21 Maximum in-plane principal stress distribution along Path II in PCBN grains with different volume fractions: (a) brazing-induced stress, (b) resultant stress in grinding, (c) the largest brazing-induced stress in the range from 0 to 320 μm, from 320 to 400 μm, and at the vertex, (d) the largest resultant stress in the range from 0 to 320 μm, from 320 to 400 μm and at the vertex.

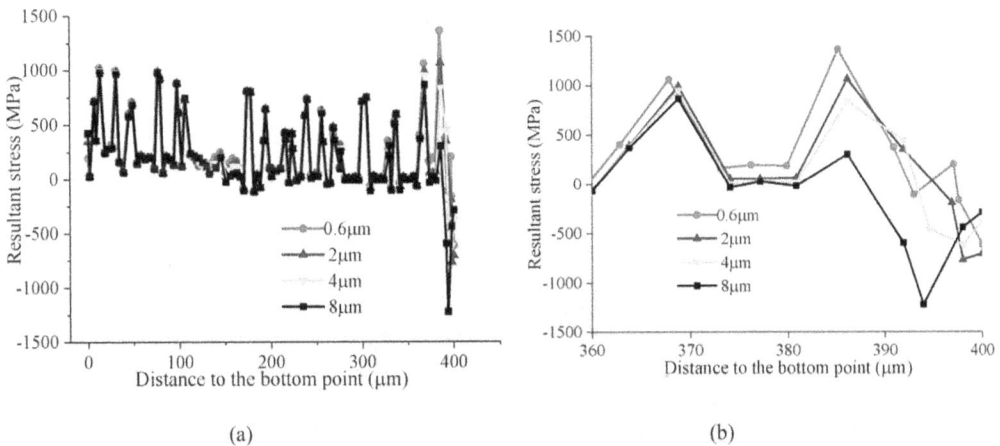

(a)

(b)

FIGURE 8.22 Resultant stress in the PCBN grain along Path II with different a_{gmax}: (a) full range, (b) narrow range.

working life of the brazed grinding tools, the brazing-induced residual stress should be as low as possible to maintain the fracture strength of the grain. However, during attrition and fracture wear, the microfracture is preferred to occur on the grain-top surface to maintain sharp edges. Therefore, the resultant stress should be large enough on the grain-top surface. In conclusion, the most suitable stress distribution was obtained in the case of an embedding depth of 50%, a volume fraction of 80%, and an uncut chip thickness of 0.6 μm.

8.3.8 Experimental Verification of Grain Wear Topography Evolution in Grinding

The measurement of stress distribution within polycrystalline CBN grains is very difficult and has low accuracy at present. However, the topology evolution can provide valuable information during the grinding process. Grinding tests have been designed and conducted with the monolayer brazed polycrystalline CBN grinding wheel; as such, the stress simulation results could be indirectly verified based on the wear topography evolution of the polycrystalline CBN grains. The grinding equipment is displayed in Figure 8.23. The nickel-based superalloy Inconel 718 is used as the workpiece material. The wheel speed is 120 m/s with a feed rate of 1.7 m/min, and the cutting depth is 10 μm; as such, the uncut chip thickness is 0.6 μm. Three grinding passes are conducted with grinding length of 60 mm. The wear topography of PCBN grain was collected using the 3D optical microscope (Hirox KH-7700), and the grains' height was also measured. Moreover, the 3D profiles of the grains were reconstructed based on the Matlab software, which made it easy to characterize the grain topography evolution. Also, to quantitatively evaluate the grain fracture wear during the grinding process, the surface fractal dimension D_s was calculated. In general, if microfracture wear happens on the grain surface, the microcutting edges can be formed, which usually enlarges the surface fractal dimension D_s. On the contrary, when large fracture wear occurs, the microcutting edges reduce, and the surface fractal dimension D_s decreases to a certain extent.

Figure 8.24 provides the wear topography evolution of the polycrystalline CBN grain during the grinding process and the evolution of the grain height. The surface fractal dimension is also displayed in Figure 8.25. The original grain has several cutting edges,

FIGURE 8.23 Equipment for the grinding experiment with a brazed PCBN grinding wheel.

Wear topography	3D reconstructed profile of wear topography
Original	
Pass I	
Pass II	
Pass III	

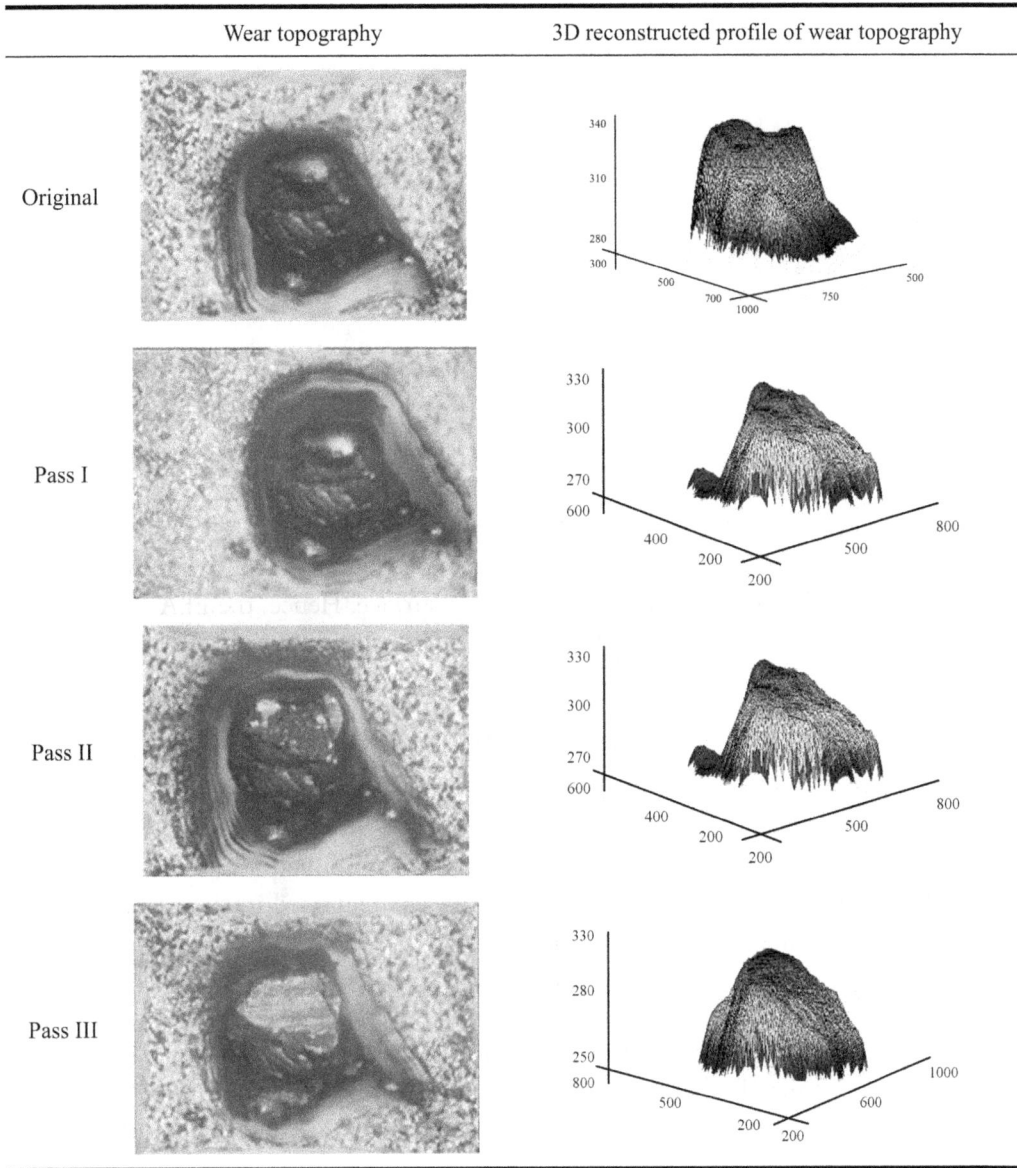

FIGURE 8.24 Typical wear topography evolution of the brazed PCBN grain during grinding.

and the grain height h and fractal dimension D_s are 336.9 and 2.0563 μm, respectively. After the first grinding pass, a few parts of microcrystalline CBN particles fall off the polycrystalline CBN grain due to the large resultant stress in grinding. The height h decreases by 3.27% to 325.9 μm; however, the fractal dimension D_s increases to 2.0589, which indicates that the number of microcutting edges increases and grain self-sharpening occurs. After the second grinding pass, the height h further decreases by 3% to 316.1 μm, and the fractal dimension continuously increases. After the third grinding pass, the grain height h decreases by 4.68% to 301.4 μm, and the fractal dimension D_s decreases by 1.15% to 2.0409.

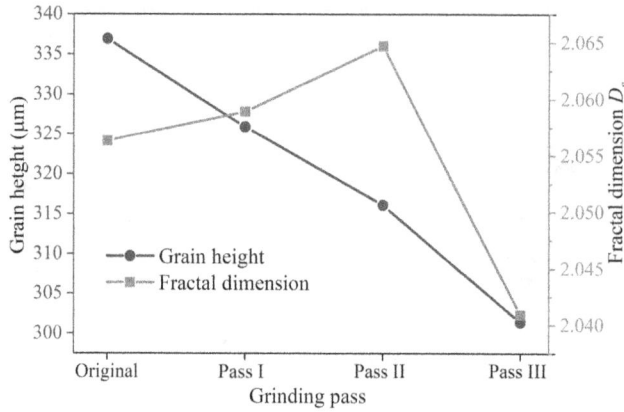

FIGURE 8.25 Evolution of the grain height and surface fractal dimension in grinding.

The wear of the brazed polycrystalline CBN grain is still dominant due the microfracture. As described in the simulation, the resultant stress on the grain top surface was so large that a microfracture might occur on the grain top surface. Hence, the FEA model in the present work is verified to be valid by the wear topography evolution of brazed polycrystalline CBN grains.

8.4 CRACK PROPAGATION OF POLYCRYSTALLINE CBN ABRASIVE GRAINS

8.4.1 Cohesive Element Model for Polycrystalline CBN Grains

8.4.1.1 Cohesive Element Theory

The crack initiation and propagation process can be simplified as a cohesive element model. The constitutive behaviour of the cohesive element used in the present work is based on the traction-separation law. The traction-separation-law consists of three basic ingredients: the initial linear elasticity stage, the fracture initiation criterion, and the fracture evolution criterion (Huang et al. 2017). Figure 8.26 shows a typical traction-separation-law with a failure mechanism. The initial elastic behaviour is described by an elastic constitutive matrix that relates the nominal stresses to the nominal strains across the interface. So the relationship between the nominal stresses and the nominal strains can be defined as:

$$t = \left\{ \begin{array}{c} t_n \\ t_s \\ t_t \end{array} \right\} = \left[\begin{array}{ccc} E_{nn} & E_{ns} & E_{nt} \\ E_{ns} & E_{ss} & E_{st} \\ E_{nt} & E_{st} & E_{tt} \end{array} \right] \left\{ \begin{array}{c} \varepsilon_n \\ \varepsilon_s \\ \varepsilon_t \end{array} \right\} = E\varepsilon \qquad (8.6)$$

where t is the nominal traction stress vector, ε is the nominal strain vector, and E is the elastic constitutive matrix. When the stress and/or strains satisfy certain damage initiation criteria, the process of degradation of the response of a material point begins. In this work, the damage is assumed to initiate when a quadratic interaction function involving the nominal stress ratios reaches a value of one. This criterion is called the quadratic nominal stress criterion, which is represented as:

$$\left\{ \frac{\langle t_n \rangle}{t_n^o} \right\}^2 + \left\{ \frac{t_s}{t_s^o} \right\}^2 + \left\{ \frac{t_t}{t_t^o} \right\}^2 = 1 \qquad (8.7)$$

where t_n^o, t_s^o, and t_t^o represent the peak values of the nominal stress when the deformation is either purely normal to the interface or purely in the first or second shear direction, respectively. Once the damage initiation criterion is reached, the rate at which the material stiffness is degraded is described by the damage evolution law. A scalar damage variable D is used in the damage evolution law to represent the overall damage to the material. Upon further loading after the initiation of damage, D monotonically evolves from 0 to 1. The effects of the damage on the stress components are described as:

$$\begin{cases} t_n = \begin{cases} (1-D)\bar{t}_n, & \bar{t}_n \geq 0 \\ \bar{t}_n, & \text{otherwise (no damage to compressive stiffness);} \end{cases} \\ t_s = (1-D)\bar{t}_s, \\ t_t = (1-D)\bar{t}_t. \end{cases} \qquad (8.8)$$

where $\bar{t}_n$, $\bar{t}_s$, and $\bar{t}_t$ are the stress components predicted by the elastic traction-separation behaviour for the current strains without damage. When the damage variable D reaches 1 at all of its material points, the cohesive element will fail and be removed. The energy dissipated due to the element failure is equal to the area enclosed by the traction-separation curve and the horizontal axis.

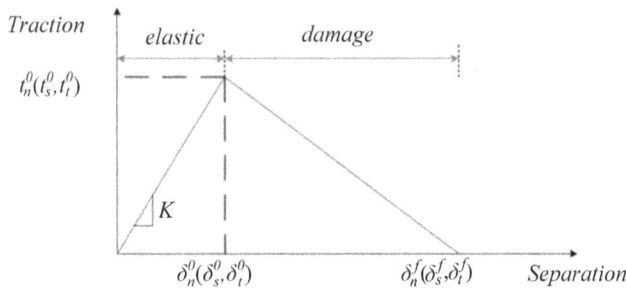

FIGURE 8.26   Typical traction separation law.

8.4.1.2 Embedding Cohesive Elements

During the fracture procedure of polycrystalline CBN materials, cracks may cause propagation both along grain boundaries and into grains. Therefore, to simulate the intergranular fracture and transgranular fracture of polycrystalline CBN materials, cohesive elements will be embedded as potential paths for crack propagation both in the grains and along the grain boundaries.

The entire model first meshes with linear triangular elements, and then cohesive elements are embedded along with each triangular element as the potential crack paths. The nodes and element data of the triangular mesh require special treatment to generate the desired cohesive elements. The basic procedure for the treatment is as follows:

i. generating new nodes according to the number of repeats of a node. The new node has the same coordinates as the old one.

ii. building new edges where two triangular elements share a common edge. The nodes of a new edge are composed of the new nodes generated in the first step.

iii. combining the new edges with the old edges into cohesive elements. Figure 8.27 shows a simple example of such a procedure.

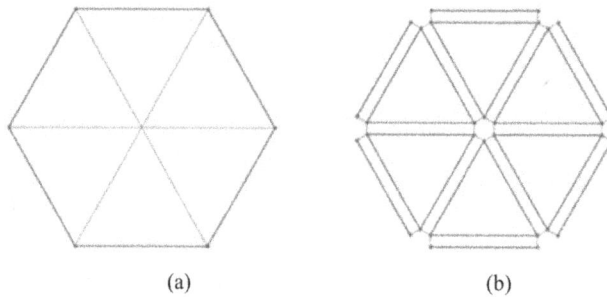

(a)   (b)

FIGURE 8.27 Cohesive elements are inserted in the initial mesh: (a) initial triangular mesh and (b) new mesh with cohesive elements.

After the treatment of the triangular mesh, the cohesive elements along the grain boundaries and those in the grains are distinguished according to the position of the vertices and convex edges of the Voronoi diagram to assign material properties to them, respectively. The whole procedure for accomplishing these tasks is done through a script written in Python. The generated mesh of the polycrystalline CBN microstructural model is shown in Figure 8.28.

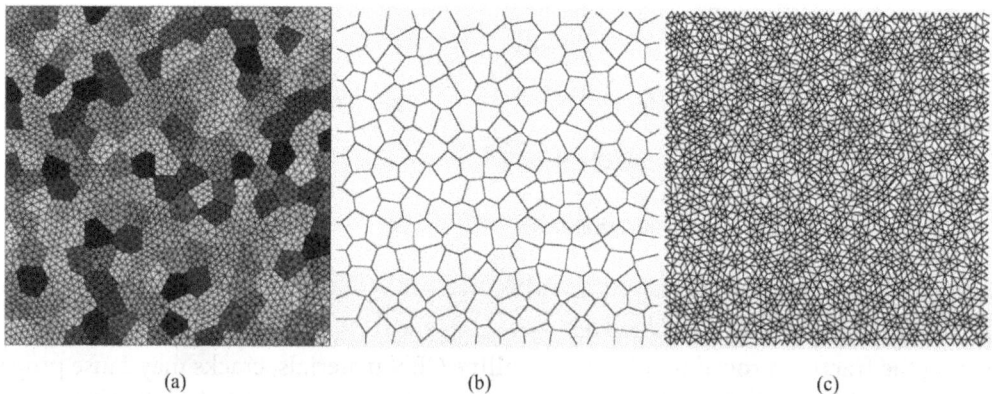

(a)   (b)   (c)

FIGURE 8.28 The mesh of the polycrystalline CBN microstructural model: (a) the triangular mesh; (b) the cohesive elements along grain boundaries; (c) the cohesive elements with microcrystalline CBN particles.

8.4.2 Boundary Conditions and Simulation Parameters for Polycrystalline CBN Grains

In an implicit solution procedure, the modelling of progressive damage involves softening in the material response, which will lead to convergence difficulties. Therefore, the simulation will be conducted on the explicit module of Abaqus. As depicted in Figure 8.29, the simulation is performed on a square model. Because the size of the abrasive grain is usually on the micron scale, the side length of the model is chosen at 100 μm. To observe the crack propagation process, a predefined crack of 4 μm is introduced on the left side of the model. Due to the explicit module of Abaqus used here, a symmetric velocity load instead of a displacement load is imposed at the top and bottom edges of the model. The imposed velocity v is chosen as 100 mm/s, which yields the effective strain rate of the model $\dot{\varepsilon} = v / l = 2 \times 10^3$ (l is the side length of the model). Thus, the dynamic crack propagation process of the model under an effective strain rate of 2,000 can be simulated.

As mentioned previously, this work mainly deals with the effects of microstructural characteristics on the fracture toughness of polycrystalline CBN materials. To this end, three microstructural models of different grain sizes are established, as shown in Figure 8.30, to study the effects of grain size. The material properties for the microstructures are listed in Table 8.2. The strength parameters of cohesive elements along

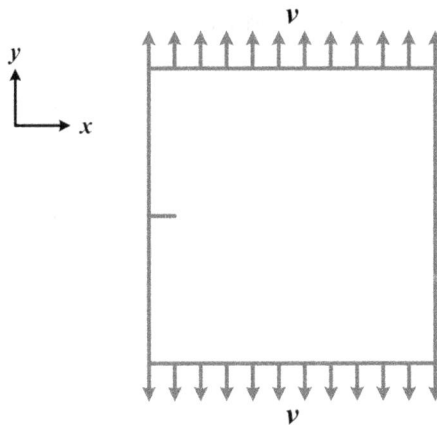

FIGURE 8.29   Geometric model and boundary conditions.

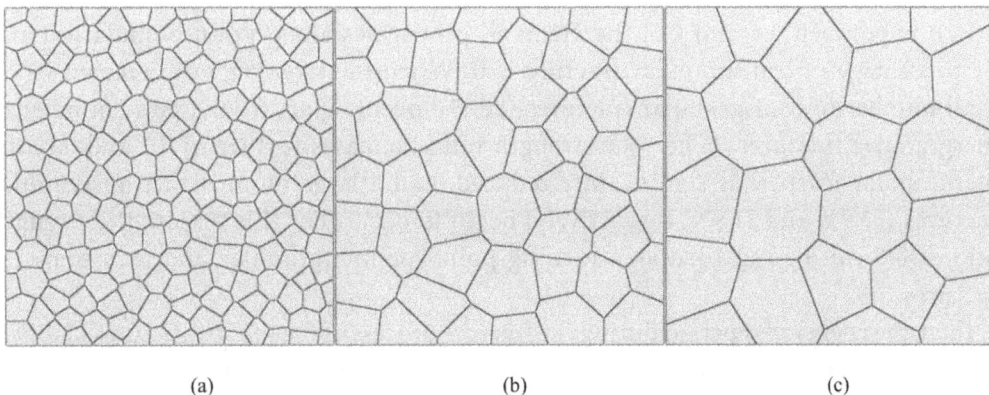

FIGURE 8.30   Microstructural models of PCBN materials: (a) $r=4$ μm; (b) $r=8$ μm; (c) $r=12$ μm.

TABLE 8.2   Materials Parameters for PCBN Microstructures

Element Type	E (Gpa)	υ	T_{max} (Mpa)	Γ (J/m²)
Triangular element (CBN)	909	0.12	-	-
Cohesive element in grains	909	0.12	800	0.06
Cohesive element along grain boundaries	400/500/600/700/800	0.12	320	0.024
	400	0.12	400	0.030
	400	0.12	480	0.036
	400	0.12	560	0.042
	400/500/600/700/800	0.12	640	0.048

grain boundaries are determined based on the ratio to the strength parameters of cohesive elements in grains. In particular, the ratios chosen in this simulation work are 0.4, 0.5, 0.6, 0.7, and 0.8, respectively, to study the effects of the bonding strength of the grain boundary. The stiffness of the cohesive elements along grain boundaries is chosen at 400, 500, 600, 700, and 800 GPa, respectively, to study the effects of grain boundary stiffness.

8.4.3 The Effects of Bonding Strength on Grain Boundaries

For brittle polycrystalline materials, the ratio between the bonding strength of grain boundaries and the bonding strength of crystal planes within microcrystalline CBN particles can have a significant effect on the fracture toughness of the material. The ratio is a function of material chemistry and sintering conditions. To investigate the effect of the bonding strengths of grain boundaries on the fracture toughness of polycrystalline CBN materials, a set of strength ratios are designed, which represent the relative bonding strength of grain boundaries and crystal planes. The other material parameters and the grain size of 8 μm are kept constant. The crack propagation of the microstructures under these strength ratios is then simulated, and the energy dissipation due to the fracture process is calculated, respectively.

The crack patterns of the microstructures of different strength ratios are displayed in Figure 8.31. The strength ratio is defined as ρ. The length of the crack path and the percentage of transgranular fracture are illustrated in Figure 8.32. When the strength ratio ρ is between 0.4 and 0.5, the crack all propagates along grain boundaries, and the percentage of transgranular fracture is 0. When the strength ratio reaches 0.6, a small number of transgranular fractures (26.1%) occur, apart from a large number of intergranular fractures. When the strength ratio increases further to 0.7 and 0.8, the intergranular fracture is significantly reduced, and a large number of transgranular fractures (73.9% and 81.9%, respectively) begin to occur. For this reason, the strength ratio of 0.6 can be regarded as a turning point for intergranular and transgranular fractures.

Then, the energy dissipation during the fracture process of the microstructures of different strength ratios is analyzed. The time histories of the energy dissipation for the microstructures and the overall energy dissipation due to total fracture of the microstructures

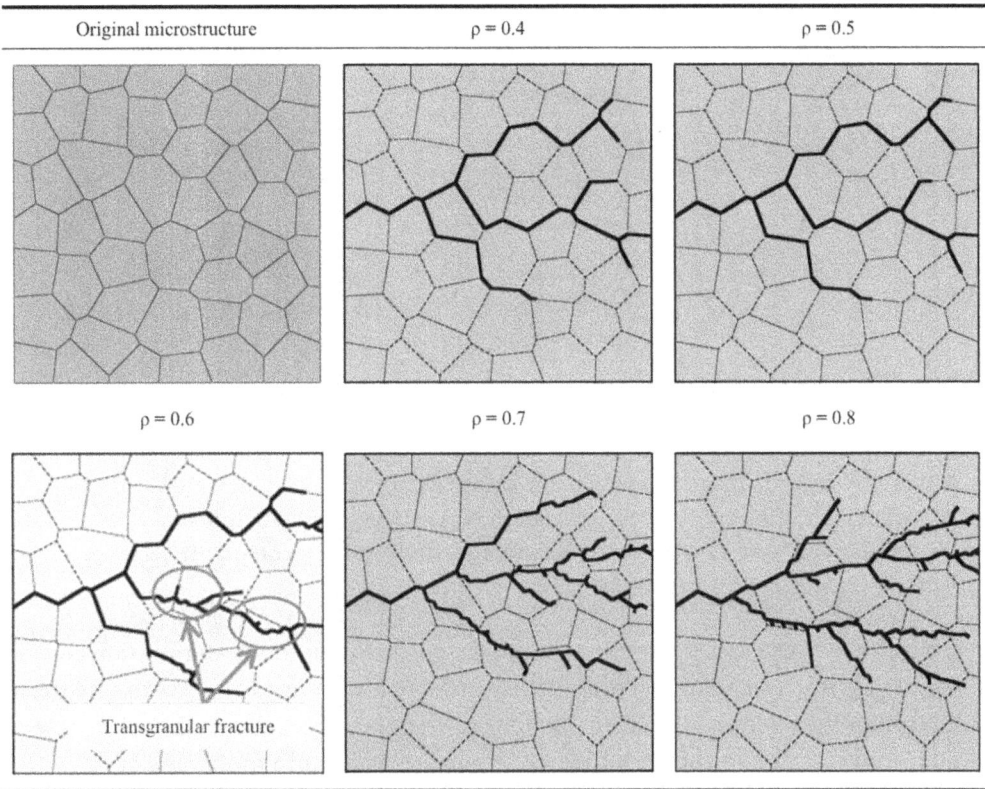

FIGURE 8.31   The crack patterns of the microstructures of different strength ratios (ρ).

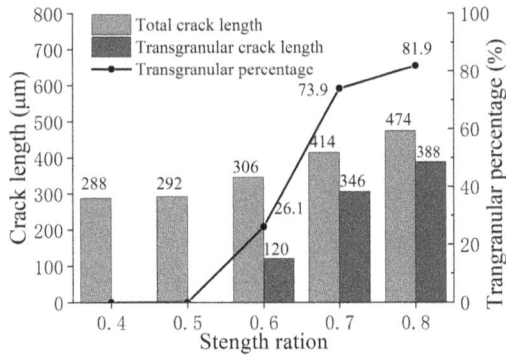

FIGURE 8.32   The length of the crack path and the percentage of transgranular fracture in terms of strength ratios.

in terms of the strength ratios are plotted in Figure 8.33. With the increase in strength ratios, the overall energy dissipation increases approximately linearly from 0.04 mJ at a strength ratio of 0.4–0.17 mJ at a strength ratio of 0.8. The reason is that, on the one hand, with the increase in strength of the grain boundary, the intergranular fracture will dissipate more energy. On the other hand, as mentioned above, with the increase in strength of the grain boundary, the crack propagation pattern will change from a mode with only

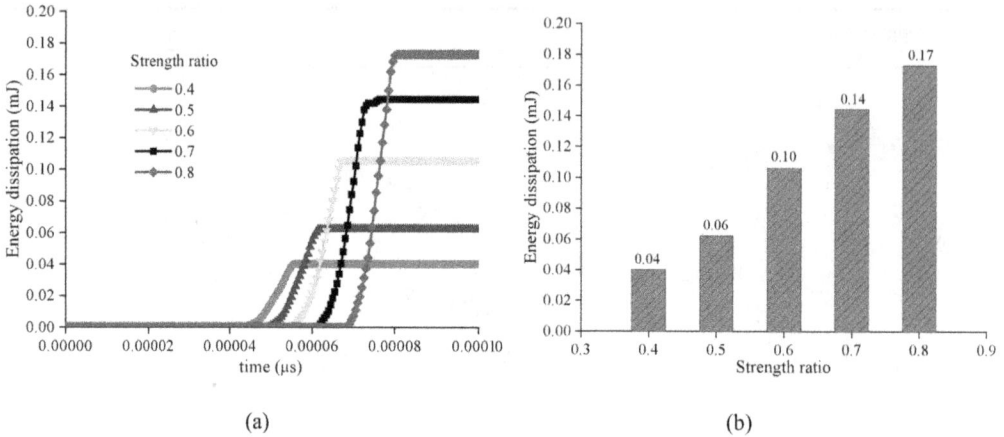

(a) (b)

FIGURE 8.33 Energy dissipation results in the case of a grain size of 8 μm: (a) time histories of the energy dissipation during crack propagation of the microstructures with different strength ratios, (b) the overall energy dissipation due to total fracture of the microstructures in terms of strength ratios.

intergranular fracture to a mode with the coexistence of intergranular and transgranular fracture. Since transgranular fracture dissipates more energy compared with intergranular fracture, improving the bonding strength of the grain boundary would increase the energy dissipation due to total fracture and the relative fracture toughness of the microstructures. Liu (2007) studied the effect of sintering temperature on the fracture toughness of polycrystalline CBN materials and found that in the low-temperature sintering zone, the bonding strength between CBN grains was weak and the fracture pattern was dominated by intergranular fracture, leading to small fracture toughness. However, in the high-temperature sintering zone, the bonding strength between CBN grains was relatively strong, and large amounts of transgranular fracture occurred in the fracture pattern, leading to a large fracture toughness. The basic trend obtained through the simulation is in agreement with the experimental results.

8.4.4 The Effects of Grain Size on Polycrystalline CBN Grain Fracture

Grain size is one of the important features that affect the material properties of polycrystalline brittle materials. It is mainly determined by sintering conditions, powder compositions, and proportions. To investigate the influence of grain size on the fracture toughness of polycrystalline CBN materials, three different grain sizes are selected in this simulation, which are 4, 8, and 12 μm, respectively. Two typical grain boundary strength ratios, i.e., 0.4 and 0.8, are chosen here to study the effect of grain size under intergranular fracture mode as well as intergranular and transgranular coexistence mode.

8.4.4.1 Intergranular Fracture Mode (Here Strength Ratio is 0.4)

Firstly, the characteristics of the crack patterns of the microstructures with different grain sizes are analyzed. Figure 8.34 shows the crack patterns for the microstructures with different grain sizes. It can be seen that for microstructures with smaller grain sizes, such as 4 μm, the crack path travels through more grain boundaries than that of microstructures with larger grain sizes, such as 12 μm. Thus, the crack path is more tortuous for micro-structures with smaller grain sizes. As plotted in Figure 8.35, the longest crack path of 390 μm is achieved with the smallest grain size of 4 μm, while the shortest crack path of 200 μm is achieved with the largest grain size of 12 μm, and the crack path of 288 μm with a grain size of 8 μm stays in the middle. Thus, for microstructures with a smaller grain size, the crack path tends to be longer.

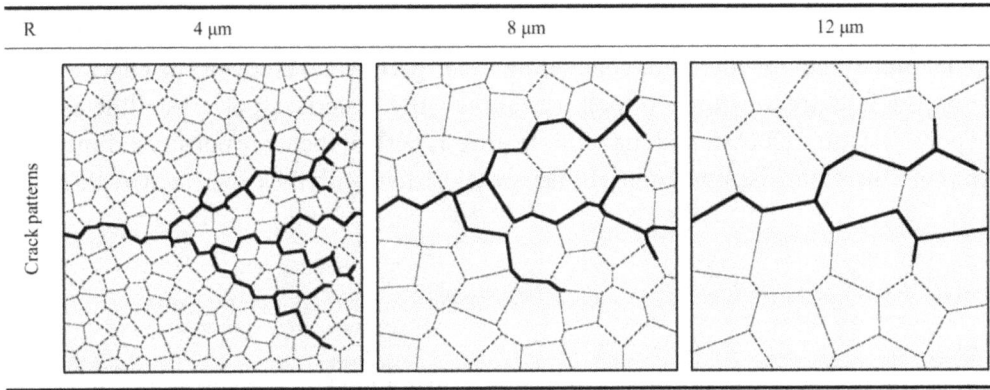

FIGURE 8.34 The crack patterns for the microstructures with different grain sizes (R) when the strength ratio is 0.4.

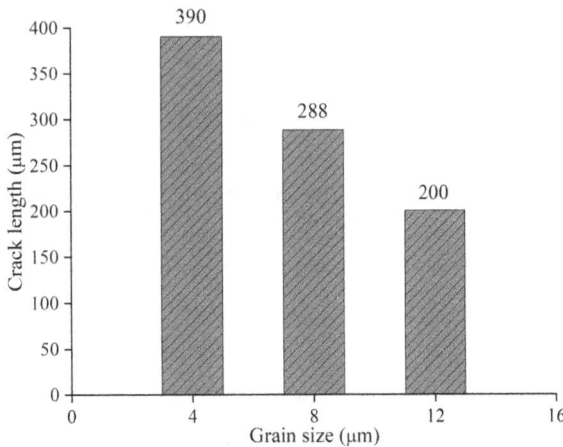

FIGURE 8.35 The length of the crack path in terms of grain size.

Secondly, the energy dissipations of the microstructures with different grain sizes due to the crack propagation process are analyzed. The time histories of the energy dissipation for the microstructures and the overall energy dissipation due to total fracture of the microstructures in terms of the grain size are plotted in Figure 8.36. It can be found that the overall energy dissipation decreases approximately linearly from 0.05 mJ at a grain size of 4 µm to 0.03 mJ at a grain size of 12 µm. Microstructures with smaller grain sizes tend to be tougher compared with those with larger grain sizes. According to the characteristics of the crack patterns mentioned above, under intergranular fracture mode, the crack path of the microstructures with smaller grain sizes is longer and more torturous than that of the microstructures with larger grain sizes. Since a longer and more torturous crack path will dissipate more energy, the material with a smaller grain size has a higher fracture toughness. The effect of grain size on the fracture toughness of polycrystalline alumina was studied (Schlacher et al. 2023; Yao et al. 2011). The crack pattern of alumina is dominated by intergranular fracture, and the fracture energy decreased as grain size increased. Although the authors didn't find any literature concerning the experimental study of the effect of grain size on the fracture toughness of polycrystalline CBN materials, polycrystalline alumina and polycrystalline CBN should share this feature as both of them are polycrystalline brittle materials. Therefore, the simulation results are consistent with the experimental results.

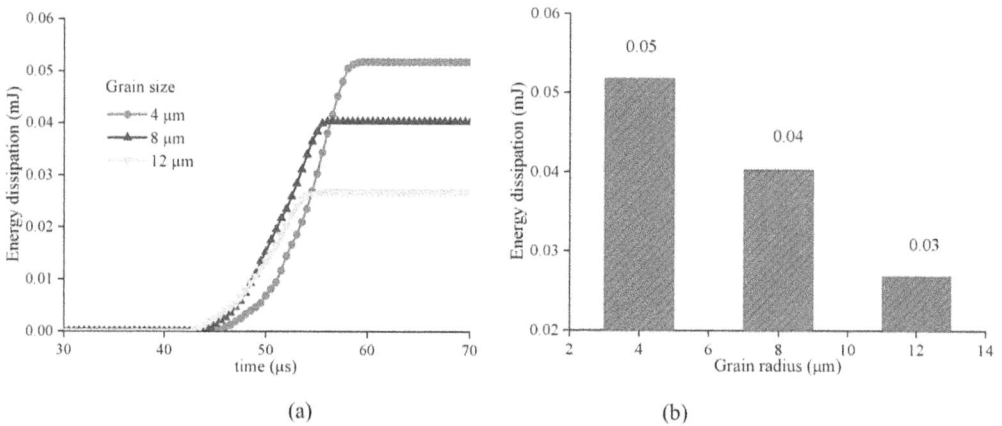

FIGURE 8.36 Energy dissipation results in the case of a strength ratio of 0.4: (a) time histories of the energy dissipation during crack propagation of the microstructures with different grain sizes, (b) the overall energy dissipation due to total fracture of the microstructures in terms of grain size.

8.4.4.2 Intergranular and Transgranular Coexistence Mode (Here Strength Ratio is 0.8)

The characteristics of the crack patterns of the microstructures with different grain sizes are also firstly analyzed. As depicted in Figure 8.37, the crack path of the microstructures consists of both intergranular and transgranular fractures. The length of the crack path and the percentage of transgranular fracture are plotted in Figure 8.38. The length of the total crack path increases from 418 µm at a grain size of 4 to 492 µm at a grain size of 12 µm, while the percentage of transgranular fracture increases from 67% to 91%. Microstructures with a larger grain size tend to have a longer crack path and a higher percentage of transgranular fracture.

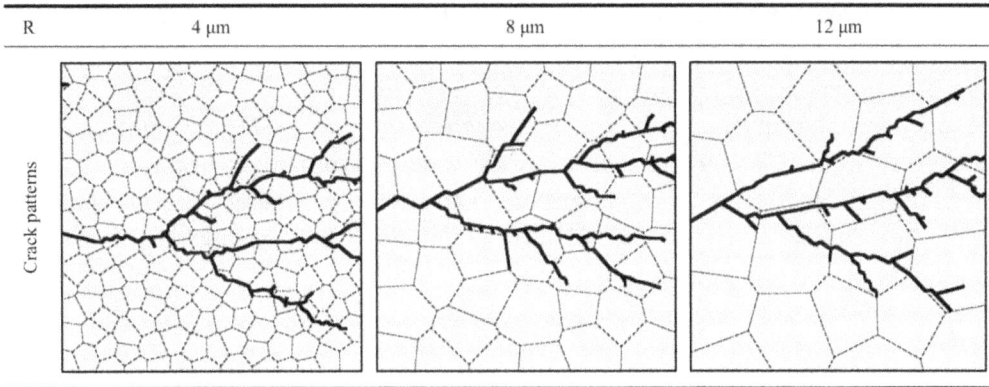

FIGURE 8.37   The crack patterns for the microstructures of different grain sizes (R) in the case of the strength ratio are 0.8.

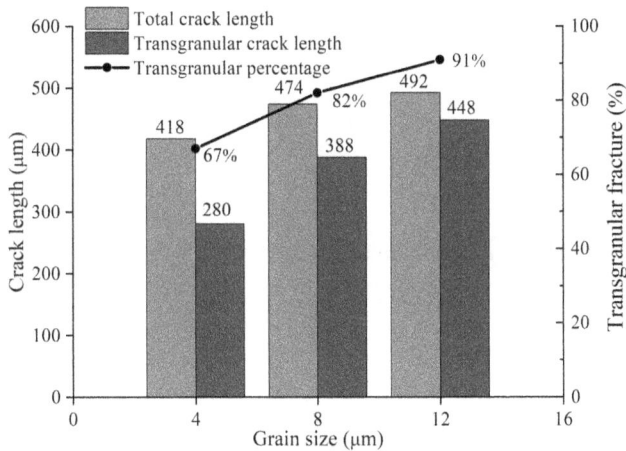

FIGURE 8.38   The length of the crack path and the percentage of transgranular fracture in terms of grain size in the case of a strength ratio of 0.8.

Then, the energy dissipations of the microstructures with different grain sizes are discussed. The time histories of the energy dissipation during the crack propagation of the microstructures and the overall energy dissipation due to the complete fracture of the microstructures are shown in Figure 8.39. As the grain size increases, the overall energy dissipation increases approximately linearly from 0.16 mJ at a grain size of 4 μm to 0.18 mJ at a grain size of 12 μm. The change tendency of the energy dissipation when the grain boundary strength ratio is 0.8 is contrary to that when the grain boundary strength ratio is 0.4. As mentioned above, with the increase in grain size, the total crack length will increase, and more importantly, the percentage of transgranular fracture will also increase. What is more, the bonding strength inside the grain is greater than that of the grain boundary, so the increase in the percentage of transgranular fracture due to the increase in grain size will lead to higher energy dissipation, i.e., higher fracture toughness. The influence of grain size on the fracture toughness of polycrystalline alumina materials was also studied. The fracture patterns are composed of both intergranular fracture and transgranular fracture. It was found that materials with larger grain sizes had higher fracture toughness (Wang and Li, 2017). The simulation results exhibit the same change tendency as the experimental results.

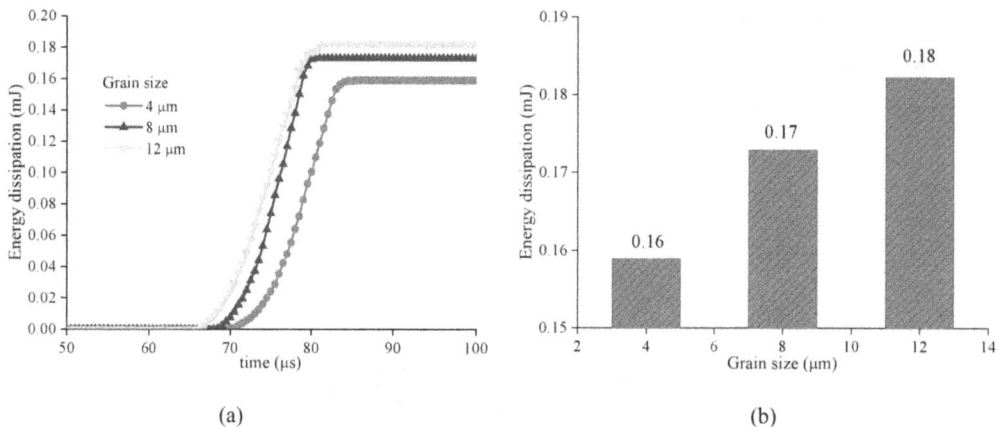

FIGURE 8.39 Energy dissipation results in the case of a strength ratio of 0.8: (a) time histories of the energy dissipation during crack propagation of the microstructures with different grain size, (b) the overall energy dissipation due to total fracture concerning grain size.

According to the analysis above, the influence of grain size on fracture toughness consists of two parts. When the crack pattern is dominated by intergranular fracture, the fracture toughness will decrease with an increase in grain size. However, when the crack pattern consists of both intergranular and transgranular fractures, the fracture toughness increases with the increase in grain size.

8.4.5 The Effect of the Grain Boundary Stiffness

The stiffness of the grain boundary is mainly determined by the type of binder materials used and the sintering conditions. To investigate the effect of grain boundary stiffness on the fracture toughness of polycrystalline CBN materials, five different values of grain boundary stiffness are selected in the simulation, which are 400, 500, 600, 700, and 800 GPa, respectively. The grain size is chosen as 6 μm. The grain boundary strength ratios are set at 0.4 and 0.8, respectively, to study the effect of grain boundary stiffness under intergranular fracture mode as well as intergranular and transgranular fracture coexistence mode.

8.4.5.1 Intergranular Fracture Mode (Here Strength Ratio is 0.4)

Firstly, the characteristics of the crack patterns of the microstructures with different grain boundary stiffness are analyzed. As depicted in Figure 8.40, the crack patterns of the microstructures with different grain boundary stiffness are almost the same. The grain boundary stiffness has little effect on the crack pattern. The length of the crack path is plotted in Figure 8.41. With the increase in grain boundary stiffness, the length of the crack path increases by 9.6%, from 354 μm at a grain boundary stiffness of 400 GPa to 388 μm at a grain boundary stiffness of 800 GPa. The change in the length of the crack path is generally small.

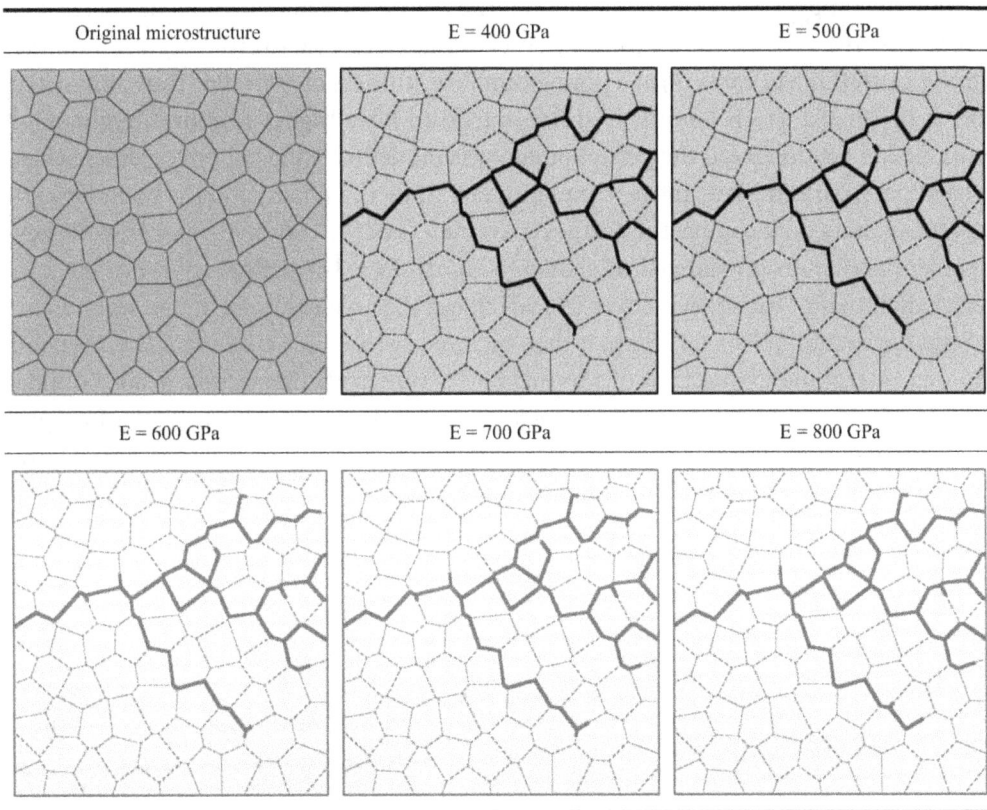

FIGURE 8.40 The crack patterns of the microstructures with different grain boundary stiffness when the strength ratio is 0.4.

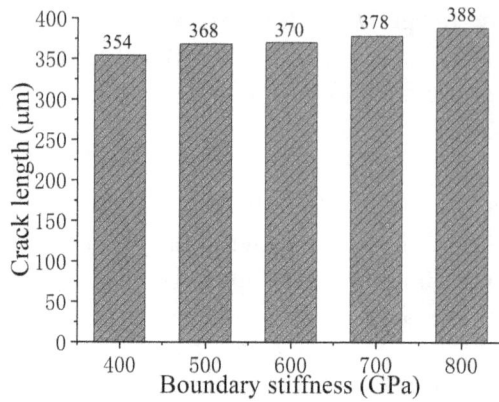

FIGURE 8.41   The length of the crack path in terms of grain boundary stiffness.

Then, the energy dissipations of the microstructures with different grain boundary stiffness are analysed. The time histories of the energy dissipation during crack propagation of the microstructures and the overall energy dissipation concerning grain boundary stiffness due to complete fracture of the microstructures are displayed in Figure 8.42. It can be seen that the overall energy dissipation decreases from 0.046 mJ at a grain boundary stiffness of 400 GPa to 0.025 mJ at a grain boundary stiffness of 800 GPa. Thus, microstructures with smaller grain boundary stiffness tend to have higher fracture toughness. As mentioned in Section 8.4.1, the grain boundary is modelled using cohesive elements based on traction-separation-law. In the simulation, to study the effect of grain boundary stiffness, the authors only change the stiffness parameters of the cohesive element while keeping other parameters unchanged, as shown in Figure 8.43. The stiffness of the first curve is greater than that of the second curve ($K1 > K2$), but the area enclosed by the second curve and the horizontal axis is larger than that enclosed by the first curve and the horizontal axis, i.e., intergranular fracture of the microstructure with the second grain boundary stiffness will dissipate more energy than that of the microstructure with the first grain boundary stiffness. Therefore, under intergranular fracture mode, the smaller the grain boundary stiffness, the higher the fracture toughness.

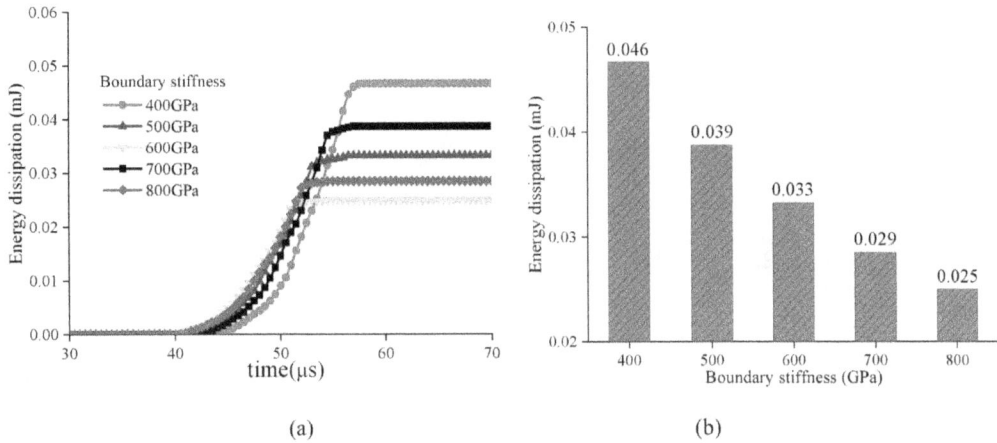

FIGURE 8.42 Energy dissipation results in the case of a grain size of 6 μm: (a) time histories of the energy dissipation during crack propagation of the microstructures with different grain boundary stiffness, (b) the overall energy dissipation due to total fracture in terms of grain boundary stiffness.

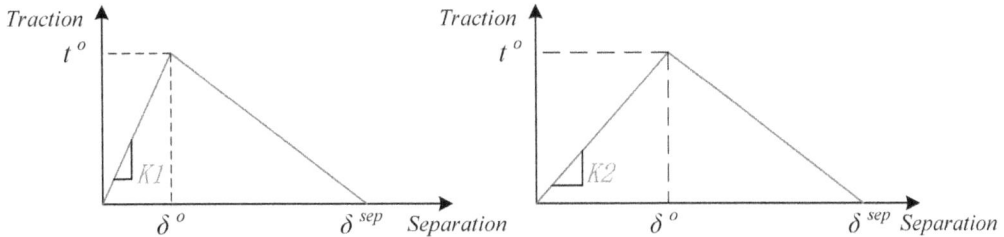

FIGURE 8.43 Traction separation law with different stiffness parameters ($K1 > K2$).

8.4.5.2 Intergranular and Transgranular Fracture Coexistence Mode

First, the characteristics of the crack patterns of the microstructures with different grain boundary stiffness are analyzed. As displayed in Figure 8.44, the crack patterns of the microstructures with different grain boundary stiffness show a significant difference from each other. The length of the crack path and the percentage of the transgranular fracture are plotted in Figure 8.45. It can be seen that when grain boundary stiffness increases from 400 to 500 GPa, the length of the total crack path increases from 402 to 492 μm, and the percentage of transgranular fracture increases from 74.1% to 85.4%. But when the grain boundary stiffness increases from 500 to 800 GPa, the length of the total crack path decreases from 492 to 364 μm, and the percentage of transgranular fracture decreases from 85.4% to 55.5%. At the grain boundary stiffness of 500 GPa, the length of the total crack path and the percentage of transgranular fracture all reach their peak values.

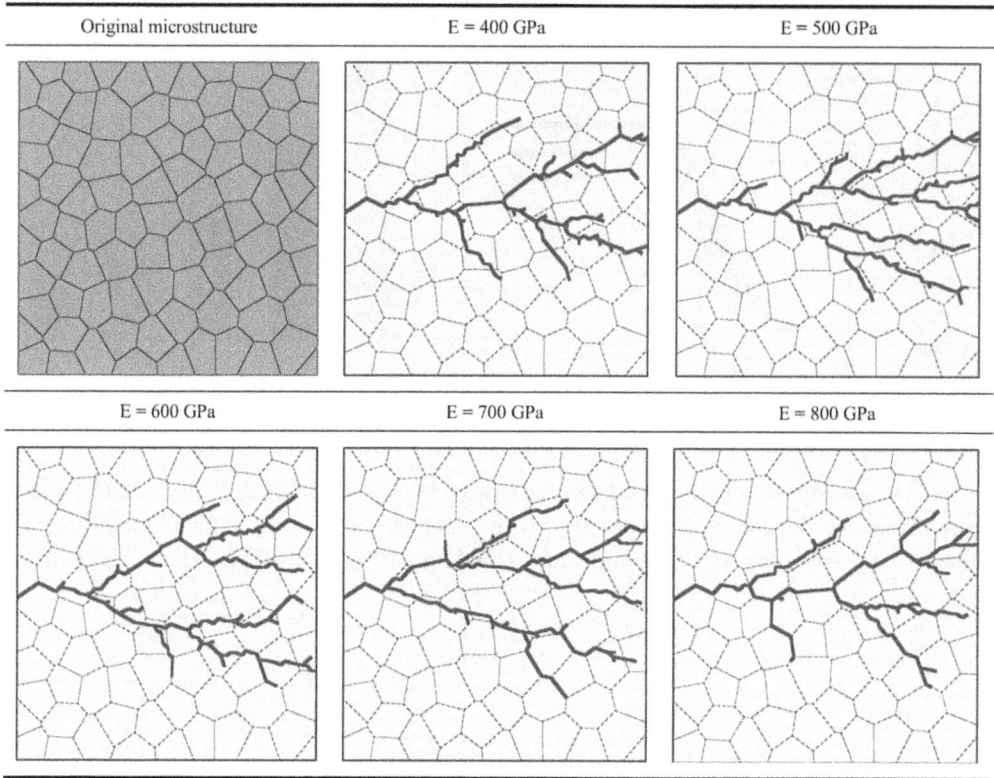

FIGURE 8.44 The crack patterns of the microstructures have different grain boundary stiffness when the strength ratio is 0.8.

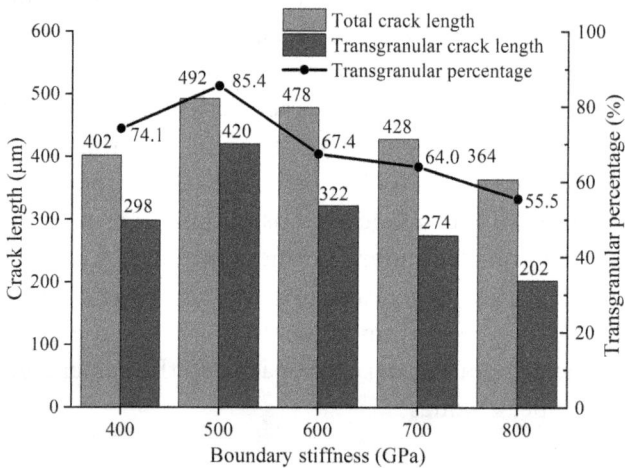

FIGURE 8.45 The length of the crack path and the percentage of transgranular fracture in terms of grain boundary stiffness.

Then, the energy dissipations of the microstructures with different grain boundary stiffness are analysed. The energy dissipation concerning time during crack propagation of the microstructures and the overall energy dissipation concerning grain boundary stiffness due to complete fracture of the microstructures are plotted in Figure 8.46. It is observed that the overall energy dissipation increases from 0.15 to 0.17 mJ with the increase of the stiffness of the grain boundary from 400 to 500 GPa, while the overall energy dissipation decreases from 0.17 to 0.11 mJ with the increase of the stiffness of the grain boundary from 500 to 800 GPa. The overall energy dissipation reaches a peak value of 0.17 mJ at a grain boundary stiffness of 500 GPa. Considering the relationship between the characteristics of the crack patterns and the overall energy dissipation, it can be found that the change tendency of the length of the crack path, especially the transgranular crack path, is consistent with the change tendency of the overall energy dissipation. Therefore, under intergranular and transgranular fracture modes, the influence of the stiffness of the grain boundary on the fracture toughness is mainly realized by changing the percentage of transgranular fracture.

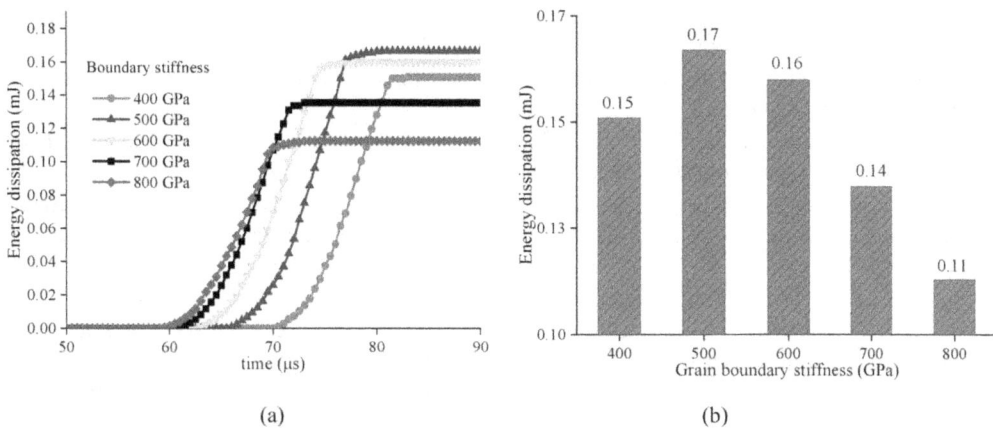

(a)　(b)

FIGURE 8.46 Energy dissipation results in the case of a grain size of 6 μm: (a) time histories of the energy dissipation during crack propagation of the microstructures with different grain boundary stiffness, (b) the overall energy dissipation due to total fracture of the microstructures in terms of grain boundary stiffness.

REFERENCES

Alveen, P., D. Carolan, D. McNamara, et al. 2013. Micromechanical modelling of ceramic based composites with statistically representative synthetic microstructures. *Computational Materials Science*, 79: 960–970.

Ding, W. F., J. H. Xu, Z. Z. Chen, et al. 2011. Grain wear of brazed polycrystalline CBN abrasive tools during constant-force grinding Ti-6Al-4V alloy. *International Journal of Advanced Manufacturing Technology*, 52: 969–976.

Huang, X., W. F. Ding, Y. J. Zhu, et al. 2017. Crack propagation simulation of polycrystalline cubic boron nitride abrasive materials based on cohesive element method. *Computational Materials Science*, 138: 302–314.

Kraft, R. H., and J. F. Molinari. 2008. A statistical investigation of the effects of grain boundary properties on transgranular fracture. *Acta Materialia*, 56: 4739–4749.

Li, H., K. Li, G. Subhash, et al. 2006. Micromechanical modeling of tungsten-based bulk metallic glass matrix composites. *Materials Science and Engineering A*, 429: 115–123.

Liu, J. 2007. *Study on the sintering of cBN-AlN composites at the HPHT. Dissertation (in Chinese)*, Sichuan University.

Liu, Y. J., D. W. He, P. Wang, et al. 2016. Microstructural and mechanical properties of cBN-Si composites prepared from the high pressure infiltration method. *International Journal of Refractory Metals & Hard Materials*, 61:1–5.

McKie, A., J. Winzer, L. Sigalas, et al. 2011. Mechanical properties of cBN-Al composite materials. *Ceramics International*, 37:1–8.

Schlacher, J., T. Csanádi, M. Vojtko, et al. 2023. Micro-scale fracture toughness of textured alumina ceramics. *Journal of the European Ceramic Society*, 43(7): 2943–2950.

Sfantos, G. K., and M. H. Aliabadi. 2007. A boundary cohesive grain element formulation for modelling intergranular microfracture in polycrystalline brittle materials. *International Journal for Numerical Methods in Engineering*, 69: 1590–1626.

Souza, A. M., E. de Silva. 2019. Global strategy of grinding wheel performance evaluation applied to grinding of superalloys. *Precision Engineering*, 57: 113–126.

Wang, Z., P. Li. 2017. Voronoi cell finite element modelling of the intergranular fracture mechanism in polycrystalline alumina. *Ceramics International*, 43(9): 6967–6975.

Warner, D. H., and J. F. Molinari. 2006. Micromechanical finite element modeling of compressive fracture in confined alumina ceramic. *Acta Materialia*, 2006, 54: 5135–5145.

Yao, W., J. Liu, T. Holland, et al. 2011. Grain size dependence of fracture toughness for fine grained alumina. *Scripta Materialia*, 65(2): 143–146.

Yin, A. Y., X. H. Yang, H. Gao, et al. 2012. Tensile fracture simulation of random heterogeneous asphalt mixture with cohesive crack model. *Engineering Fracture Mechanics*, 92: 40–55.

Yue, Z. M., L. M. Yang, J. H. Gong, et al. 2016. Experimental investigation on microstructure and mechanical properties of cBN-Ti3SiC2 composites. *Advanced Engineering Materials*, 18: 1568–1573.

Zhang, X. H., Z. H. Deng, G. Y. Chen, et al. 2016. A theoretical and experimental study on laser-induced deterioration in wet grinding of Al2O3 engineering ceramic. *International Journal of Advanced Manufacturing Technology*, 82(9): 1949–1957.

Zhou, T. T., C. Z. Huang, H. L. Liu, et al. 2012. Crack propagation simulation in microstructure of ceramic tool materials. *Computational Materials Science*, 54: 150–156.

Zhu, Y. J., W. F. Ding, T. Y. Yu, et al. 2017. Investigation on stress distribution and wear behavior of brazed polycrystalline cubic boron nitride superabrasive grains: Numerical simulation and experimental study. *Wear*, 376-377: 1234–1244.

Zhao, Y. C., and M. Z. Wang. 2009. Effect of sintering temperature on the structure and properties of polycrystalline cubic boron nitride prepared by SPS. *Journal of Materials Processing Technology*, 2009, 209: 355–359.

Zheng, G. M., J. Zhao, Z. J. Gao, et al. 2012. Cutting performance and wear mechanisms of Sialon-Si3N4 graded nano-composite ceramic cutting tools. *International Journal of Advanced Manufacturing Technology*, 58(1): 19–28.

CHAPTER 9

Fractal Analysis of CBN Grain Wear Morphology in Grinding

9.1 INTRODUCTION

The abrasive wheel surface topography, which works as an important input variable, is one of the key factors influencing the grinding performance of brazed cubic boron nitride (CBN) abrasive wheels (Dai et al. 2022; Li et al. 2019; Linke and Klocke, 2010; Yan et al. 2011). Detailed knowledge of the topography of an abrasive wheel highlights the results of wheel preparation and working state. Topographical changes during grinding contribute to accounting for the effects of grinding parameters on grinding performance. Also, analysis of the nature of topographical changes can help improve the control of the grinding process. For the further advancement of grinding technology, it is therefore helpful to establish procedures for measurement and monitoring of CBN grain topography in grinding. It has been asserted that once the topography of the abrasive grain is identified, the grinding process behaviour of the abrasive wheels can be predicted with greater reliability (Bredthauer et al. 2023). Nevertheless, the precise description of the tool surface of the abrasive wheels is still scarce and difficult because the tool surface is not only mainly composed of hundreds of irregular CBN abrasive grains but is also strongly influenced by grinding conditions. It is therefore very important to describe the tool surface by quantitatively analysing the topography of the abrasive grains.

In the grinding process, the grain topography not only consists of geometrically undefined cutting edges in statistical distribution but is also influenced by wear behaviour and the preceding profiling and sharpening processes (Jimoto et al. 2006; Stachowiak and Podsiadlo, 2001). However, in traditional statistical approaches, the topography of CBN grains is always postulated as stationary. The majority of methods used in three-dimensional (3D) topography characterization provide functions or parameters that strongly depend on the scale at which they are calculated. This means that these statistical parameters, i.e., variance of height, slope and curvature, are not unique for some particular surface. The characterization provided by these functions and parameters is not a reliable

DOI: 10.1201/9781032678047-11

description method for the 3D features of the grain topography and the multi-scale nature of tribological surfaces (Podsiadlo and Stachowiak, 2000). Therefore, being well known for the topography of CBN grains as a non-stationary random process implies that some intrinsic parameters have to be introduced to better characterize the grain surface.

Given the random distribution and shape of the cutting edges of brazed CBN grains, the statistical method is applied to analyse the grain wear mechanism during grinding. Fractals are regarded as 'self-similar or self-affinity', meaning that they exhibit similar fine-scale features at many magnifications (Mandelbrot, 1967). Though the grain surface is not strictly a fractal, the surface topography is statistically self-affinity. Moreover, the surface topography resembles, at least in some scale regimes, a fractal so closely that it makes good sense to model a fractal using the grain surface statistically. Also, fractal theory has been widely applied in geomorphology, physics, and biology. For surface engineering, fractal theory is used to study contact mechanics, wear, crack analysis, and characterization of surface topography, and some achievements have been acquired accordingly. Therefore, fractal theory could be used to analyse the wear behaviour of the CBN grains during grinding. That is, the fractal dimension could be utilized as a descriptor of the grain surface. In general, higher fractal dimensions express a more complicated wear topography of CBN grains.

The main purpose of this work is to discuss the feasibility of measuring abrasive grain topography through reconstruction models based on fractal theory. Surface grinding experiments were carried out using a single-layer brazed abrasive wheel. The wear topography of brazed abrasive grains was observed using the 3D optical video microscope with a resolution of $1,600 \times 1,200$ pixels. Accordingly, the 3D topography of the grains that have complicated and random shapes was extracted and reconstructed based on fractal theory. In particular, limited to either vertical or horizontal resolution of the present optical equipment, the images obtained are not clear and sharp, so it is rather difficult to evaluate the grain directly and quantitatively. However, the complicated change in grain wear topography could be analysed quantitatively using fractal theory. As a consequence, the relationship between fractal dimension and grain wear topography was investigated, which could provide a guide to establishing the correlation between the fractal dimension and the grinding performance of abrasive wheels.

9.2 FRACTAL ANALYSIS OF GRAIN WEAR DURING GRINDING NICKEL-BASED SUPERALLOY

9.2.1 Details of the Grinding Experiment

Before the grinding test, the polycrystalline CBN grains were carefully filtered for similarity in mesh size and shape and then arranged on the surface of filler power to ensure the wear behaviour of every grain was uniform as possible. After the brazing test was carried out, surface grinding experiments were conducted subsequently. The experimental conditions are listed in Table 9.1 (Miao et al. 2013). The topography of the brazed polycrystalline CBN abrasive wheel is displayed in Figure 9.1.

After grinding, the brazed polycrystalline CBN abrasive wheel was cleared first with the ultrasonic cleaning method. Then, the optical video microscope (Hirox KH-7700) was

TABLE 9.1   Grinding Experimental Condition

Types	Contents
Grinding machine	Precise horizontal spindle grinding machine modelled HZ-Y150
Abrasive wheels	Brazed polycrystalline CBN abrasive wheel
Grain mesh size	#40/50
Grinding fluid	Water-based emulsion
Ground material	Nickel-based superalloy GH4169 (similar to Inconel718)
Peripheral wheel speed ν_s	30 m/s
Workpiece infeed speed ν_w	6 m/min
Depth of cut a_p	10 μm

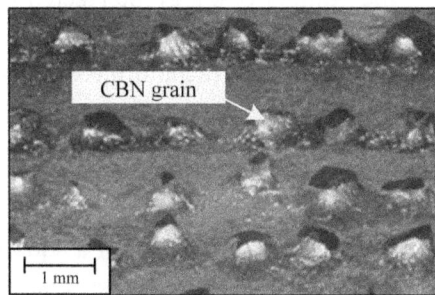

FIGURE 9.1   Topography of the brazed polycrystalline CBN abrasive wheel.

utilized for 3D topography observation of polycrystalline CBN grains on the wheel surface. As is well known, though all the polycrystalline CBN grains of the brazed abrasive wheel have worked under a nominally identical grinding condition, the wear behaviour and topography of every grain are perhaps quite different from each other. For this reason, four polycrystalline CBN grains with complete wear topography were chosen randomly with the serial numbers polycrystalline cubic boron nitride-1 (PcBN-1), PcBN-2, PcBN-3, and PcBN-4 in this work.

9.2.2 Basic Principles and Analysis Methods of Three-Dimensional Fractal Theory
9.2.2.1 The Basic Principles of 3D Fractal Theory
Fractal dimension is a quantitative value that can express a complicated degree in the geometrical shape of an object (James and Davide, 2010; Liang et al. 2012). A higher fractal dimension always corresponds to a more complicated shape and richer microstructure details. In this work, the fractal dimension is applied to evaluate the complicated change in the wear topography of brazed polycrystalline CBN abrasive grains during grinding. The 3D analysis method is based on the idea of the box dimension, which could be described as follows:

Generally, the dimension p of a set F is subject to the relationship of the exponential law when $\delta \to 0$. That is,

$$M_\delta(F) \sim c\delta^{-p} \tag{9.1}$$

where c and p are constant numbers. When Eq. (9.1) is logarithmized, the following formula can be calculated:

$$\log M_\delta(F) \sim \log c - p\log\delta \tag{9.2}$$

when $\delta \to 0$, the equation is given as follows:

$$p = \lim_{\delta \to 0} \frac{M_\delta(F)}{-\log\delta} \tag{9.3}$$

According to the above equations and theory, a 3D profile is divided by a cub grid with a mesh size of δ. Subsequently, the number of cubes intersected with the 3D profile $N(\delta)$ is counted. If there is a fractal nature in the 3D profile, when the mesh size δ ($\delta > 0$) is scaled down, the relationship between $N(\delta)$, δ, and fractal dimension Ds can be expressed by the following equation:

$$N(\delta) = c\delta^{-D_s} \tag{9.4}$$

where c is a constant number. Assume that the area of a square divided by mesh size δ is $S(\delta)$, then the surface area of a 3D profile based on $N(\delta)$ is illustrated by Fujimoto and Ichida (2008).

$$S(\delta) = \delta^2 N(\delta) = c\delta^{2-D_s} \tag{9.5}$$

Then, take the logarithm of both sides; Eq. (9.5) is rewritten in the form as follows:

$$\log S(\delta) = \log c + (2 - D_s)\log\delta \tag{9.6}$$

Finally, if the value of δ scale is in a proper scope, when the log δ differential on both sides of Eq. (9.6) is taken, the fractal dimension D_s could be estimated using the slope of a straight line, that is,

$$D_s = 2 - \frac{d\log S(\delta)}{d\log\delta} \tag{9.7}$$

9.2.2.2 The Analysis Method of 3D Fractal Theory

In this work, 3D fractal analysis is conducted according to the following procedures: First of all, every 3D coordinate with an interval of 1 μm in the selected area is extracted using 3D optical video microscopy. The polycrystalline CBN grain surface could exhibit topographical features with a measuring accuracy of microscale of 1 μm. Afterwards, the raw data is re-extracted from the obtained 3D coordinates with different mesh sizes. As seen in Figure 9.2a, when the coordinates are re-extracted with an interval of 16 μm, a cub grid with a mesh size $\delta = 16$ μm is set on the surface of the polycrystalline CBN grain. The top surface of the grain is divided into many triangles with the re-extracted 3D coordinates, and the area of each triangle could be calculated using the re-extracted coordinates. Finally, the grain topography is decided by the sum of these small triangle surface areas. With the mesh size δ scaling down, as it turns out in Figure 9.2b–d, the topography of brazed polycrystalline CBN grain divided by small triangles is performed more and more clearly. Measurement of these triangles area is taken in many passes with different μm steps. The smaller the scale value δ, the bigger the triangle number is. Moreover, the richer the detail of the surface topography, the closer it is to the true surface.

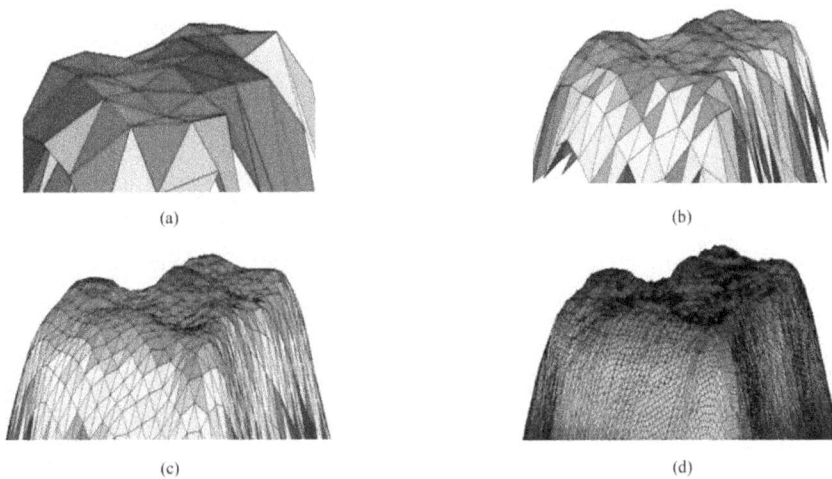

(a)

(b)

(c)

(d)

FIGURE 9.2   Method of reconstruction of grain topography based on 3D fractal analysis (Sample: surface profile of PcBN-1 grain): (a) $\delta = 16$ μm, (b) $\delta = 8$ μm, (c) $\delta = 4$ μm, (d) $\delta = 2$ μm.

Based on the above analysis and equations, $\log \delta$ is taken on the horizontal axis, and $\log S(\delta)$ is taken on the vertical axis, as displayed in Figure 9.3. When the data points are fitted into a straight line in the rectilinear coordinate, fractal dimension D_s is obtained according to the slope of the line. As stated previously, the grain topography is not strictly a fractal and resembles, in some scale regimes, a fractal. As seen in Figure 9.3, the area values of dots A and B on the grain surface are so close that they do not meet the requirements of fractal nature anymore. Consequently, the fractal dimension is 2.03609 calculated within a range of 2 μm $< \delta < 10$ μm.

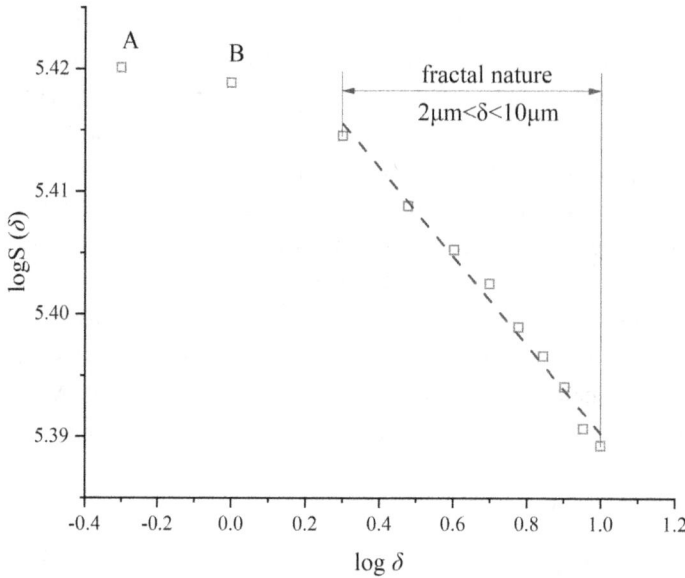

FIGURE 9.3   The calculation of the 3D fractal dimension of the surface profile sample.

9.2.3 Reconstruction of the Polycrystalline CBN Grains Topography

After the grinding test, the raw surface profiles of brazed polycrystalline CBN abrasive grains were extracted and then imported into MATLAB software. For different mesh sizes δ (unit: μm), when 3D coordinates were extracted, the topography of the brazed polycrystalline CBN grains was accordingly reconstructed by estimating the fractal dimension using the basic principle of 3D fractal analysis.

Take the PcBN-1 grain as an example. Just as described above, because of the limitations in the resolution of the present optical equipment, it is found from Figure 9.4a that the obtained image of the brazed PcBN grain is rough, dark, and obscure, which makes it difficult to observe the topographical detail. In contrast, the reconstruction model of the grain topography is clear and sharp-edged, as displayed in Figure 9.4b.

Figure 9.5 shows a magnified image in the vicinity of the grain top surface with different mesh sizes. Microfracture is apparent in these images. When mesh size δ is 1 μm, as shown in Figure 9.5a, the surface profile of the grain is fine and the wear flat, which is the typical characteristic of abrasion wear, occurs on the top surface of the brazed polycrystalline CBN grain. The image clearly shows a concentration of microfracture in the region around the wear flat. When the microfracture behaviour of the brazed polycrystalline CBN grains happens due to the self-sharpening effect during grinding, the space taken by the grain profile is complicated, large, and irregular. Also, as displayed in Figure 9.5a, the wear-flat region takes a small proportion in comparison to the whole surface of the PcBN grain. That is, microfracture and abrasion wear take place simultaneously in different regions with different proportions during grinding. When mesh size δ is chosen as 2 μm, as shown in Figure 9.5b, the comparison of 3D profile sharpness is obvious. Meanwhile, the flat region on the grain surface exhibits visibly. However, compared to the image in Figure 9.5a, the complexity and irregularity of the grain shape are slightly less. Figure 9.5c demonstrates

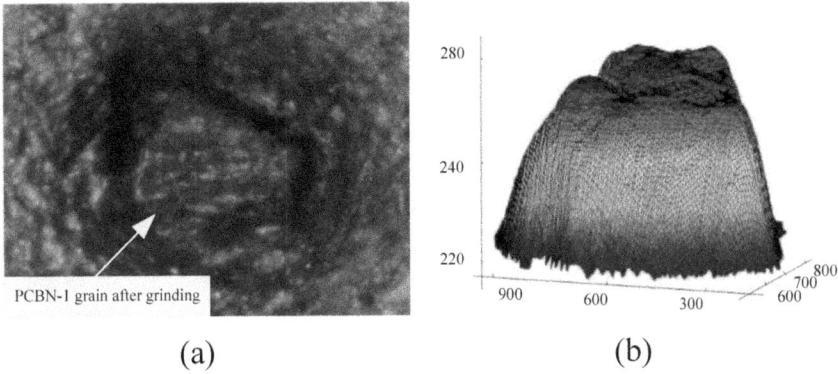

FIGURE 9.4   Topography of PcBN-1 grain after grinding: (a) 3D video microscope image, (b) 3D reconstruction image (unit: μm).

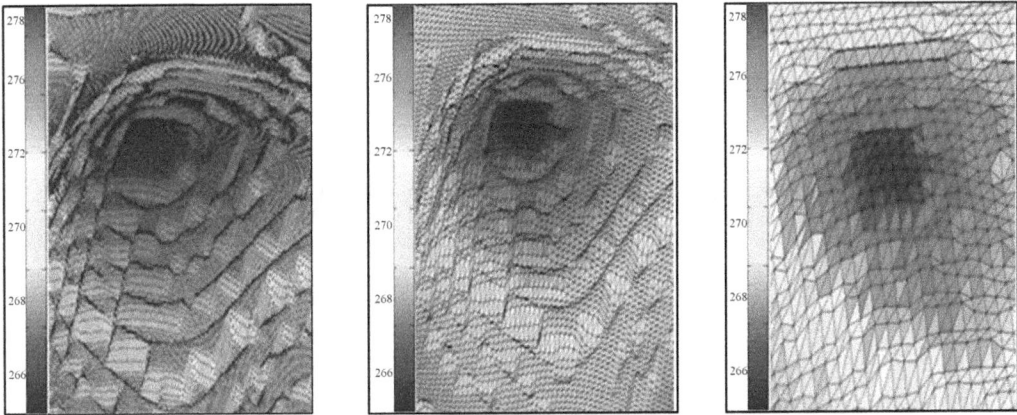

FIGURE 9.5   Regional 3D reconstruction topography of PcBN-1 grain surface with different mesh size δ: (a) $\delta=1\,\mu m$, (b) $\delta=2\,\mu m$, (c) $\delta=4\,\mu m$.

merely the broad grain profile when mesh size δ is 4 μm. Under such circumstances, it is difficult to grasp the detailed microstructure on the top surface of the brazed polycrystalline cBN grain. Therefore, the establishment of the reconstruction models is favourable and necessary to figure out the rich detail of the grain topography.

9.2.4 Calculation of the 3D Fractal Dimension

The wear topography of polycrystalline CBN abrasive grain after grinding is extremely like a small 'island'. Therefore, it can be regarded as the natural surface topography of a downsized mountain. As mentioned above, the actual surface topography of the polycrystalline CBN abrasive grain should be treated as a non-stationary random process. The grinding process is too complicated to exhaustively evaluate the grain wear process using a limited grain region. Consequently, the results of 3D fractal dimensions obtained only for grain contour lines are limited, and the fracture could take place in multiple places on the surface of one grain simultaneously. Figure 9.6 shows the contour line of $h=220\,\mu m$

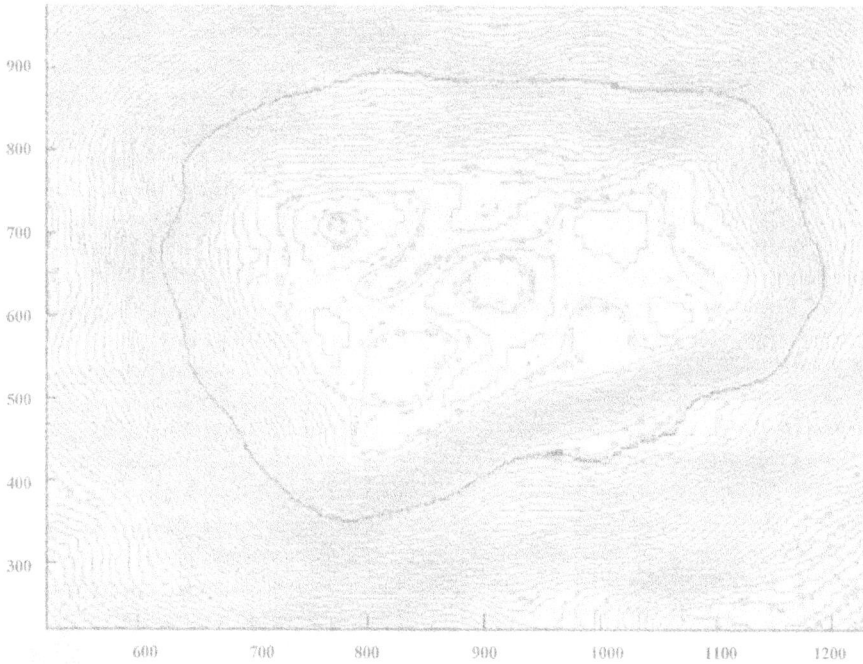

FIGURE 9.6   The contour map of the PcBN-1 grain with an interval of 2 μm.

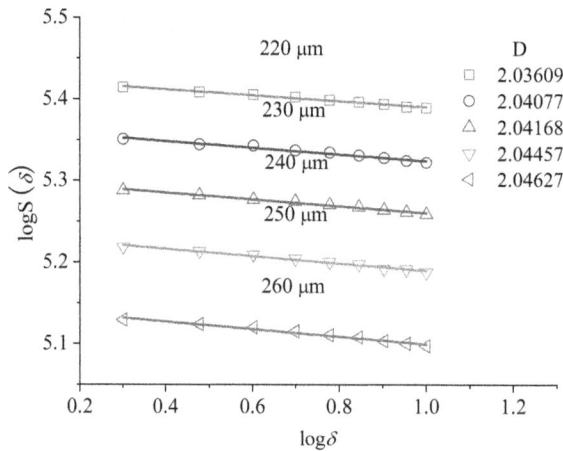

FIGURE 9.7   The fractal dimension of the 3D topography of the PcBN-1 grain with different contour lines.

enclosed in the red frame. As seen in Figure 9.7, even if the calculated 3D surface is chosen above the contour lines, the region still closely resembles a fractal.

Considering those factors mentioned above and the influence of the actual exposing height of the brazed grain (grain size: #40/50), the 3D fractal dimension in the vicinity of the grain top surface is calculated above the contour line of $h = 220\,\mu m$. Under such conditions, more protrusion of the whole brazed PcBN grain is included. It can be observed from

Figure 9.7 that a correlation exists between the fractal dimension and the cutting contour line h in the case of the PcBN-1 grain.

Also, it has been approved that fractal nature is in a region of $2\,\mu m < \delta < 10\,\mu m$ for PcBN-1 grain. The fractal dimension is accordingly calculated. The result is 2.03609 with a standard error of 0.00105. The fractal dimension values of the other three grains, PcBN-2, PcBN-3, and PcBN-4, are obtained and displayed in Table 9.2.

9.2.5 Fractal Analysis of Wear Topography of Brazed PcBN Grains

As mentioned above, using the reconstruction of grain wear topography, the surface microstructure formed through the microfracture behaviour can be grasped clearly. During grinding, though the grinding parameters are identical, the shape of every grain is complicated and different from each other. Therefore, every brazed polycrystalline CBN⁻ grain of the abrasive wheel shows a difference in fractal dimension and microstructure of the grain surface. The topography of grains composed of microfracture, large-fracture, and wear flat is shown in Figure 9.8 for PcBN-1, PcBN-3, and PcBN-4 grains. Compared with the image of the PcBN-2 grain, as displayed in Figure 9.8a, the topography of the top surface of the other three grains has relatively simple and less detailed microstructures, and it appears that many wear flats and large fractures on the surfaces of the grains. The formation of wear flats on polycrystalline CBN grains leads to a loss of grain sharpness and a tendency of dull polycrystalline CBN grains. Also, as seen in Figure 9.8b and c, the topography of the brazed polycrystalline CBN grains has changed significantly. The fracture range connected with the change in topography of polycrystalline CBN grains results in a different level of wear resistance for the grains. This is confirmed comparatively in Figure 9.8a, which illustrates that the space taken by the profile of the PcBN-2 grain

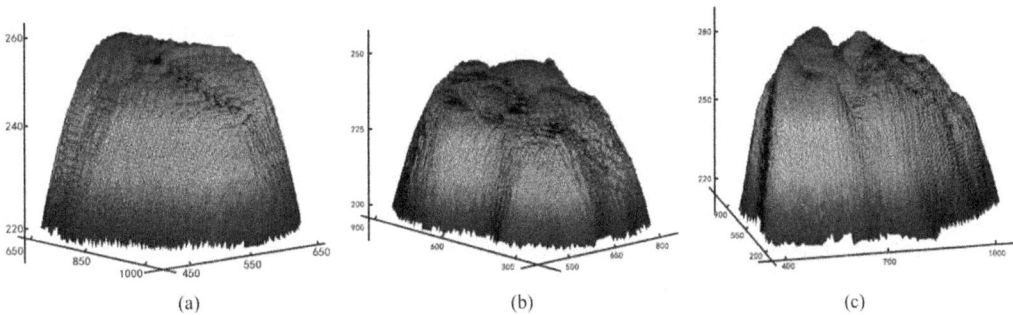

(a)   (b)   (c)

FIGURE 9.8 The 3D reconstruction topography of three polycrystalline CBN grains (unit: μm): (a) PcBN-2, (b) PcBN-3, (c) PcBN-4.

TABLE 9.2 Fractal Dimension of Different PcBN Grains

Grain No.	The Cutting Contour Line (μm)	Fractal Dimension Ds	Standard Error
PcBN-1	220	2.03609	0.00150
PcBN-2	220	2.06408	0.00498
PcBN-3	220	2.04019	0.00174
PcBN-4	220	2.03850	0.00147

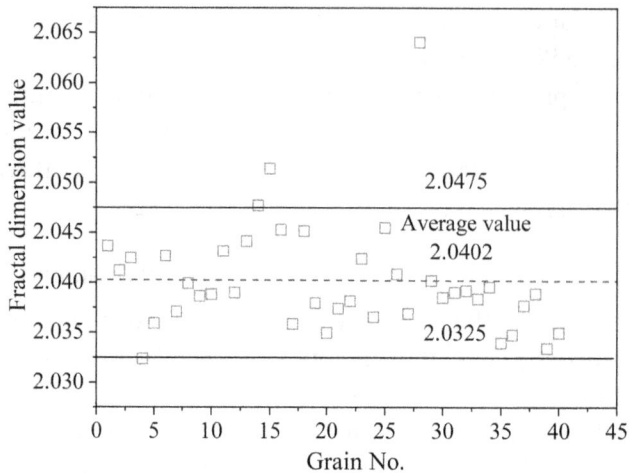

FIGURE 9.9 The fractal dimension of 40 polycrystalline CBN grains of the brazed abrasive wheel.

is complex and irregular. The top surface of the PcBN-2 grain is distinguished by a fine microcrystalline structure without wear flat. The fractal dimension of the PcBN-2 grain is the highest among the four grains. That is to say, the fractal dimension of the complicated topography of the brazed polycrystalline CBN grains with microfracture behaviour due to the self-sharpening effect is higher than that of the rough topography of the other three grains due to large fractures and wear flat.

On the other hand, the fractal dimensions of 40 grain samples in the brazed polycrystalline CBN abrasive wheel were also calculated and analyzed in the present investigation. The result is displayed in Figure 9.9. For the vast majority of the forty polycrystalline CBN grains, the achievable fractal dimension range of the wear topography is limited to 2.0325–2.0475 with a concentrated value of 2.04. It is very close to the average value of 2.0402 for all the grains. Under such conditions, the wear topography of these different polycrystalline CBN grains in the brazed abrasive wheel could be quantified using an approximate value. At the same time, this value could be defined within a certain standard error.

The measurable results suggest the existence of a common correlation between the fractal dimension and the wear behaviour of brazed polycrystalline CBN grains. Therefore, the determination of the relationship between fractal dimension and the grinding performance of polycrystalline CBN grains, as well as the identification of the wear topography of a superabrasive wheel in the grinding process, will be the key work in the subsequent work.

9.3 FRACTAL ANALYSIS OF GRAIN WEAR DURING GRINDING TITANIUM ALLOY

9.3.1 Details of the Grinding Experiment

In this work, experiments were conducted on the PROFIMAT MT-408 high-speed precision grinding machine (Wang et al. 2018). During the high-speed grinding process, the grinding wheel usually expands so that it is not precious to set the grinding wheel in a stationary state. In this work, an acoustic emission device was used to set the grinding wheel dynamically. When the grinding wheel rotates at the working speed and moves close to the workpiece, the acoustic emission device would receive a signal if the grain came into contact with the workpiece. Then the grinding parameters can be set precisely.

The grinding wheel was composed of a wheel substrate and two balancing weights. The diameter of the grinding wheel is 400 mm. Two balancing weights were located on the opposite sides of the grinding wheel's diameter line to maintain dynamic balance in the high-speed grinding process. Both monocrystalline CBN grains and polycrystalline CBN grains were tested. The grains were selected through a 3D optical video microscope to ensure similar grain shape and size. The maximum grain size was 300–400 μm, corresponding to the standard 40/50 mesh size. The grain sample was fixed by the induction brazing method on one of these balancing weights, as illustrated in Figure 9.10. The grain was embedded in Ag-Cu-Ti alloy filler metal. As a result, a strong metallurgy-chemical combination was generated between the single grain and the brazing filler metal so that the single grain could be held firmly on the balancing weight. The workpiece was made of Ti-6Al-4V titanium alloy material and shaped into a cuboid with a 100×50 mm top surface and an 8 mm height. Before experiments, the top surface should be ground to ensure the surface roughness Ra less than 0.4 μm.

The parameters of the single-grain grinding process are listed in Table 9.3. The workpiece infeed speed and the grinding depth were constant at 80 mm/min and 0.02 mm, respectively. Because there was only single grain on the wheel, the interval between two adjacent grains λ was exactly the perimeter of the grinding wheel. Meanwhile, the diameter of the grinding wheel was 400 mm. The value of a_{gmax} was calculated and determined by the rotating speed of the grinding wheel. A stroke was set to be 50 mm to ensure grains were changing obviously. In this work, each grain should complete at least 10 strokes to get fractured. When completing a stroke, the balancing weight with the single grain was taken down without moving the grinding wheel to take photographs of the grain surface topography, and the topography data was collected with a 3D optical video microscope.

TABLE 9.3 Parameters of Single Grain Grinding Process

Single Grain Grinding Wheel	Parameters
Grinding wheel speed	20–120 m/s
Workpiece infeed speed	80 mm/min
Grinding depth	0.02 mm
a_{gmax}	0.2–1 μm
Grinding stroke	50 mm

FIGURE 9.10   Grinding wheel with a single grain.

Therefore, the installation precision of the grinding wheel will not change, while the precision of the balancing weight can be compensated with the acoustic emission device. From these photographs, the fracture situation of grains can be clearly observed. Then the grinding region was changed, and the grain was reset to continue the next stroke so that the surface topography of the grinding region could be protected. When finishing ten strokes, the topography data of all the grinding regions was collected with a Sensofar 3D optical profiler. With the collected data on surface topography, the profile of grains and the grinding scratches on the workpiece can be digitally reconstructed. The fractal dimension can also be calculated.

9.3.2 Comparative Fractal Analysis of Polycrystalline and Monocrystalline Grains

Some comparison grinding experiments with monocrystalline CBN grains were conducted when a_{gmax} was 0.6 μm. Both monocrystalline and polycrystalline CBN grains had to complete 10 strokes. After each stroke, photographs of surface morphology were taken with a 3D optical video microscope. Among these grains, two monocrystalline CBN grains and two polycrystalline CBN grains were selected to make comparisons quantitatively. With the topography data collected, the fractal dimensions of grains after each stroke were calculated with Matlab software. In general, it is known that the value of fractal dimensions will rise when the surface gets sharper and decrease otherwise. Four fitting curves were then drawn based on these fractal dimensions to approximately display the change in grain sharpness and topography profile. The slope of the fitting curve indicated the falling speed of grain sharpness, which also reflected the average wear speed of the grain surface. Considering that not all the fractal dimension nodes were located on the fitting curves, the standard deviation was therefore calculated to describe the dispersion degree, which reflected the stability of topography change.

Figure 9.11 displays four fitting curves for the fractal dimensions. It is obvious that all fitting curves fall down, but with different slopes, listed in Table 9.4, which was because the four grains got worn at different speeds. The slopes of polycrystalline CBN grains were −0.00108 and −0.00262, respectively, while the slopes of monocrystalline CBN

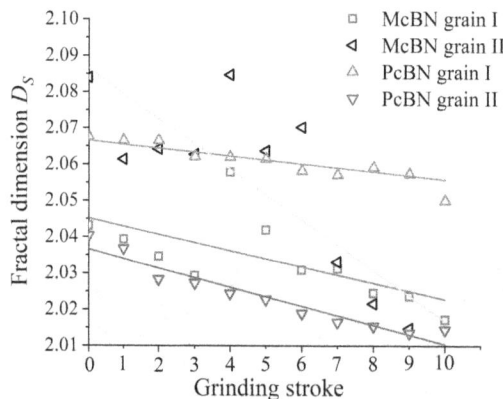

FIGURE 9.11 Fractal dimension fitting curves of monocrystalline and polycrystalline CBN grains.

TABLE 9.4 Slope and Standard Deviation of Fractal Dimension Fitting Curves

Grains	Slope	Standard Deviation
McBN grain I	−0.00224	0.00502
McBN grain II	−0.00698	0.00872
PcBN grain I	−0.00108	0.00217
PcBN grain II	−0.00262	0.00157

grains were −0.00224 and −0.00698, respectively. It means that polycrystalline CBN grains (PcBN) I and II got worn at a similar speed. But for monocrystalline CBN grains, the grain I was worn slowly, but the slope of grain II was much larger than the former, which means the wear speeds of monocrystalline CBN grains were much different from each other. By comparison, the wear speed of polycrystalline CBN grains seemed lower and more stable. Generally, it is easier for polycrystalline CBN grains to keep their initial sharpness.

On the other hand, the standard deviations are listed in Table 9.4. The values of monocrystalline CBN grains were 0.00502 and 0.00872, which were much larger than polycrystalline CBN grains. The fractal dimension nodes of monocrystalline CBN grains were much farther from the fitting curves than polycrystalline CBN grains in Figure 9.11. It means that there was little change occurring on polycrystalline CBN grains. For monocrystalline CBN grains, the fractal dimension of grain I rose rapidly after the fourth stroke and fell sharply after the fifth stroke and the sixth stroke. Like the same, the value of the grain II rose at the fourth stroke and fell after the fifth stroke and the seventh stroke. It indicated that the grains were large and fractured, so the change process was so rapid. But for polycrystalline CBN grains, there was no such rapid change in all ten strokes. According to this, it is inferred that monocrystalline CBN (McBN) grains are easier to get seriously fractured than polycrystalline CBN grains, which makes monocrystalline CBN grains fail earlier.

Some rapid changes in fractal dimension were selected to make comparisons. It is known that the fractal dimension of McBN grain I rise rapidly after the fourth stroke, and that of McBN grain II falls sharply after the seventh stroke. Therefore, the photographs of grain surface morphology, 3D profiles of reconstruction models, and corresponding contour maps are shown in Figure 9.12.

Figure 9.12a and b illustrate the different topography profiles of McBN grain I before and after the fourth stroke. From the 3D profiles and contour maps, it is easy to see that there was a wide cutting edge on the top of the grain, but there was only a peak after the fourth stroke. A large fracture occurred on the left side of the grain, so that the left part of the original cutting edge disappeared. The original cutting edge got short, but many new microcutting edges were generated in the fractured region, making the topography profile more complicated. As a result, the fractal dimension rose. Figure 9.12c and d show the different topography profiles of McBN grain II before and after the seventh stroke. It can be observed that there was a peak at the cutting edge in Figure 9.12c. However,

	Surface morphology	3D profile	Contour map
MCBN grain I (a) 3rd stroke			
(b) 4th stroke			
MCBN grain II (c) 6th stroke			
(d) 7th stroke			
PCBN grain I (e) The initial			
(f) 8th stroke			

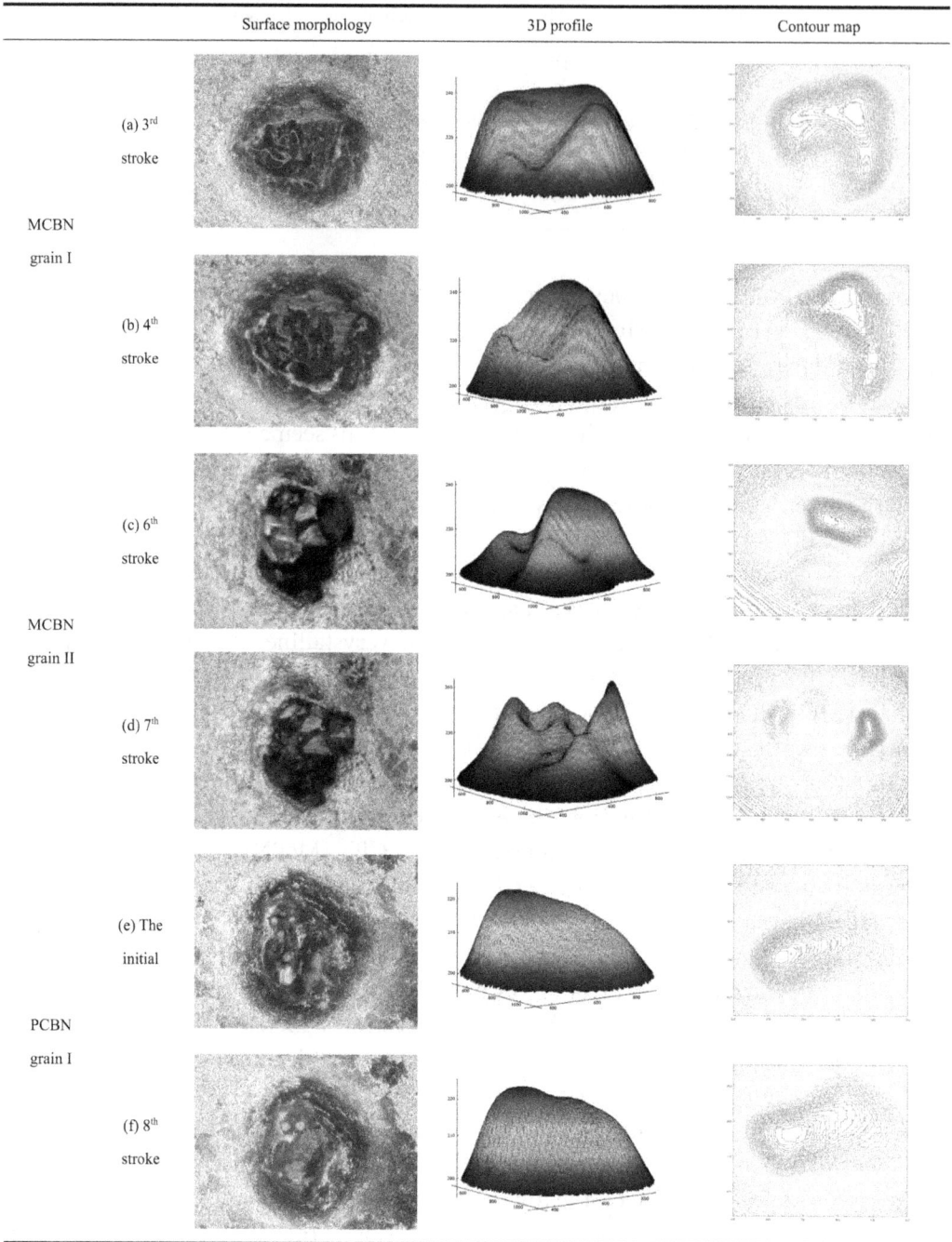

FIGURE 9.12 Surface morphology, 3D profiles, and contour maps of typically worn cutting edges.

after the seventh stroke, a large fracture occurred, and the whole peak was replaced by several smaller peaks. From the contour map in Figure 9.12d, the newly formed surfaces in fractured regions seemed smooth, which means the large fracture brought just a few microcutting edges. The protruding height also decreased sharply owing to the disappearance of the original peak. A large volume was consumed so that the grinding performance would get worse after this stroke. Figure 9.12e and f display the topography profiles of PcBN grain I at the beginning and after the eighth stroke. It can be seen that the cutting edge didn't change a lot after eight strokes, except for some microfractures. There were also some abrasion platforms generated on the top of the grain, but no large fracture occurred. By comparison, it is easy to find that polycrystalline CBN grains can sharpen themselves better but cost less volume, which means polycrystalline CBN grains perform better than monocrystalline CBN grains in the high-speed grinding process. According to the microstructure, McBN grains are non-isotropic, so there are ordered, weak interfaces inside the grains. In the grinding process, tensile stress and compressive stress are generated inside the grain. Once the stresses exceed their strength, they will destroy these interfaces, making the grain fractured.

On the contrary, polycrystalline CBN grains are composed of ultrafine microcrystal grains and an AlN binder. When grinding, the binder in the region suffering the load will fail, and only a few ultrafine microcrystal grains in this region will fall down so that the microfracture occurs. It won't break the whole grain so that the cutting edges can remain and the height decreases smoothly. According to this, it is proven that polycrystalline CBN grains perform better than monocrystalline CBN grains in high-speed grinding processes.

9.3.3 Comparison with Different Uncut Chip Thickness

Usually, the polycrystalline CBN grains perform better than the monocrystalline CBN grains under the same conditions. The fracture behaviour is influenced by the load, while the load is determined by grinding parameters. In this work, the grinding force is determined by the undeformed chip thickness (a_{gmax}). In this section, more experiments were conducted to investigate the effect of grinding depth on fracture behaviour.

9.3.3.1 Fractal Analysis

Figure 9.13a shows the fractal dimension fitting curves of polycrystalline CBN grains with different cutting thicknesses. The fitting curves all fell at different speeds. The slope tendency is illustrated in Figure 9.13b. The smallest slope was -1.46×10^{-3} when a_{gmax} was 0.2 μm. The largest slope was -3.96×10^{-3} when a_{gmax} was 1.0 μm, which was nearly three times the smallest one. Generally, the slope will get steeper when the a_{gmax} gets larger, which indicates the fracture is more serious.

Furthermore, Figure 9.13b illustrates the different standard deviations of polycrystalline CBN grains with different a_{gmax}. The smallest standard deviation was 0.671×10^{-3} when a_{gmax} was 0.2 μm, while the largest standard deviation was 1.57×10^{-3} when a_{gmax} was 0.6 μm, increasing by 134%. It seemed that the largest standard deviation was twice larger than the least one, but the maximum was still much less than that of the monocrystalline CBN grains.

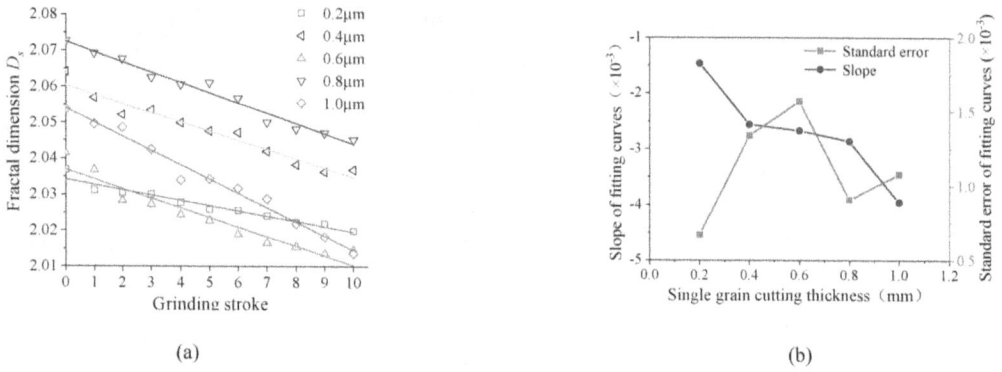

(a)

(b)

FIGURE 9.13 Fractal dimension fitting curves of polycrystalline CBN grains with different a_{gmax}: (a) Fitting curves of each grain, (b) Change of the slope and the standard deviation.

9.3.3.2 Change of Topography Profile

Photographs of grain surface morphology were taken with a 3D optical video microscope after finishing a stroke. The changes in grain surface morphology under the condition of different a_{gmax} are relatively illustrated in Figures 9.14–9.16. To protect the former grinding scratch, the grain was reset and moved to a new grinding region after each stroke. When completing 10 strokes, the surface data of all grinding scratches was collected with the Sensofar 3D optical profiler. Figure 9.17 shows the process of grinding a scratch profile in the cut-in region on the Ti-6Al-4V workpiece surface. According to this scratch profile, the corresponding grain topography in each stroke can be speculated.

Figure 9.14 shows the change in grain surface morphology when a_{gmax} is 0.2 μm. After the previous two strokes, the cutting edge was worn and an abrasion platform was generated on the top of the grain so that the cut-in scratch profile in the third stroke seemed almost plane, as displayed in Figure 9.17a. The large grinding force caused by the abrasion platform made the grain fracture. Cracks were generated on the top of the abrasion platform after the third stroke, as shown in Figure 9.17c. Therefore, the cut-in scratch profile in the fourth stroke became complicated. In the following strokes, the top surface of the grain was gradually worn. There were evident wear traces on the top surface, which can be seen from Figure 9.14d and h. Due to the grinding force, the cracks were also extended and enlarged, making the scratch profile more complicated.

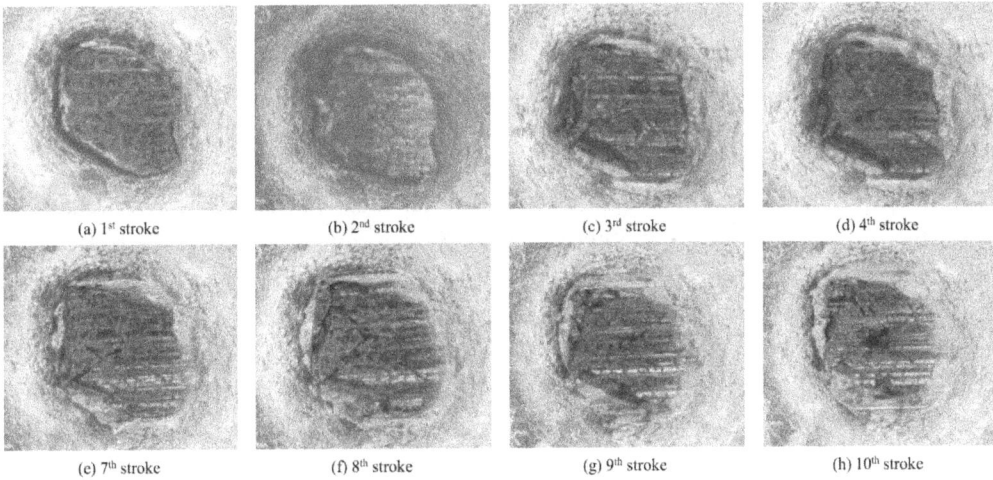

FIGURE 9.14 Surface morphology of PcBN grains when a_{gmax} is 0.2 μm.

Figure 9.15 shows the change in grain surface morphology when a_{gmax} is 0.4 μm. It can be seen that some Ti-6Al-4V chips cling to the right side of the grain. Considering that these chips would not affect the observation, they were left to get involved in grinding, making the process more realistic. According to Figure 9.15a and b, the grain was worn after the previous two strokes. From Figure 9.15c–9.15h, there was no large fracture occurring except for some microfractures. Some little cracks and wear traces were generated on the grain top in Figure 9.15. Generally, no dramatic change occurred on the topography profile of the grain under this condition, which can be proved from Figure 9.17b. It means that the grain was consumed at a very slow speed, with the cutting-edge self-sharpened compared with monocrystalline CBN grains.

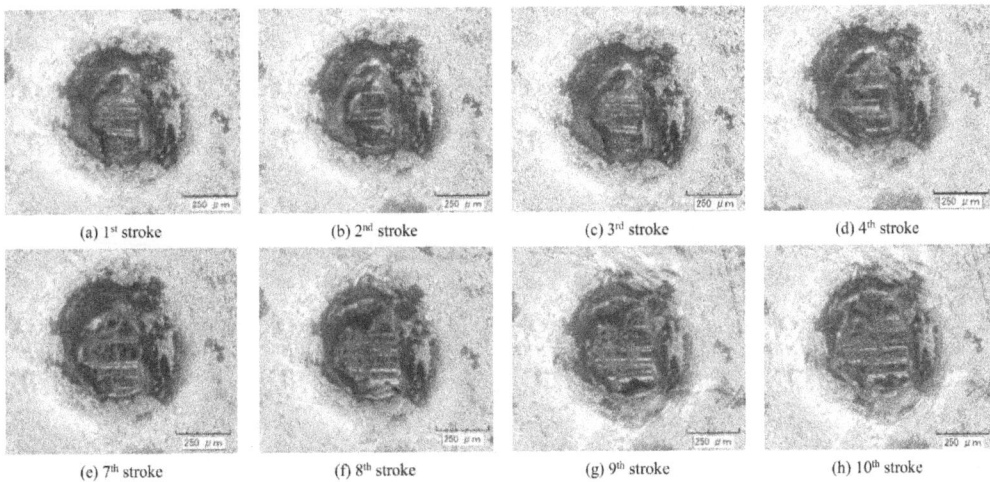

FIGURE 9.15 Surface morphology of polycrystalline CBN grains when a_{gmax} is 0.4 μm.

Figure 9.16 displays the change in grain surface morphology when a_{gmax} is 0.6 μm. In the previous three strokes, the top of the grain was worn, and the cutting edge got smooth, as shown from Figure 9.16a–c. In Figure 9.16d, a little crack appeared in the middle of the grain. From Figure 9.16e–h, it can be seen that many microfractures occurred on the top surface and changed the grain topography. In Figure 9.17c, the cut-in scratch profile got complicated after the fourth stroke. In the beginning, the deepest region on the scratch profile was located at 300 μm on the X-axis, which indicated the peak of the cutting edge. Then, after two strokes, the peak was fractured and turned into an abrasion platform. The abrasion platform led to a large grinding force, so the grain was slightly broken after the third stroke. Then, the grain was gradually fractured, and a peak appeared on the scratch profile located at 250 μm on the X-axis, corresponding to the cracks on the grain. In Figure 9.16g and h, there were some scratches on the filler metal, indicating that the protruding height of the grain decreased, which means the fracture consumed several volumes of the grain under this condition. Figure 9.18 shows the change in grinding depth on the workpiece. It is obvious that the grinding depth decreased sharply after the seventh stroke due to the fracture.

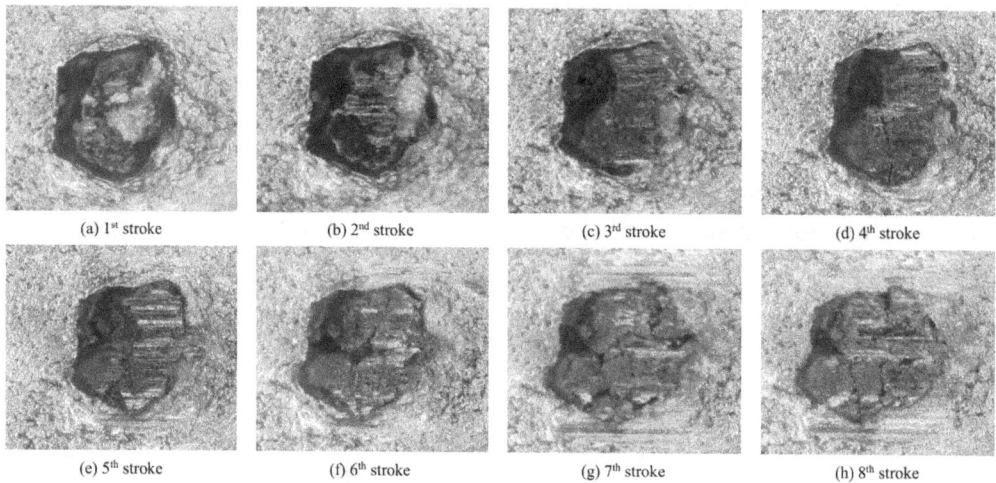

(a) 1st stroke   (b) 2nd stroke   (c) 3rd stroke   (d) 4th stroke

(e) 5th stroke   (f) 6th stroke   (g) 7th stroke   (h) 8th stroke

FIGURE 9.16   Surface morphology of PcBN grains when a_{gmax} is 0.6 μm.

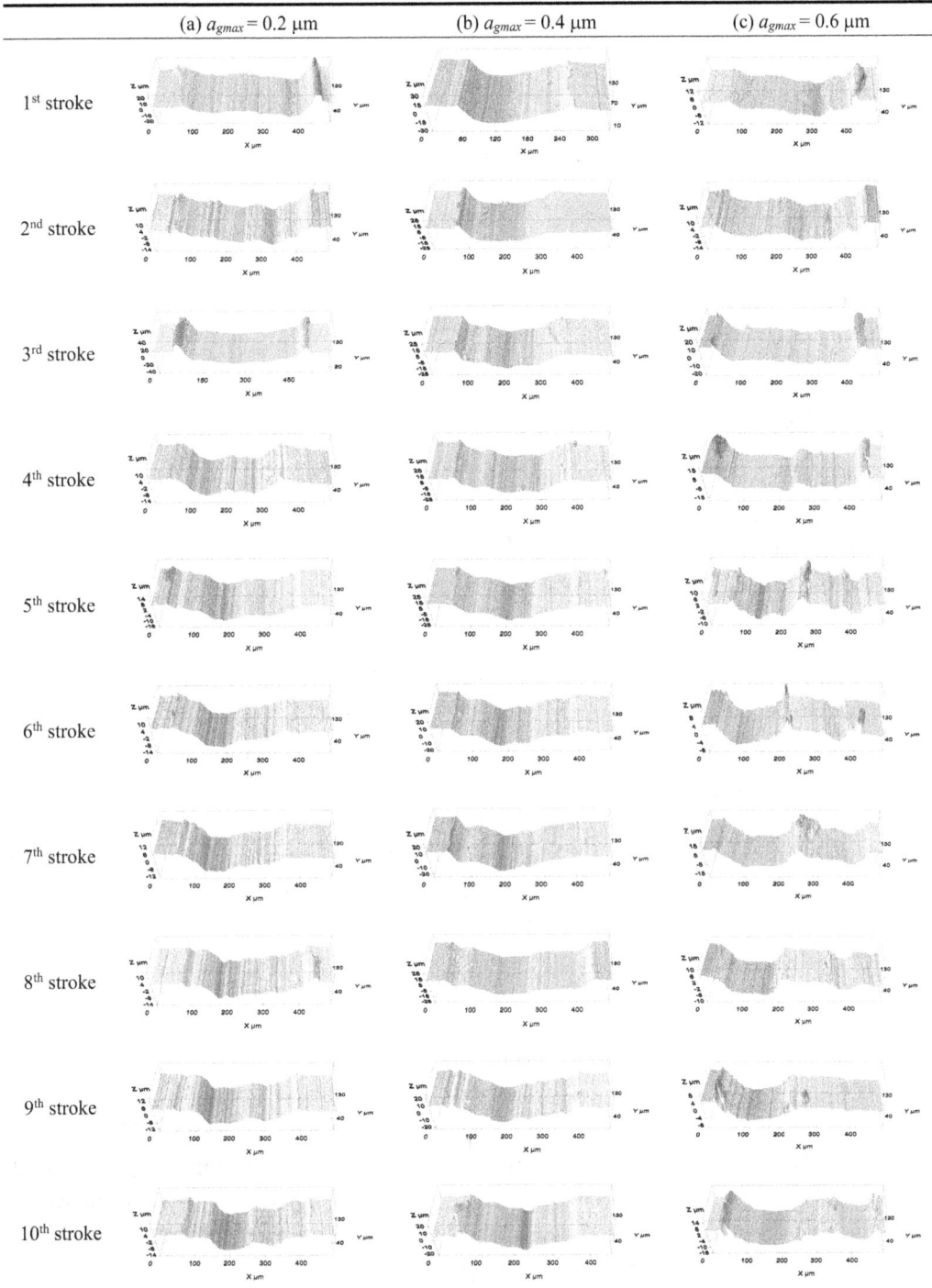

FIGURE 9.17 Change process of cut-in scratch profile with different a_{gmax}.

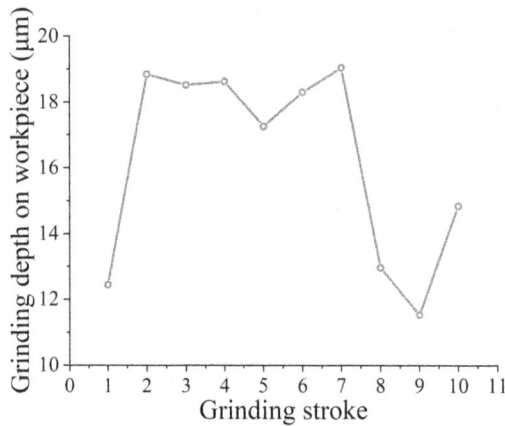

FIGURE 9.18   Change of grinding depth on the workpiece when a_{gmax} is 0.6 μm.

By comparison, when a_{gmax} is 0.6 μm, the grain is fractured more seriously than the other two grains. It indicates that the grinding depth or grinding force will influence the fracture behaviour of polycrystalline CBN grains. That is to say, the grain will get worn more seriously with a larger grinding depth.

9.4 COMPARATIVE FRACTAL ANALYSIS OF ABRASIVE WHEEL WEAR DURING GRINDING

9.4.1 Experimental Details

In this work, a single-layer brazed wheel containing monocrystalline CBN grains and polycrystalline CBN grains, respectively, was fabricated, with which a high-speed grinding experiment was conducted on nickel-based superalloy. The wear behaviour and self-sharpening phenomenon of monocrystalline CBN grains and polycrystalline CBN grains versus the accumulated volume of removal material were analysed quantitatively. The influence of the tool self-sharpening phenomenon on radial wheel wear, grinding force, and force ratio were discussed.

The size of both the monocrystalline CBN grains and polycrystalline CBN grains was 40/50 mesh (355–425 μm in diameter). The main properties of the monocrystalline CBN grains and polycrystalline CBN grains are listed in Table 9.5. The particular wheel structure containing a groove was designed to braze monocrystalline CBN grains and polycrystalline CBN grains into the different parts of one grinding wheel (Figure 9.19). The external diameter of the wheel was 400 mm, and the inner diameter was 127 mm. The pre-alloyed $(Cu_{80}Sn_{20})_{90}Ti_{10}$ (wt.%) powders were utilized as the connecting layer of the grinding wheel to join the abrasive grains and the metallic wheel matrix at a brazing temperature of 900°C. The brazing method and mechanism were reported in our previous work. The grain distribution pattern with a row distance of 1.2 mm is demonstrated in Figure 9.19c, which ensures sufficient space to store chips.

Straight surface grinding experiments were performed on an instrument, BLOHM PROFIMAT MT-408 grinding machine, while applying a 5% solution of a commercial

TABLE 9.5 Main Properties of Monocrystalline and Polycrystalline CBN Grains

Contents	McBN	PcBN
Chemical composition	BN	CBN-Al
Strength (MPa)	210–700	355–454
Knoop hardness (GPa)	47	15–40
Fracture toughness (MPa·m^{-2})	2.5	6.4–8.0

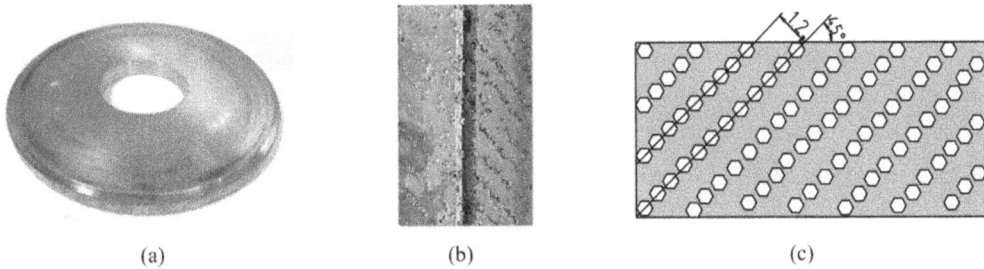

(a) (b) (c)

FIGURE 9.19 Monolayer brazed grinding wheels with McBN and PcBN grains: (a) monolayer brazed wheel; (b) brazed McBN (left) and PcBN (right) grains; (c) grain distribution patterns.

soluble water-based coolant with a pressure of 15 MPa. The maximum rotational speed of the spindle was 8,000 rpm, and the output power was 45 kW. The ground material was nickel-based superalloy Inconel718, a typical difficult-to-cut material. During grinding, the part with monocrystalline CBN grains and that with polycrystalline CBN grains of the particular wheel worked, respectively.

In the present day, high-speed grinding of difficult-to-cut material is a popular and interesting topic. Based on the potential industrial application and the research status of high-speed grinding technology, grinding experiments were carried out in the up mode with fixed grinding parameters, namely, wheel speed v_s of 120 m/s (spindle speed of 5,732 rpm), workpiece infeed speed v_w of 1,676 mm/min, and depth of cut a_p of 10 μm in the present investigation. Under such conditions, the maximum undeformed chip thickness a_{gmax} is 0.58 μm and the material removal rate (MRR) is about 0.28 mm^3/ (mm.s). It is noted that the MRR is defined as the product of the workpiece infeed speed and the depth of cut during grinding.

Grinding experiments were terminated at the end of the useful working life of the brazed monocrystalline CBN abrasive wheel, which was readily identified by a sudden increase in the grinding force and destruction of the grain layer. The grinding forces were measured using a quart piezoelectric-type dynamometer (Kistler 9272), attached to a 5070A10100 multichannel charge amplifier and computer data acquisition software.

To keep track of the wear behaviour and self-sharpening phenomenon of the grinding wheel, 30 monocrystalline CBN grains and 30 polycrystalline CBN grains, respectively, were selected randomly from eight symmetrical regions of the single-layer brazed abrasive wheel. Three-dimensional optical microscopy modelled by Hirox KH-7700 was periodically applied to capture the grain topography in the different wear stages during grinding. All the images and corresponding data were obtained under identical conditions, including the magnification level (350×), the light source, and the microscope parameter settings.

9.4.2 Fractal Dimension of Typical Grain Cutting Edges Morphology

A statistical analysis of the fractal dimension of the cutting edge morphology was conducted for the sample grains. Figure 9.20 displays the average value of the fractal dimension for different wear patterns of the grain cutting edge morphology. In general, different wear patterns correspond to different average values of the fractal dimension. Furthermore, the fractal dimension value of polycrystalline CBN grain cutting edges is always larger than that of their monocrystalline CBN counterparts, which indicates that a more complicated microstructure is formed on the tip of the worn polycrystalline CBN grain. For instance, the average value of the fractal dimension for the monocrystalline CBN grain cutting edges formed due to attrition wear on the tip of the grain is 2.042, while that for the polycrystalline CBN grain is 2.047. When the monocrystalline CBN grain cutting edges endure microfracture, the average value of the fractal dimension is 2.046, while at this time it is 2.051 for the polycrystalline CBN grain cutting edges. In particular, the average fractal dimension is 2.043 for the cutting edges of the worn monocrystalline CBN grain due to the large fracture. However, large fractures seldom take place for the polycrystalline CBN grain cutting edges, and the corresponding fractal dimension is not obtained here.

9.4.3 Wear Behaviour of Monocrystalline and Polycrystalline CBN Grains

Figure 9.21 displays some typical sequential optical images, 3D profiles, and counter maps of a monocrystalline CBN grain with an increase in the accumulated volume of material removed per unit grinding width V' in high-speed grinding. Generally, the grain-cutting edges changed their shape in different forms with the advance of the wear behaviour. As seen in Figure 9.21a, a ductile attritious wear flat was formed in the central part (white region) on the top surface of the grain cutting edges when the stock removal V' was 100 mm³/mm. Furthermore, the wear flat surface became larger with the increasing stock removal, as displayed in the comparison among Figure 9.21a–d, for example, the attrition

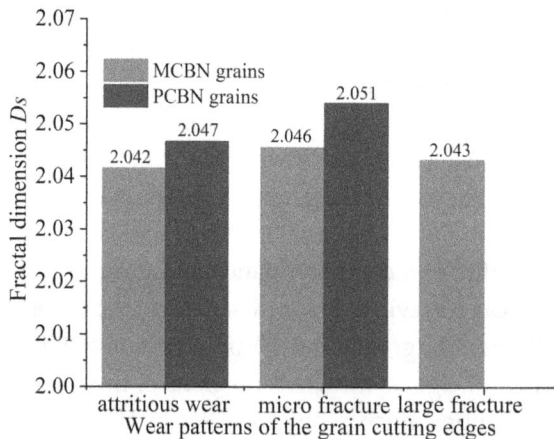

FIGURE 9.20 Relationship between fractal dimension and wear patterns of the grain cutting edges.

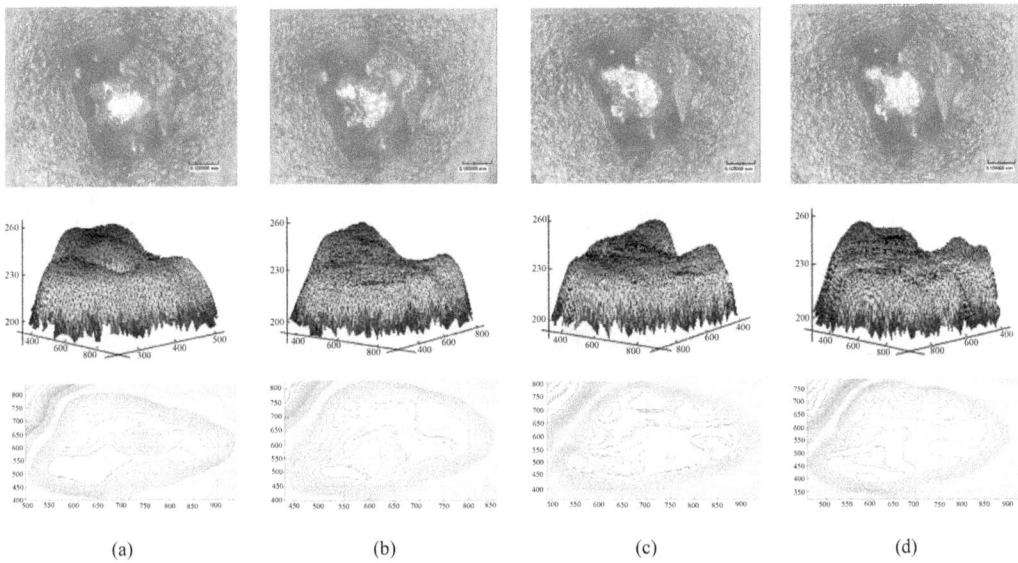

(a)   (b)   (c)   (d)

FIGURE 9.21 Surface morphology, 3D profiles, and counter maps of typical worn monocrystalline CBN grain cutting edges: (a) $V' = 100$ mm3/mm; (b) $V' = 400$ mm3/mm; (c) $V' = 600$ mm3/mm; (d) $V' = 1,100$ mm3/mm.

wear flat area at $V'=400$ mm^3/mm and $V'=600$ mm^3/mm was significantly larger than that at $V'=100$ mm^3/mm. When the accumulated volume of the material removed V' reached 1,100 mm^3/mm, the largest value of the attritious wear flat area of the grain cutting edges was obtained. Meanwhile, some fracture behaviour also happened at the top surface of the grain. Under such conditions, although the grain cutting edges may become dull because of the attritious wear, they could reproduce and maintain their sharpness due to the fracture that occurred repeatedly on the top surface of the grain. That is to say, the self-sharpening phenomenon of the monocrystalline CBN grain cutting edges was exhibited due to grain microfracture. It should be noted that, compared to the microfracture behavior, an attritious wear and large fracture took the lead in the wear behaviour and grinding performance of the monocrystalline CBN grain cutting edges.

The morphology of the polycrystalline CBN grain cutting edges was also detected sequentially with the accumulated volume of the removal material, as displayed in Figure 9.22. As seen in Figure 9.22a, some microfracture and ductile attritious wear happened at $V'=100$ mm^3/mm. New sharp edges were formed on the top surface of the grain cutting edges. Afterwards, according to Figure 9.22b–d, though the flat surface due to the attritious wear increased gradually, a large fracture never took place. High sharpness of the grain cutting edges was always ensured because the microfracture of the polycrystalline CBN grain occurred repeatedly in high-speed grinding. Because the fracture area of polycrystalline CBN grain was smaller than that of monocrystalline CBN grain under the identical condition, a lower tool wear rate and longer tool life were obtained for the polycrystalline CBN wheel compared to the monocrystalline CBN wheel in high-speed grinding.

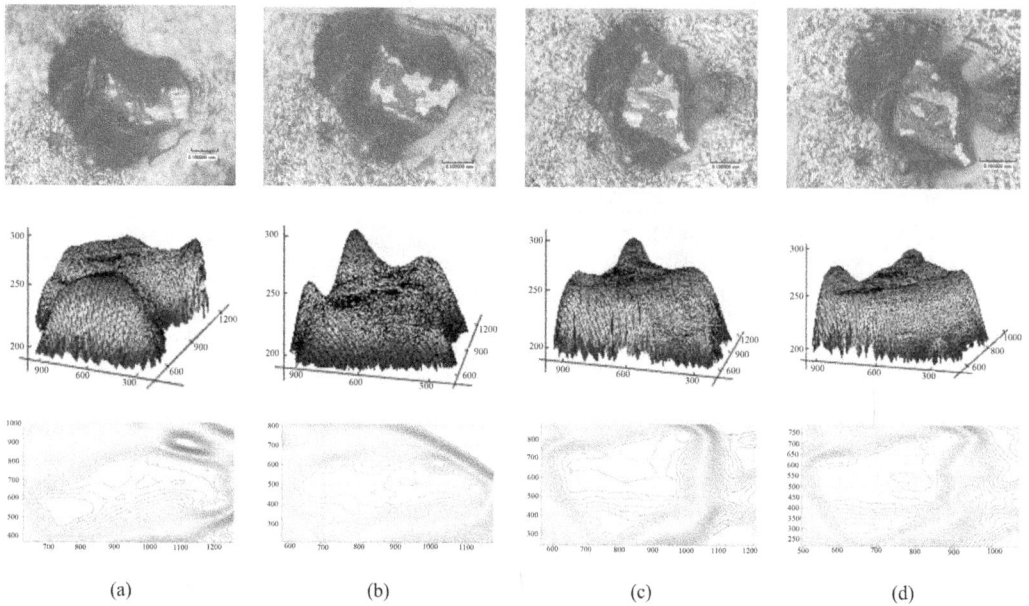

(a)	(b)	(c)	(d)

FIGURE 9.22 Surface morphology, 3D profiles, and counter maps of typical worn polycrystalline CBN grain cutting edges: (a) $V'=100\,mm^3/mm$; (b) $V'=400\,mm^3/mm$; (c) $V'=600\,mm^3/mm$; (d) $V'=1{,}100\,mm^3/mm$.

9.4.4 Fractal Analysis of Typical Monocrystalline and Polycrystalline CBN Grains

Figure 9.23 shows the changes in the fractal dimension of the grain cutting edges versus the accumulated volume of the removal material in the high-speed grinding process. Seen from Figure 9.23a, the monocrystalline CBN grain firstly took the low values of fractal dimension, i.e., 2.032–2.035, due to an increase in the attrition wear area when the stock removals were between 50 and $100\,mm^3/mm$. Afterwards, because the attrition wear area decreased and new sharp cutting edges were formed due to the large fracture and microfracture, the fractal dimension of monocrystalline CBN cutting edges tended to increase to the maximum value, $D_s=2.044$, between the stock removals from 100 to $200\,mm^3/mm$. However, the fractal dimension value went down due to the significant large fracture behaviour over a range of stock removals from 200 to $400\,mm^3/mm$. Furthermore, between the stock removals from 400 to $600\,mm^3/mm$, an increase in the fractal dimension was found due to the increasing microfracture. Finally, the fractal dimension of the monocrystalline CBN grain cutting edges decreased gradually due to large fractures over a range of stock removals from 600 to $1{,}100\,mm^3/mm$.

According to Figure 9.23b, compared to the fractal dimension of the monocrystalline CBN grain cutting edges, the different changes in the fractal dimension of the polycrystalline CBN grain cutting edges were formed during high-speed grinding, even though identical grinding parameters were applied. The maximum value of the fractal dimension, i.e., $D_s=2.060$, was obtained when the stock removal was $50\,mm^3/mm$. Afterwards, the fractal dimension decreased significantly due to an increase in the attrition wear and a decrease in the microfracture between the stock removals from 50 to $200\,mm^3/mm$. Moreover, between the stock removals from 200 to $400\,mm^3/mm$, a stable value of the fractal dimension, i.e.,

$D_s=2.048$, was taken due to the attritious wear. And then, over a range of stock removals from 400 to 600 mm³/mm, the fractal dimension increased because the microfracture occurred. Finally, the fractal dimension was almost kept constant at 2.053 due to the repeating microfracture of the polycrystalline CBN grain when the stock removal material was between 600 and 1,100 mm³/mm.

In brief, according to Figure 9.23, it is known that the wear behaviour of the monocrystalline CBN grain cutting edges was, in order, attritious wear→large fracture→microfracture→large fracture→attritious wear, while that of the polycrystalline CBN grain cutting edges was microfracture→attritious wear→microfracture in the high speed grinding process.

9.4.5 Fractal Analysis of Monocrystalline and Polycrystalline CBN Abrasive Wheels

The self-sharpening phenomenon of the abrasive wheels could be understood based on the fractal dimension of large quantities of grain-cutting edges. In the current investigation, 30 monocrystalline CBN grains and 30 polycrystalline CBN grains were selected randomly from the eight symmetrical regions of the abrasive wheel, respectively. The fractal dimension of the cutting edges of these grains was calculated at different stock removals. Figure 9.24 displays the correspondent results obtained at a stock removal of 50 mm³/mm. Furthermore, the statistical information on the fractal dimension of the monocrystalline CBN wheel and polycrystalline CBN wheel was accordingly obtained after several grains with particularly large or small values of fractal dimension were discarded.

As seen in Figure 9.24, for the monocrystalline CBN abrasive wheel, the fractal dimension of grain cutting edges is between 2.033 and 2.062, with an average value of 2.047. For the polycrystalline CBN abrasive wheel, however, the fractal dimension ranges from 2.035 to 2.068, with an average value of 2.050.

The average value of the fractal dimension of the abrasive wheels versus the accumulated volume of material removed per unit grinding width V' is illustrated in Figure 9.25. The average value of the fractal dimension is 2.040–2.047 for the monocrystalline CBN wheel and 2.049–2.054 for the polycrystalline CBN counterpart in high-speed grinding.

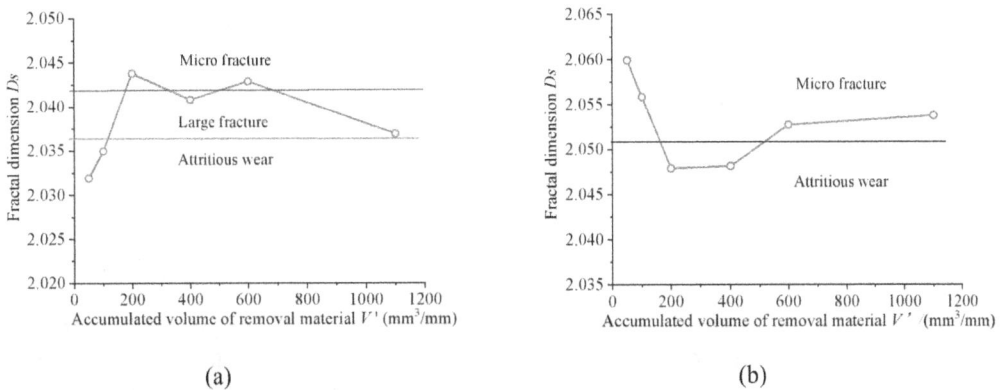

(a) (b)

FIGURE 9.23 Changes in fractal dimension of the grain cutting edges: (a) monocrystalline CBN grain, (b) polycrystalline CBN grain.

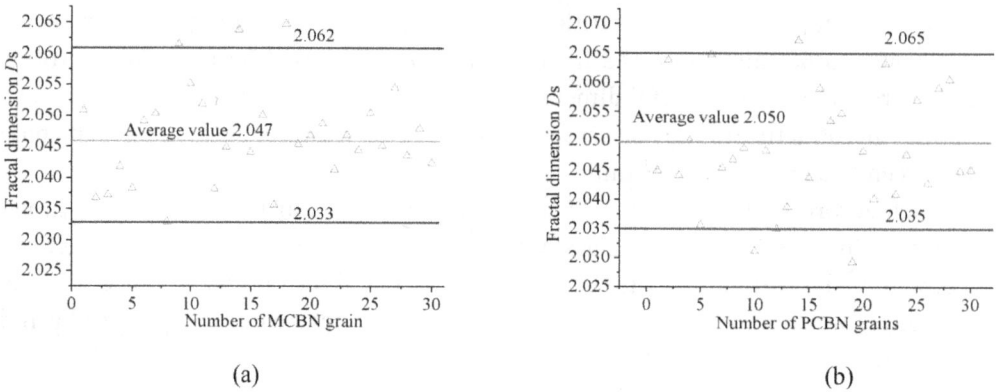

(a)

(b)

FIGURE 9.24 Fractal dimension of different wheels for the accumulated volume of removal material of 50 mm³/mm: (a) monocrystalline CBN wheel, (b) polycrystalline CBN wheel.

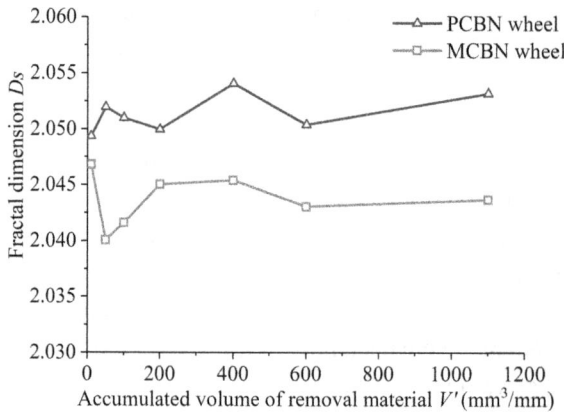

FIGURE 9.25 Changes in the average fractal dimension of the monocrystalline and polycrystalline CBN abrasive wheels.

Moreover, the former is always lower than the latter. Usually, the fractal dimension of the fine cutting edges formed due to microfracture is higher than that of the rough cutting edges formed due to large fractures and attritious wear. Therefore, it is known that the polycrystalline CBN grain cutting edges are finer than those of their monocrystalline CBN counterparts, which could enhance the self-sharpening phenomenon and improve the grinding performance of the polycrystalline CBN abrasive wheel.

9.4.6 Radial Wear of Monocrystalline and Polycrystalline CBN Abrasive Wheels in High-Speed Grinding

The radial wheel wear is another important parameter to quantitatively characterize the wear-resistance ability of the grinding wheels. In this work, the sequential reduction of the protrusion height of 30 grains for each wheel was measured, the average value of which was calculated and analysed to reflect the radial wear of the single-layer brazed wheels. The results are provided in Figure 9.26 as plots of radial wheel wear versus accumulated volume of removal material per unit width of grinding.

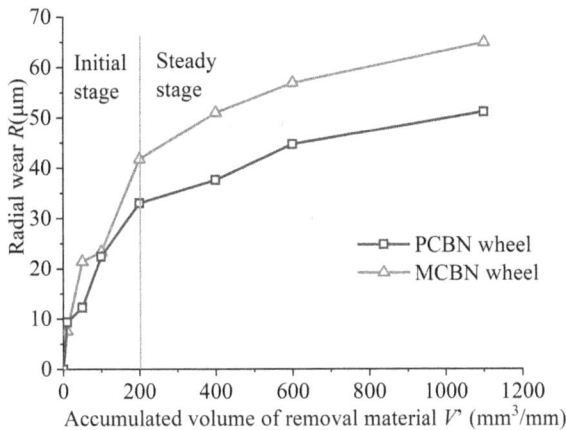

FIGURE 9.26   Radial wear of monocrystalline and polycrystalline CBN abrasive wheels.

In general, there is an initial transient run-in wear stage with radial wheel wear of 30–40 μm at a decreasing rate to a steady-state wear stage at a lower wear rate for each wheel. The slope in the steady-state wear region tends to be smaller. On the other hand, the wear-resistance ability of the polycrystalline CBN wheel is always better than that of the monocrystalline CBN wheel in the grinding process. The radial wheel wear at the termination of testing arrived at approximately 65 μm for monocrystalline CBN wheels and 50 μm for polycrystalline CBN wheels, respectively.

9.4.7 Influence of the Self-sharpening Phenomenon of Abrasive Wheels on Grinding Forces and Forces Ratio

Grinding forces and force ratios could reflect the effects of the self-sharpening phenomenon of the abrasive wheels. The good self-sharpening phenomenon of abrasive wheels corresponds to high wheel sharpness, which could produce a low-grinding force and force ratio. Figure 9.27 displays the grinding forces and forces ratios versus the accumulated volume of removal material per unit width during grinding Inconel718 nickel-based superalloy.

As seen in Figure 9.27a, for the monocrystalline CBN wheel, the normal force F_n' increased rapidly in the initial grinding stage. When the stock removal reached 200 mm³/ mm, F_n' kept at a high value of 5.5 N/mm. Afterwards, the normal forces were increased slowly to 7 N/mm when the stock removal arrived at 1,100 mm³/mm. For the polycrystalline CBN wheel, not only the absolute value but also the change range of the normal force, 2.5–3.0 N/mm, was smaller than that of the monocrystalline CBN wheel in the initial grinding stage.

Furthermore, the normal force kept an increasing tendency at a very low level of 4.0 N in the steady grinding stage. In general, the normal force of the monocrystalline CBN was almost as 1.4–2 times as high as that of the polycrystalline CBN wheel. Taking into account the tangential grinding forces, it was kept at 1.2–1.5 N/mm for the monocrystalline CBN wheel and 0.9–1.3 N/mm for the polycrystalline CBN wheel.

The grinding force ratio of normal force F_n to tangential force F_t is correlated to the change in the sharpness degree of the grinding wheel. A small value of the ratio of force always corresponds to a good ability to maintain high sharpness. Figure 9.27b shows the

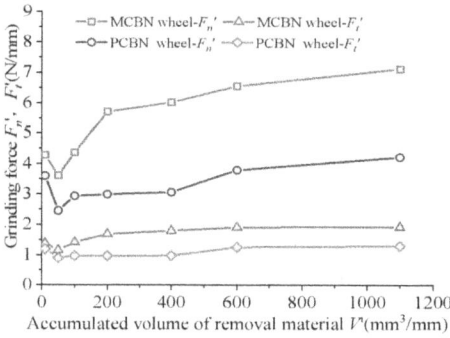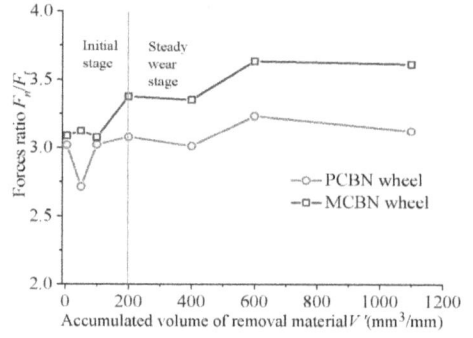

(a)   (b)

FIGURE 9.27   Grinding force and forces ratio versus accumulated volume of removal material: (a) grinding force, (b) forces ratio.

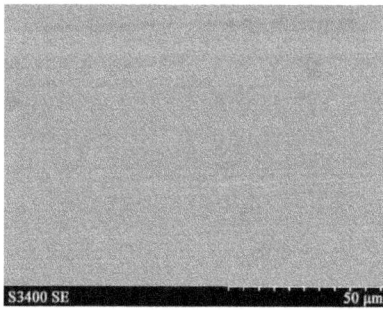

(a)   (b)

FIGURE 9.28   Ground surface with (a) monocrystalline CBN grains, (b) polycrystalline CBN grains.

ratio of the force versus the stock removal material. The force ratio of both the monocrystalline CBN and polycrystalline CBN abrasive wheels did not keep constant in the initial grinding stage until the stock removal reached 200 mm³/mm. In the steady grinding stage, the ratio of force was increased from 3.1 to 3.6 for the monocrystalline CBN abrasive wheel, while it was increased from 2.8 to 3.2 for the polycrystalline CBN wheel. The polycrystalline CBN grinding wheel had a better capacity than the monocrystalline CBN wheel to maintain high sharpness. The excellent wear-resistance ability and self-sharpening phenomenon were obtained due to the main microfracture behaviour of polycrystalline CBN grains during grinding. However, for the monocrystalline CBN grains, the attrition wear and the large fracture were the primary wear patterns, and the microfracture behaviour and self-sharpening capacity were generally bad.

Additionally, the ground surface obtained in the steady wear stage of the grinding wheels is also compared. The surface roughness Ra of the ground surface is 0.35 μm with monocrystalline CBN grains and 0.28 μm with polycrystalline CBN grains, respectively. Figure 9.28 shows the morphology of the ground surface, from which it is found that the flaws in the surface ground with polycrystalline CBN grains are much smaller than those with monocrystalline CBN grains. Accordingly, lower surface roughness and smaller surface defects are obtained due to the high sharpness degree of the polycrystalline CBN grains.

REFERENCES

Bredthauer, M., P. Snellings, P. Mattfeld, et al. 2023. Wear-related topography changes for electroplated cBN grinding wheels and their effect on thermo-mechanical load. *Wear*, 512–513: 204543.

Dai, J., Y. Li, D. Xiang, et al. 2022. The mechanism investigation of ultrasonic roller dressing vitrified bonded CBN grinding wheel. *Ceramics International*, 48(17): 24421–24430

Fujimoto, M., and Y. Ichida. 2008. Micro fracture behavior of cutting edges in grinding using single crystal cBN grains. *Diamond & Related Materials*, 17: 1759–1763.

James, G. M., and M. Davide. 2010. Fractal geometry in the nucleus. *The EMBO Journal*, 29(1): 2–3.

Jimoto, M., Y. Ichida, R. Sato, et al. 2006. Characterization of wheel surface topography in cBN grinding. *JSME International Journal*, 49: 106–113.

Li, M., W. Ding, Z. Zhao, et al. 2019. An investigation on the dressing contact behavior between vitrified bonded CBN abrasive wheel and diamond grit dresser. *Journal of Manufacturing Processes*, 58: 355–367.

Liang, X. H., B. Lin, X. S. Han, et al. 2012. Fractal analysis of engineering ceramics ground surface. *Applied Surface Science*, 2012, 258: 6406–6415.

Linke, B., and F. Klocke. 2010. Temperatures and wear mechanisms in dressing of vitrified bonded grinding wheels. *International Journal of Machine Tools & Manufacture*, 50: 552–558.

Mandelbrot, B. B. 1967. How long is the coast of Britain? Statistical self-similarity and fractional dimension. *Science*, 156: 636–638.

Miao, Q., W. F. Ding, J. H. Xu, et al. 2013. Fractal analysis of wear topography of brazed polycrystalline cBN abrasive grains during grinding nickel superalloy. *International Journal of Advanced Manufacturing Technology*, 68: 2229–2236.

Podsiadlo, P., and G. W. Stachowiak. 2000. Scale-invariant analysis of tribological surfaces, thinning films and tribological interfaces. *Tribology Series*, 38: 546–557.

Stachowiak, G. W., and P. Podsiadlo. 2001. Characterization and classification of wear particles and surfaces. *Wear*, 249: 194–200.

Wang, J. W., W. F. Ding, Y. J. Zhu, et al. 2018. Fracture mechanism of polycrystalline cubic boron nitride abrasive grains during single-grain grinding of Ti-6Al-4V titanium alloy. *International Journal of Advanced Manufacturing Technology*, 94: 281–291.

Yan L., Y. M. Rong, F. Jiang, Z. X. Zhou. 2011. Three-dimension surface characterization of grinding wheel using white light interferometer. *International Journal of Advanced Manufacturing Technology*, 55: 133–141.

CHAPTER 10

Stress Distribution Effects on Grain Wear Evolution in Grinding

10.1 INTRODUCTION

Many studies have utilized single cubic boron nitride (CBN) grain grinding to delineate the distinct features of the CBN grinding mechanism. Considering the grain size, Buhl et al. (2013) found that small grains mainly fail due to grain pullout, while large grains tend to be fractured during the grinding process. Zhao et al. (2016) investigated the grain topography after various grinding passes and found that the grain wear process evolves from attritor wear to macro and microfracture. Ding et al. (2014) found that the grain wear modes are mainly macro and microfracture in ultrasonic vibration-assisted grinding. However, attrition wear mainly occurs in conventional grinding, wherein multiple microcutting edges are generated due to the grain wear process. Fujimoto and Ichida (2008) calculated the fractal dimension of the geometry profile of polycrystalline CBN grain, which can quantitatively characterize the grain wear state. Yu et al. (2017) have shown that grain wear mainly has three types: grain pullout, grain fracture, and grain attritious wear, and they affect the wheel topology evolution and grinding performance. In this work, grain fracture and attrition wear are combined to derive the wear flat area on the grain top surface, which contributes to the frictional force/power during the grinding process. A sparse amount of work has been conducted on grain fracture and grain morphology evolution at the microscale.

The grain-scale mechanisms have also been studied numerically. Jackson (2004) estimated the stress distribution inside the alumina grain and CBN grain under constant tangential and normal grinding forces with finite element analysis (FEA). The tensile or compressive stress-induced fracture region was assessed based on a Griffiths fracture criterion. It was found that grain fracture and pull-out processes were mainly caused by mechanical tensile stress. Zhu et al. (2016) used the Voronoi method to investigate the

DOI: 10.1201/9781032678047-12

polycrystalline CBN grain wear model and estimated the stress distribution under grinding force. Alveen et al. (2015) also used the Voronoi model to simulate the crack propagation process with a predetermined crack path and constant tensile load. Mei et al. (2017) predicted the wear evolution of the grain profile during grinding by using numerical simulation. The grain-level grinding forces were estimated from a single-grain model. These forces were then utilized to estimate the stress distribution and the corresponding wear region, utilizing the enforcement of static equilibrium. Through the iterative repeatability of this two-step process, grain wear evolution was simulated. However, the process kinematics were not captured in such a wear model. A grain-level wear model should include the role of chip formation in the high wheel speed.

This work presents the developed modelling framework for single-CBN grains to simulate the experimentally observed morphological evolution of both single grains. The model employs a full kinematic simulation of the high-speed grinding of Inconel 718 nickel-based superalloy with a single-CBN grain. The model encompassed both the friction and the chip formation process during the grinding process, thereby facilitating the investigation of the actual stress distribution during multiple grinding passes. The model also employs the details of the fracture processes at the microscale of the grit, thereby facilitating the simulation of the wear morphology evolution for each grinding pass and the associated internal stresses. The simulation results provide insights on grain chipping and its effect on grain morphology evolution, grinding force, and the formation of new cutting edges. The study helps to develop a better understanding of the CBN grain fracture behaviour during the grinding process. This framework will facilitate the development of the grinding wheel life expectancy model in future work.

10.2 FINITE ELEMENT MODEL OF SINGLE-GRAIN WEAR EVOLUTION DURING GRINDING

10.2.1 Modelling Framework

A commercial FEA package Abaqus Dynamic/Explicit was used in this study. The wear evolution of a single-CBN grain and the material removal process during high-speed grinding of Inconel 718 were simulated. A 2D FEA model was utilized to reduce the computational complexity and to show the general trend of the process under different machining conditions (Wang et al. 2018). The modelling steps are shown in Figure 10.1, and the implemented framework can be divided into four stages:

a. Material removal simulation: the cutting model was performed under predefined grinding conditions.

b. Post-processing: the stress distribution, wear regions within the grain, material removal situation after grinding, and grinding force were estimated to support the next step.

c. Grain morphology update: according to the simulation result achieved in (b), the fractured regions within the grain were removed, and the grain geometry profile was updated accordingly.

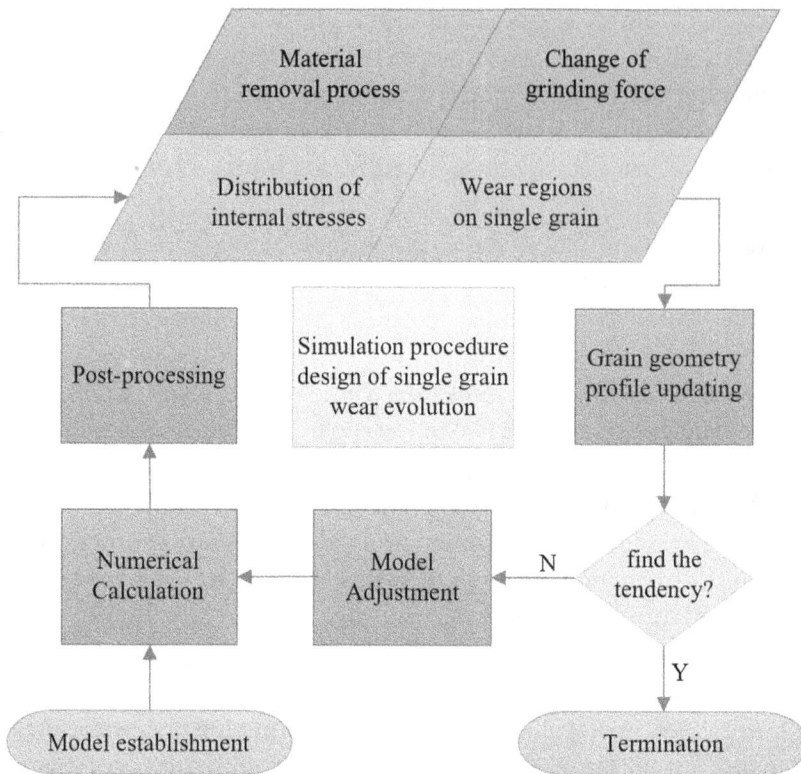

FIGURE 10.1 Flow diagram for the simulation framework of single-grain wear evolution during high-speed grinding.

    d. Model adjustment: re-define the contact condition between grain and workpiece. If the bottom cutting edge of a single grain was removed, the coordinate of the newly formed cutting edge was adjusted to ensure a constant cutting depth.

This sequence of steps was repeated for each grinding pass so that the accumulated grain wear could be tracked. Additionally, the chip formation has also been concurrently simulated with the grain wear evolution.

10.2.2 Model Geometry and Governing Equations

The initial grain shape selection is significantly important to properly model the grain wear evolution. The grain shape is usually modelled in 3D as a sphere, pyramid, irregular polyhedron, or hexahedron (3D shape) (Eder et al. 2015; Liu et al. 2013). In 2D modelling, a circle, triangle, or hexagon is typically employed (Mcdonald et al. 2017; Liu et al. 2015; Ding et al. 2014). Among these, the sphere and half-sphere have no defined cutting edges; the pyramid has sharp cutting edges, and the irregular polyhedron has a high degree of variation. In this study, a hexagonal grain of 0.02 mm with an edge length of 0.01 mm has been utilized, as shown in Figure 10.2a. The grain has a flank angle of five to avoid numerical

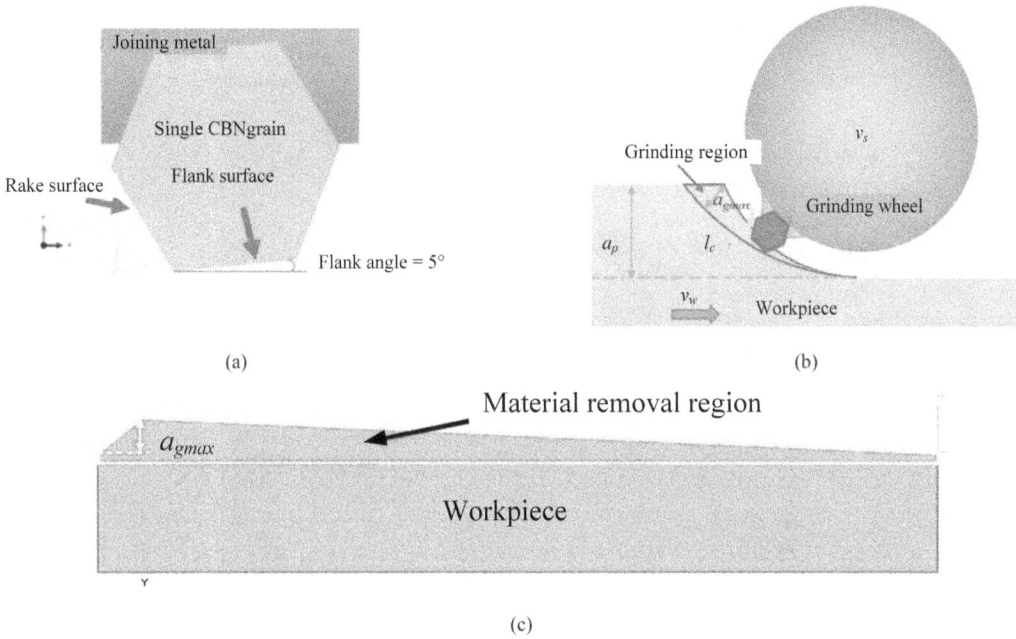

FIGURE 10.2 Geometric details of the FEA model: (a) single-CBN grain model, (b) scheme of single-grain grinding process, (c) Inconel 718 nickel-based superalloy workpiece model.

issues. The grain meshed into 53,998 elements with the type of CPE3T (a three-node plane strain thermally coupled triangle with linear displacement and temperature interpolation).

In a high-speed surface grinding process, the chip has a non-uniform thickness. As shown in Figure 10.2b, v_w is the workpiece infeed speed; v_s is the wheel speed; a_p is the grinding depth; and a_{gmax} is the maximum undeformed chip thickness in single-grain grinding. In this study, the material removal region of the workpiece was chosen as a triangle with a height of a_{gmax}, as illustrated in Figure 10.2c. The workpiece was meshed into 36,596 elements with the mixed type of CPE3T and CPE4RT (a four-node plane strain thermally coupled quadrilateral with bilinear displacement and temperature interpolation, reduced integration, and hourglass control).

The material properties of CBN grain and Inconel 718 nickel-based superalloy are listed in Table 10.1 (Akhtar et al. 2016).

The material constitutive relation for the workpiece was simulated by the Johnson and Cook constitutive model to account for large strain, high strain rate, and high temperature. The model has the form of:

$$\sigma = \left[A + B\varepsilon_p{}^n\right]\left[1 + C\left(\ln\frac{\dot{\varepsilon}_p}{\dot{\varepsilon}_0}\right)\right]\left[1 - \left(\frac{T - T_{\text{tran}}}{T_{\text{melt}} - T_{\text{tran}}}\right)^m\right] \tag{10.1}$$

where ε_p represents the equivalent plastic strain; $\dot{\varepsilon}_p$ and $\dot{\varepsilon}_0$ are the equivalent plastic strain rate and reference strain rate, respectively; T_{melt} and T_{tran} are melting temperature and transition temperature, respectively. Parameters A, B, C, n, and m used in this study are listed in Table 10.2.

TABLE 10.1   Materials Properties

Subjects	CBN Grain	Inconel 718 Nickel-Based Superalloy
Density (kg/mm³)	3.4×10^{-6}	8.25×10^{-6}
Young's modulus (GPa)	710	220
Poisson's ratio	0.15	0.3
Conductivity (W/(m·K))	80	20
Thermal expansion	2.2×10^{-6}	1.5×10^{-5}
Specific heat (J/(kg·K))	430	203
Tensile strength (MPa)	700	-
Compressive strength (MPa)	2,500	-

TABLE 10.2   Parameters of Johnson–Cook Plasticity Model

Subjects	Parameters
A (MPa)	450
B (MPa)	1,700
C	0.017
n	0.65
m	1.3
T_{melt} (K)	1,593
T_{tran} (K)	293

10.2.3 Abrasive Wheel Process Kinematics

According to the cutting path and relative motion, a_{gmax} and the contact arc length l_c of the material removal region can be calculated as follows:

$$a_{gmax} = 2\lambda \frac{v_w}{v_s} \sqrt{\frac{a_p}{d_s}} \tag{10.2}$$

$$l_c = \sqrt{d_s \times a_p} \tag{10.3}$$

where d_s is the grinding wheel diameter and λ is the interval between two adjacent grains, which can be considered the perimeter of the abrasive wheel in a single-grain grinding process. An exhaustive parametric study is beyond the scope of the current study; however, a set of simulations representative of typical process parameters are presented here, as summarized in Table 10.3. Accordingly, a_{gmax} was calculated as 1.39 μm and l_c was 2 mm, which are much larger relative to a_{gmax}. The l_c in this model was used as 0.036 mm, and the time for one pass was defined as 0.35 μs with 0.05 μs of initial engaging stage. A grinding pass was divided into 1,000 intervals to precisely collect simulation data.

Two types of contacts have been considered in this study: (i) a bond/CBN grain contact and (ii) a CBN grain/workpiece contact. For the bond/CBN grain contact, neither normal separation nor tangential sliding between the pairs was allowed. For the CBN grain/workpiece contact, separation was allowed in the normal direction, and sliding was allowed

TABLE 10.3   Parameters of Grinding Process

Process Parameters	Value
v_w (m/min)	0.8
v_s (m/s)	120
a_p (mm)	0.01
d_s (mm)	400

in the tangential direction, with coulomb frictional resistance. The frictional stress τ was calculated by:

$$\tau = \begin{cases} \mu p & (\text{if } \mu p \le \tau_{\lim}) \\ \tau_{\lim} & (\text{if } \mu p > \tau_{\lim}) \end{cases} \tag{10.4}$$

where p is the normal pressure, the friction coefficient μ was set to be 0.3, and $\tau_{\lim}$ is the maximum frictional stress. When the stress becomes greater than $\tau_{\lim}$, the tangential sliding will occur.

10.2.4 Criteria for Single-CBN Grain Wear and Workpiece Material Removal

The grain fracture was mainly caused by tensile stress. Though compressive stress-induced fractures have also been observed. In this study, the stress distribution within the grain was used to predict its wear evolution. Whenever the principal stresses in an element exceed a critical limit (tensile strength σ_t or compressive strength σ_c) according to Eq. (10.5), it will be a candidate for removal in the following step of the calculation.

$$\begin{cases} \sigma_{\max} > \sigma_t \\ \sigma_{\min} < \sigma_c \end{cases} \tag{10.5}$$

Figure 10.3 shows a typical stress distribution within the grain during grinding. The red and blue regions represent the domains primarily influenced by tensile stress and compressive stress, respectively. According to Eq. (10.5), the region with a maximum principal stress greater than 700 MPa and a minimum principal stress less than −2,500 MPa is illustrated, which represents the two possible candidates for fractured regions.

For Inconel 718 superalloy, the shear force in the primary deformation zone causes segmented chip formation. Therefore, the shear damage criterion was applied in this simulation. The damage state variable ω_s was used to numerically describe the material failure process, which increases monotonically with the plastic strain of the elements. It can be calculated as:

$$\omega_S = \int_0^{\varepsilon_p} \frac{d\varepsilon_p}{\varepsilon_S^{pl}(\theta_s, \dot{\varepsilon}_p)} \quad (0 \le \omega_S \le 1) \tag{10.6}$$

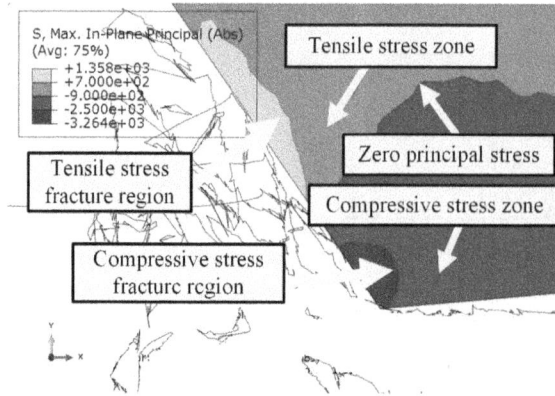

FIGURE 10.3 Failure regions of the single-grain based on the strength fracture criteria.

where $\varepsilon_s^{pl}(\theta_s, \dot{\varepsilon}_p)$ is a variable determined by shear ratio θ_s and strain rate $\dot{\varepsilon}_p$. When ω_s it reaches 1, the corresponding elements start to fail due to the large plastic strain. In post-processing, the material state is used to display the final state of each element, which is:

$$\text{state value} = \begin{cases} 1 & (\text{if } \omega_S < 1) \\ 0 & (\text{if } \omega_S = 1) \end{cases} \tag{10.7}$$

In post-processing, grain chipping can be observed. Figure 10.4a shows the Mises stress distribution during grinding with the initial grain and the shearing-induced chip formation. It can also be expressed as Figure 10.4b. The grey region (state value = 1) in Figure 10.4b represents the chips or the unremoved material, while the black region (state value = 0) represents the removed material, where the elements were selected to be deleted automatically. Figure 10.4b clearly shows shearing-induced chip formation and segmentation, with slip lines appearing periodically. Some microcracks are also generated on chips.

10.3 WEAR EVOLUTION OF SINGLE-CBN GRAIN

The evolution of CBN grain wear for a different number of grinding passes is shown in Figure 10.5. The green region represents the unaffected elements, while the red region represents the elements amenable to fracture and removal in subsequent loading increments or passes. After each grinding pass, elemental damage occurred on both the rake surface and near the cutting edge. The rake surface damage would further propagate inward in the subsequent steps. From Figure 10.5d–f, it can be speculated that this type of damage/cracking will result in a macro fracture/chipping of the CBN grain and is primarily driven by the tensile stresses during the high-speed grinding operation. The microfracture/damage near the cutting nose will drastically change the geometry of the cutting profile/nose radius of the CBN grain and thereby significantly deteriorate the cutting condition. The progression of these two effects will determine grinding performance.

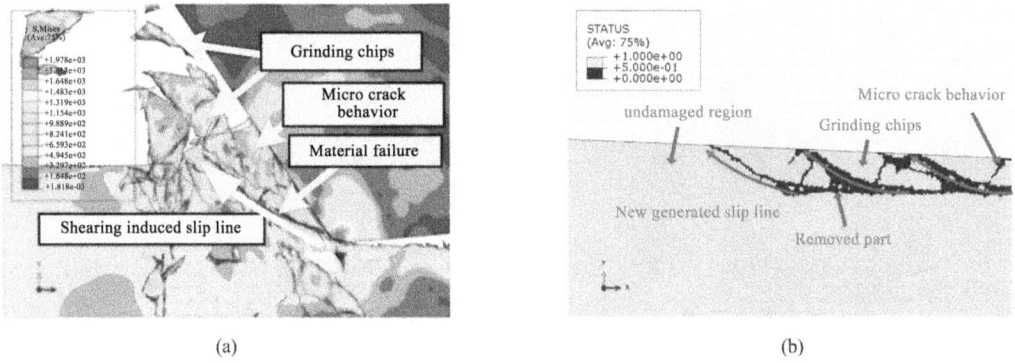

(a)

(b)

FIGURE 10.4   Chip segmentation criterion from the workpiece, based on the material state value: (a) The single-grain material removal process, (b) The corresponding material state contour.

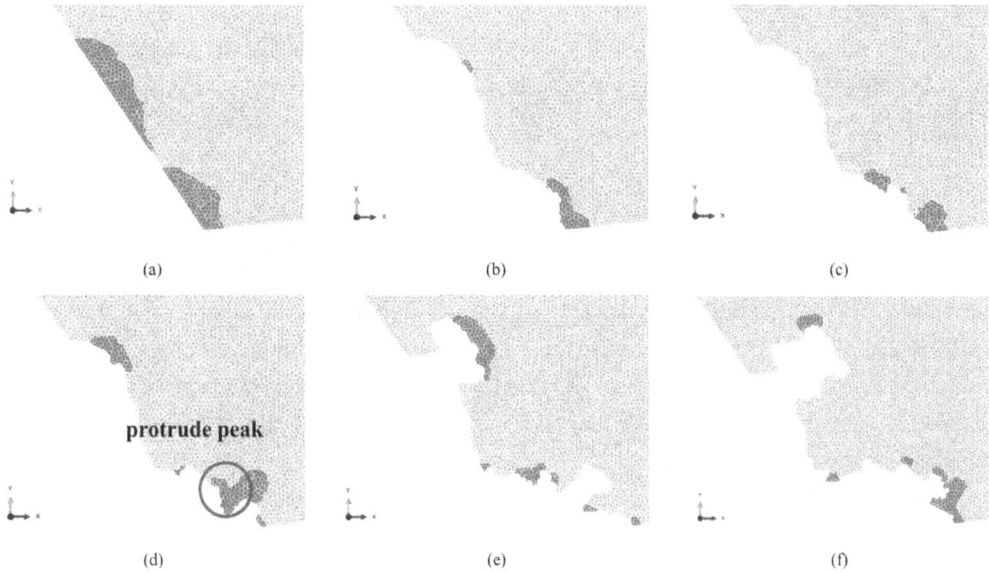

(a)

(b)

(c)

protrude peak

(d)

(e)

(f)

FIGURE 10.5   Grain wear evolution during FEA modelling: (a) Step-1, (b) Step-2, (c) Step-3, (d) Step-4, (e) Step-7, (f) Step-10.

10.4   INFLUENCE OF INTERNAL STRESSES ON GRAIN WEAR

The internal stresses during grinding modelling have also been investigated to better understand the wear process. Figures 10.6 and 10.7 show the principal stress contour for different grinding steps, taken at the instant when the largest grinding force occurred. In the contour plots, domains with principal stresses between 700 and −2,500 MPa are displayed with a false colour scale. The damaged/fractured region corresponding to exceeding the tensile strength is shaded in grey, while the one corresponding to exceeding the compressive strength is shaded in black. The maximum principal stress location in each step is also marked.

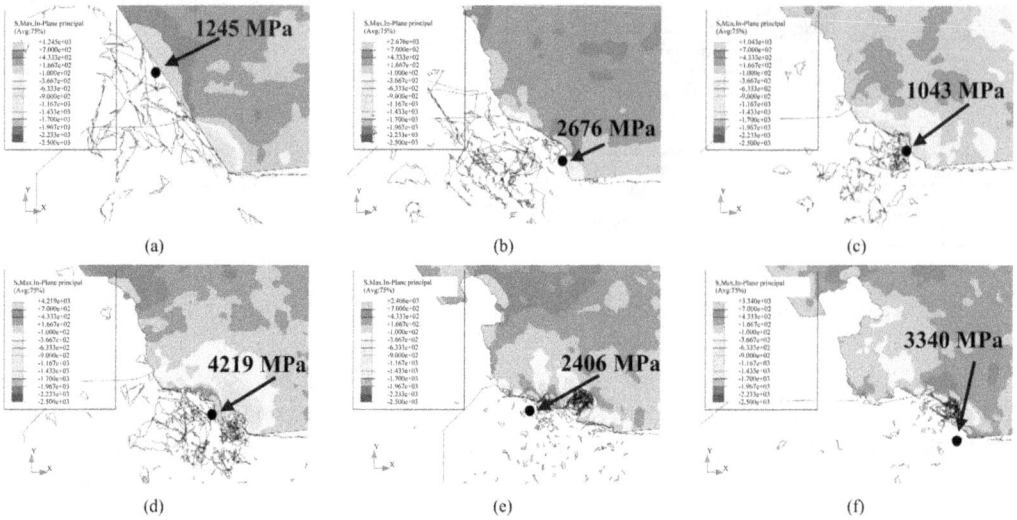

FIGURE 10.6 The maximum principal stress distribution and tensile stress-induced damage/ fractured regions at different grinding steps: (a) Step-1, (b) Step-2, (c) Step-3, (d) Step-4, (e) Step-7, (f) Step-10.

Figure 10.6 shows the effect of maximum principal stress on grain wear evolution. In Figure 10.6a, a crescent region was formed on the rake surface, leading to the crack initiation. In the following steps, the crack propagated inward due to the local tensile stress state. Additionally, there were some tensile fractures near the cutting edge corner. It stemmed from the stress concentration within the concave regions formed by grain wear. Therefore, it could be inferred that the tensile fracture might occur at both the bottom of the cutting edges and in the cracked region.

Figure 10.7 exhibits the distribution of the minimum principal stress. In Figure 10.7a Step 1, only a single-stress concentration region has been observed. In Figure 10.7c Step 3, it expanded into three regions, marked with red circles. In subsequent steps, the stress concentration regions became more dispersed and disordered. Thus, it is apparent that the compressive stress mostly influenced the wear behaviour near the cutting edges.

The evolution of the grain damage after each grinding step is further analysed by counting the number of damaged elements on the grit rake surface and near cutting edges. Damage evolution per grinding step is summarized on the bar chart in Figure 10.8. The observed cumulative damage trend indicates that the grain was damaged at a relatively steady rate. A correlation fit of the total number of damaged elements shows a liner fit of the form,

$$y = 0.213 + 0.199x \tag{10.8}$$

There were about 200 elements damaged in each step before a macrofracture occurred due to crack propagation. The ratio (represented by the blue line on Figure 10.8) of damaged

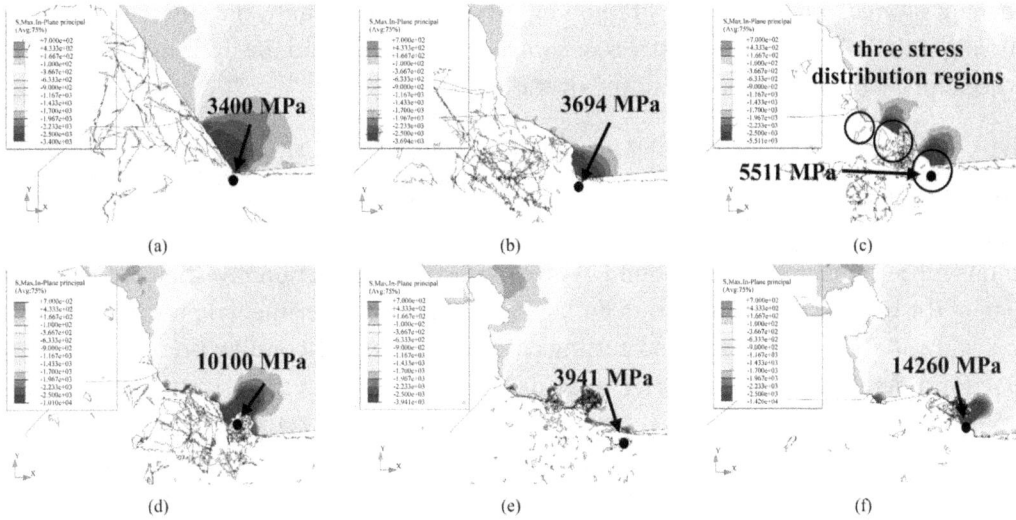

FIGURE 10.7 The minimum principal stress distribution and compressive stress-induced damage/ fractured regions at different grinding steps: (a) Step-1, (b) Step-2, (c) Step-3, (d) Step-4, (e) Step-7, (f) Step-10.

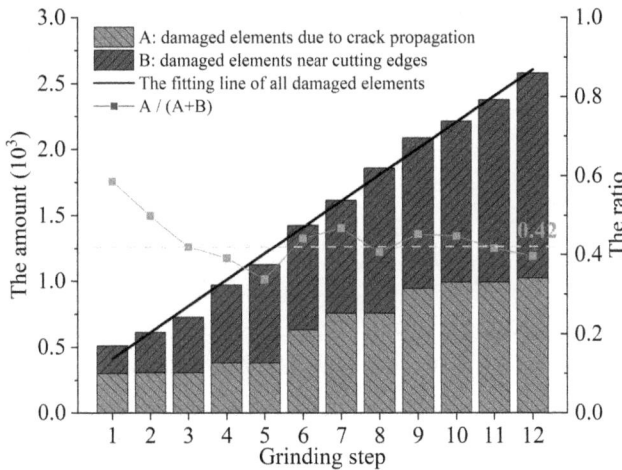

FIGURE 10.8 The number of damaged elements due to crack propagation on the gain rake surface and near cutting edges after each grinding step. The blue line represents the ratio of the damaged elements in the crack region relative to the total damaged elements.

elements due to crack propagation relative to the total number of damaged elements approached 0.42 throughout the grinding process. This implies that about 42% of damaged elements were caused by tensile stress, and the remaining 58% were influenced by a combination of tensile and compressive stresses. Moreover, the numerical results indicate that the macrofractures caused by crack propagation will consume a larger amount of the grain volume compared to grain material volume loss arising from micro and local attritions and thereby significantly reduce the lifespan of the grain. Accordingly, the simulation results highlight the role of the tensile stress in controlling the wear and life cycle of the grinding wheel during the high-speed grinding operation.

10.5 EVOLUTION OF CUTTING EDGES DUE TO GRAIN WEAR

The number of cutting edges is continuously changing during the grain wear process and thus can provide more effective cutting edges (self-sharpening). However, this phenomenon will be compensated and bound by attritous wear. This section investigates the evolution of the grain cutting edges from both a numerical and experimental perspective.

Figure 10.9 shows the extended cutting paths and the corresponding pressure distribution for different simulation steps. Figure 10.9a shows the intact single-cutting edge at the initial state. A sequence of nodal points 1–40 is highlighted, starting from the engaged front of the rake surface to the final contact point on the flank surface. The corresponding contact pressure profile changes from a tensile (−0.4 GPa) to a compressive state (~2.0 GPa) along the flank and then reaches a lower level (~0.75 GPa) at the wake of the cutting nose. At a subsequent loading step (Step-3), the cutting edge progressively evolves, as shown in Figure 10.9b, into three segments of cutting edges, marked by nodal locations 3–13, 14–26, and 26–30. A pressure redistribution commences with the splitting of the cutting edge and splits into three domains with a maximum pressure of $A = 1.123$ GPa, $B = 1.739$ GPa, and $C = 2.203$ GPa, respectively, as indicated in Figure 10.9b. The cutting edge C near the grain nose (bottom corner) carries the most of the cutting load among these three edges.

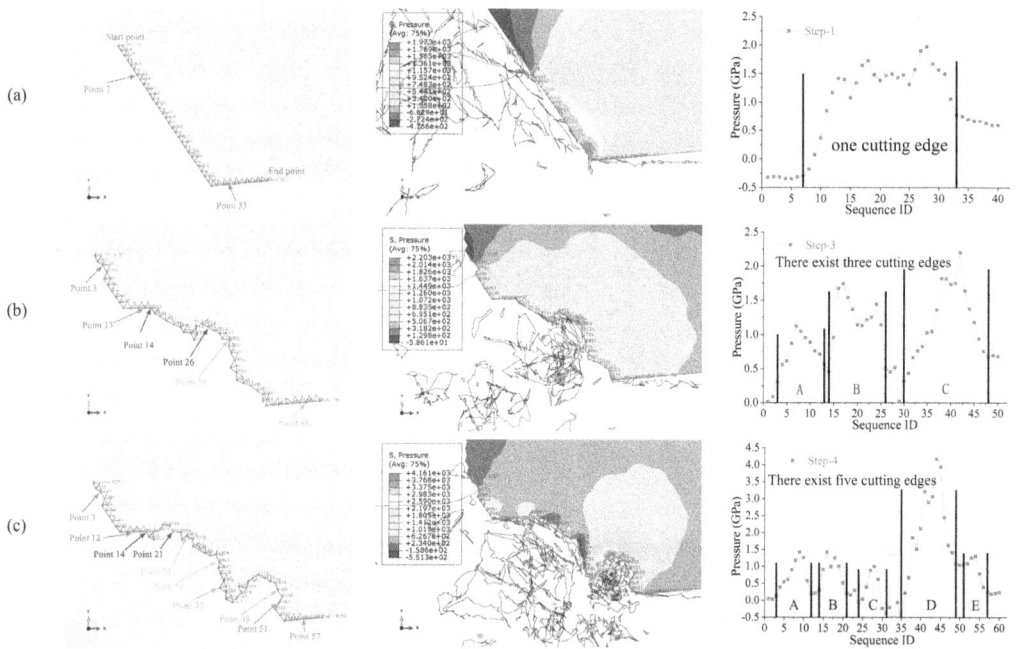

FIGURE 10.9 Evolution of the cutting edge profile and the corresponding pressure distribution at different grinding steps: (a) Step-1, (b) Step-3, (c) Step-4.

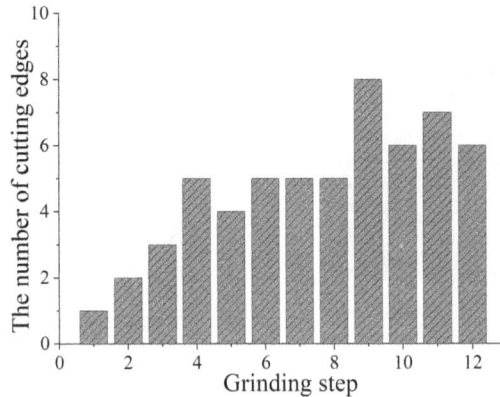

FIGURE 10.10 The number of cutting-edge segments.

Additionally, the cutting edge A under the least pressure maintained its original shape as fresh grain, while the other two cutting edges fractured due to the larger load (Figure 10.9b). The cutting edge evolved into five cutting edges in Step-4 as shown in Figure 10.9c, with maximum peak pressures of $A = 1.418\,GPa$, $B = 1.413\,GPa$, $C = 0.995\,GPa$, $D = 4.161\,GPa$, and $E = 1.298\,GPa$, respectively. The cutting edge A, which preserved its profile from the previous loading step, exhibited a similar loading level as that of edge B. Such behaviour resulted from the unique profile of the cutting edge A, wherein it has a dull obtuse angle without enough protrusion to be fractured. It is apparent that the current geometric profile at a given step will significantly influence its evolution, the corresponding grain fracture behaviour and the subsequent step. Moreover, the initial geometric mesh details might have an effect on the level of cutting-edge segmentation and fracture. Since each periodicity is represented by close to 10 nodal points, further refinement of the mesh resulted in a minor high-frequency fluctuation of the cutting edge profile, while the overall features were preserved. This observation could be substantiated by the sampling theory, wherein a signal could be reproduced at 99% fidelity when sampled at ten or more increments for a given wavelength.

The evolution of the cutting edge and its splits into individual segments are highlighted in Figure 10.10 for the different grinding steps. In general, the number of cutting-edge segments will increase during the grinding process, rendering a complex geometric profile of the cutting grain and its subsequent wear evolution. Figure 10.11 shows the experimental observation for the evolution of the CBN grain from its fresh state with a single-cutting edge (Figure 10.11a) to a complex striated surface (Figure 10.11b). The grinding process resulted in severe wear of the grain with the fracture-induced striated geometry. The fracture-induced microcutting edges promote the self-sharpening effect. While one might acknowledge the absence of full one-to-one correspondence between the experimentally observed wear profile and the numerically simulated evolution of the cutting edge surface due to the difference in the initial starting configuration and the 3D effect. However, the general features highlighted by the model remain dominant, in the form of the initial macrocracking of the grain rake surface, followed by the microcracking of the cutting nose area of the grain.

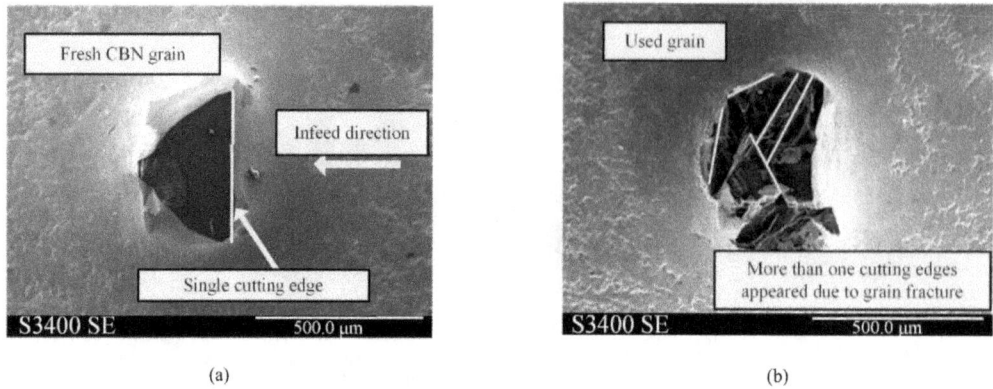

(a)   (b)

FIGURE 10.11   CBN grain and its morphological evolution after the grinding process: (a) Fresh grain, (b) Wear morphology.

REFERENCES

Akhtar, W., J. F. Sun, and W. Y. Chen. 2016. Effect of machining parameters on surface integrity in high speed milling of super alloy GH4169/Inconel 718. *Advanced Manufacturing Processes*, 31: 620–627.

Buhl, S., C. Leinenbach, R. Spolenak, et al. 2013. Failure mechanisms and cutting characteristics of brazed single diamond grains. *International Journal of Advanced Manufacturing Technology*, 66: 775–786.

Ding, K., Y. C. Fu, H. H. Su, et al. 2014a. Wear of diamond grinding wheel in ultrasonic vibration-assisted grinding of silicon carbide. *International Journal of Advanced Manufacturing Technology*, 71: 1929–1938.

Ding, W. F., Y. J. Zhu, L. C. Zhang, et al. 2014b. Stress characteristics and fracture wear of brazed CBN grains in monolayer grinding wheels. *Wear*, 332–333: 800–809.

Eder, S. J., U. Cihak-Bayr, and A. Pauschitz. 2015. Nanotribological simulations of multi-grit polishing and grinding. *Wear*, 340–341: 25–30.

Fujimoto, M., and Y. Ichida. 2008. Micro fracture behavior of cutting edges in grinding using single crystal cBN grains. *Diamond & Related Materials*, 17: 1759–1763.

Jackson, M. J. 2004. Fracture dominated wear of sharp abrasive grains and grinding wheels. *Proceedings of the Institution of Mechanical Engineers, Part J: Journal of Engineering Tribology*, 218: 225–235.

Liu, Y., B. Z. Li, C. J. Wu, et al. 2015. Simulation-based evaluation of surface micro-cracks and fracture toughness in high-speed grinding of silicon carbide ceramics. *International Journal of Advanced Manufacturing Technology*, 86: 799–808.

Liu, Y. M., A. Warkentin, R. J. Bauer, et al. 2013. Investigation of different grain shapes and dressing to predict surface roughness in grinding using kinematic simulations. *Precision Engineering*, 37: 758–764.

Mcdonald, A., A. O. Mohamed, A. Warkentin, et al. 2017. Kinematic simulation of the uncut chip thickness and surface finish using a reduced set of 3D grinding wheel measurements. *Precision Engineering*, 49: 169–178.

Mei, Y. M., Z. H. Yu, and Z. S. Yang. 2017. Numerical investigation of the evolution of grit fracture and the impact on its cutting performance in single grit grinding. *International Journal of Advanced Manufacturing Technology*, 89: 3271–3284.

Wang, J. W., T. Y. Yu, W. F. Ding, et al. 2018. Wear evolution and stress distribution of single CBN superabrasive grain in high-speed grinding. *Precision Engineering*, 54: 70–80.

Yu, T. Y., A. F. Bastawros, and A. Chandra. 2017. Experimental and modeling characterization of wear and life expectancy of electroplated CBN grinding wheels. *International Journal of Machine Tools and Manufacture*, 121: 70–80.

Zhao, Z. C., Y. C. Fu, J. H. Xu, et al. 2016. Behavior and quantitative characterization of CBN wheel wear in high-speed grinding of nickel-based superalloy. *International Journal of Advanced Manufacturing Technology*, 87: 3545–3555.

Zhu, Y. J., W. F. Ding, T. Y. Yu, et al. 2016. Investigation on stress distribution and wear behavior of brazed polycrystalline cubic boron nitride superabrasive grains: numerical simulation and experimental study. *Wear*, 376–377: 1234–1244.

Grain Wear Effect on Material Removal Behaviour during Grinding

11.1 INTRODUCTION

The wear behaviour of abrasive wheels can greatly affect the machined surface quality and integrity, the reason for which is that the different wear statuses of the abrasive tools could produce different material removal behaviours during grinding (Dai et al. 2017). Generally, the classification of wear behaviour in grinding includes the wear of abrasive tools and the wear of abrasive grains. In recent years, more attention has been paid to the research on the wear behaviour of abrasive tools from a macro-perspective. For example, the wear behaviour of cubic boron nitride (CBN) and diamond abrasive wheels was investigated, with the radial wear investigated as an evaluation parameter for grinding performance (Pazmiño et al. 2023; Shen et al. 2015). At the same time, the effects of the wear of CBN abrasive wheels on the surface roughness during grinding were also discussed (Zhao et al. 2021; Xiao et al. 2021). It is noted that, however, further studies are still necessary on the wear behaviour of abrasive grains from the micro-perspective since several micro-cutting events happen simultaneously on thousands of active cutting grains.

Furthermore, with the significant development of grinding methods with a single-diamond or CBN grain in the present day, great efforts have been made to further understand the material removal mechanism. Besides, more and more research work has also been carried out to investigate the grain wear behaviour based on the single-grain grinding method. Particularly, Buhl et al. (2013) and Wu et al. (2016), respectively, reported that the rake angle and grain shape have a great influence on the grinding force and wear resistance of abrasive grains. It should be noted that the investigation on grain wear behavwunder different process parameters, which would affect abrasive tool wear in the grinding process. It is also a well-known fact that, according to the previous publication about grinding mechanisms, the maximum undeformed chip thickness (UCT for

DOI: 10.1201/9781032678047-13

short) greatly affects the grinding process in not only the grain grinding force but also the grinding temperature within the wheel-workpiece contact zone. Under such conditions, the grain wear behaviour is highly dynamic and hard to predict in the grinding process.

For these reasons, to improve the abrasive tool life and material removal efficiency, it is particularly important to investigate the effect of grain wear behaviour on the material removal mechanism by controlling the UCT. In the present work, nickel-based superalloy Inconel 718, as a typical difficult-to-cut material, has been chosen as workpiece material during the grinding experiments with a single grain. To control the initial cutting-edge condition more easily, the diamond grains with a regular shape are utilized rather than CBN grains with an irregular shape. The grinding force of a single grain has been measured. The pile-up ratio is also calculated to detect material removal behaviour. Meanwhile, in order to further investigate the variation of grain wear behaviour, grain morphologies and grinding force are analysed under different values of UCT. Finally, the influence of grain wear behaviour on the material removal process has been discussed. Accordingly, in future research work, the abrasive tool wear could be controlled to improve the material removal rate (MRR) based on the current experimental and theoretical findings.

11.2 EXPERIMENTAL DETAILS AND PROCEDURE

The experimental set-up for the present single-grain grinding operation is displayed in Figure 11.1. Surface grinding tests were carried out on a high-speed surface grinder (BLOHM PROFIMAT MT 408) with the capabilities of a workpiece infeed speed ranging from 15 to 25 m/min, a maximum spindle power of 45 kW, and the highest revolution of 8,000 rpm. Besides, the coolant with a pressure of 15 MPa and flowrate of 90 L/min was utilized; as such, the effect of grinding temperature on grain wear could be generally ignored due to the excellent cooling condition. The cub-octahedral diamond grains in a 35/40 mesh size with a regular shape were used to control the cutting edge more easily. The wear resistance of a diamond grain usually differs significantly when grinding with different crystal faces, where the crystal face (100) shows better performance than the crystal face (111). To keep a constant grinding condition, the diamond grains were controlled at the particular crystal face (100) from the top view. Under such conditions, each diamond grain would have better wear resistance to work for a longer time in single-grain grinding. Besides, in each grinding test, only one cutting edge for each diamond grain perpendicular to the wheel speed direction was used. Therefore, the present single-grain grinding operation is similar to the conventional cutting operation (such as milling), which uses a tool with fixed geometrical shapes and angles. As a result, in the current work, it is generally reasonable to refer to the tool wear behaviour in the micro-milling operations (such as Zhang et al. 2016; Kuram and Ozcelik 2016; Liu et al. 2015; Zhan et al. 2015) to discuss the grain wear behaviour during grinding.

In this section, a single-diamond grain was brazed with Ag-Cu-Ti alloy on a steel holder, which was mounted to a steel block with screws. The assembled components were then fixed in a V-shaped slot in the wheel substrate. The workpiece material was nickel-based superalloy Inconel 718. Before the single-grain grinding tests, precision surface grinding had been carried out using alumina wheels on the workpiece to obtain the required ground surface

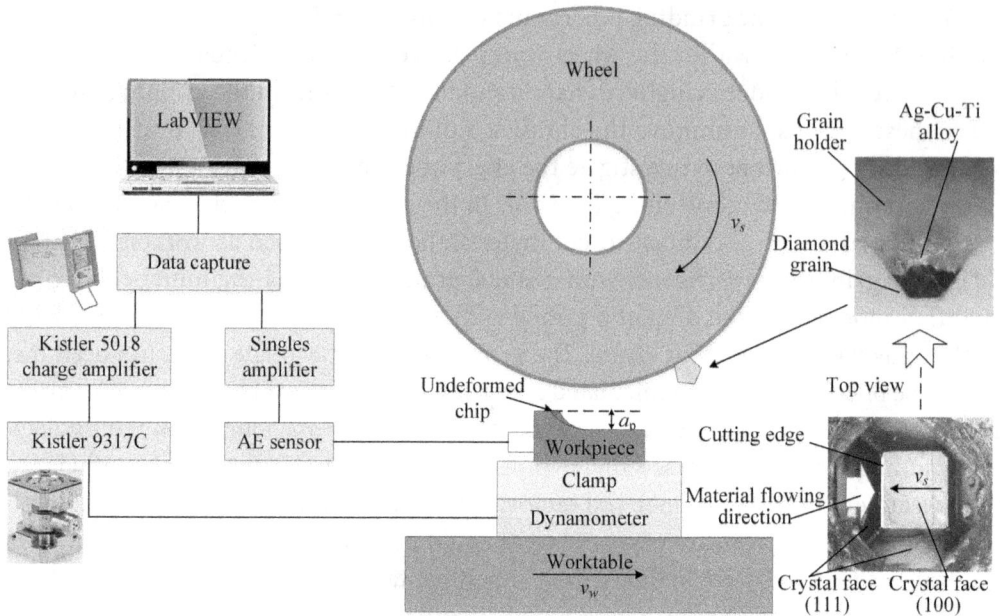

FIGURE 11.1   Schematic diagram of the present single-grain grinding operation.

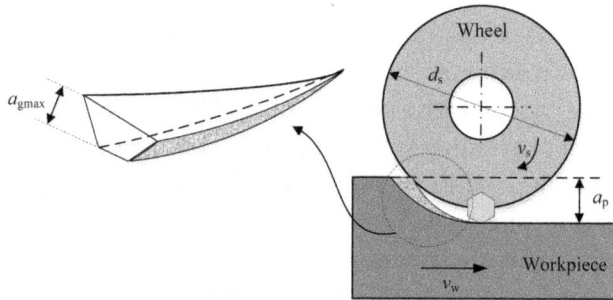

FIGURE 11.2   Schematic diagram of wedge-shaped chips.

roughness, such as Ra 0.4 μm. In the trials, the single-abrasive grain was precisely preset by using an acoustic emission instrument (DITTEL AE6000). The grinding force was measured in different grain wear stages by the three-component piezoelectric dynamometer (KISTLER 9317C) attached to a charge amplifier (KISTLER 5018).

The radius of the cutting edge circle of single-abrasive grains applied has been measured using an atomic force microscope (AFM). The dimension of the radius was nearly 0.54 μm. For this reason, tests were conducted on grain wear-free status to better understand the material removal process. A parametric study was first conducted under various UCT values, ranging from 0.16 to 1 μm. Then, to comprehensively investigate the grain wear behaviour, the UCT values were set at three different levels, such as 0.2, 0.5, and 1 μm. For a generation of wedge-shaped chips, as shown in Figure 11.2, the required workpiece infeed speed could be calculated according to the following equation:

TABLE 11.1   Grinding Process Parameters

Process Parameters	Values	
	Grain Wear-Free	Grain Wear
Undeformed chip thickness (UCT) a_{gmax} (µm)	0.16–1	0.2, 0.5, 1
Depth of cut a_p (mm)	0.02	0.02
Wheel speed v_s (m/s)	30, 50, 80	30
Workpiece infeed speed v_w (mm/min)	16.4–223.5	20.5, 51.3, 102.6
Grinding mode	Up grinding, water-based coolant	

$$a_{\text{gmax}} = 2\lambda \frac{v_w}{v_s} \sqrt{\frac{a_p}{d_s}} \tag{11.1}$$

where λ is the spacing between active grains (taking $\lambda = \pi d_s/2$ here), a_{gmax} is the UCT, v_w is the workpiece infeed speed, v_s is the wheel speed, a_p is the depth of cut, and $d_s = 390$ mm is the wheel diameter.

Table 11.1 lists the parameters applied in the present grinding tests with single-diamond grains.

During the grain wear experiment, for each group of grinding parameters, the grinding force was first collected in every two passes (each pass length is 60 mm). From the ninth pass, reading of the grinding force was carried out for every three passes. During the whole grinding tests, after each force reading, the diamond grain morphology has been recorded with the 3D optical profiler (SENSOFAR S NEOX) by disassembling the grain holder from the wheel substrate. The grain wear morphology has been characterized quantitatively in each test. Since each grain size is different in a different test, the specific material removal volume (SMRV) $\Delta V'$ is used to evaluate the amount of material removed:

$$\Delta V' = V/b = a_p \cdot L \tag{11.2}$$

where V is the material removal volume, b is the groove width, which is different for different grains, $a_p = 0.02$ mm is the depth of cut in the current experiment, and L is the length of the total grinding pass.

The pile-up ratio, which is defined as the ratio of the total pile-up area to the total groove section area, is a sound method to quantify the material removal behaviour. As illustrated in Figure 11.3, the pile-up areas are B_1 and B_2, and the groove section area is characterized by A. Thus, the pile-up ratio R_s could be expressed as:

$$R_s = (B_1 + B_2)/A \tag{11.3}$$

To obtain the values of pile-up ratio, the cross-section profiles have been measured three times along the workpiece longitude direction by using a 3D optical profiler for each test. Then, AutoCAD has been used to reconstruct the cross-section profiles, and the average value of these three pile-up ratios has been calculated.

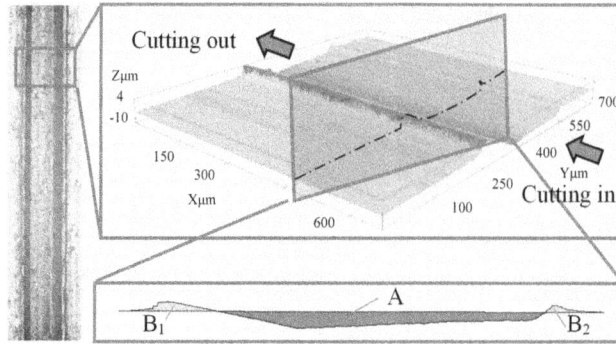

FIGURE 11.3   3D morphology and cross-section profiles of a single-grain grinding surface.

11.3 MATERIAL REMOVAL CHARACTERISTICS IN THE CASE OF GRAIN WEAR-FREE

In general, the grinding-workpiece contact could be separated into three stages: rubbing, ploughing, and cutting. In rubbing, only elastic deformation occurred, in which each grain slides on the material with a small penetration into the workpiece. When the grain penetrates deeper into the workpiece, ploughing occurs in both elastic and plastic deformation forms, where the scratch becomes more evident with ridges formed on both sides. As the grain further penetrates, the material is removed rapidly, and chip formation takes place. Since the pile-up ratio reflects the material removal ability, it is obtained at different UCT values under grain wear-free conditions to identify different material removal stages, as shown in Figure 11.4. If the UCT value is smaller than 0.3 μm, the pile-up ratio increases gradually with increasing UCT values. That is to say, more materials are left over on the workpiece surface, which indicates that the ploughing stage dominates. Once the UCT value reaches 0.3 μm, the pile-up ratio drops rapidly. Then, it remains constant as the UCT value is larger than 0.4 μm. In this case, the amount of left pile-up material has decreased for the reason that the workpiece material is mainly removed as chip formation rather than the pile-up around the scratch groove. Thus, a critical UCT value a_g, i.e., about 0.3 μm, can be identified for nickel-based superalloy Inconel 718, which can be expressed as:

$$a_{g,critical} = 0.56r_e \tag{11.4}$$

where r_e is the radius of the cutting edge.

Moreover, by comparing the measured grinding force at different values of UCT (Figure 11.5a), it could be found that the grinding force varies in two different stages due to the different material deformation characteristics in grinding. On the one hand, when the UCT value is less than 0.3 μm, the grinding process is in the rubbing and ploughing stages, where the material is friction and pulled along the front and side faces of a grain. On the other hand, in the case of the UCT larger than 0.3 μm, chip formation takes place, and the material removal increases rapidly.

The definition of critical UCT value from the variations of pile-up ratio and grinding force can be interpreted from the specific grinding energy. In the current grinding tests,

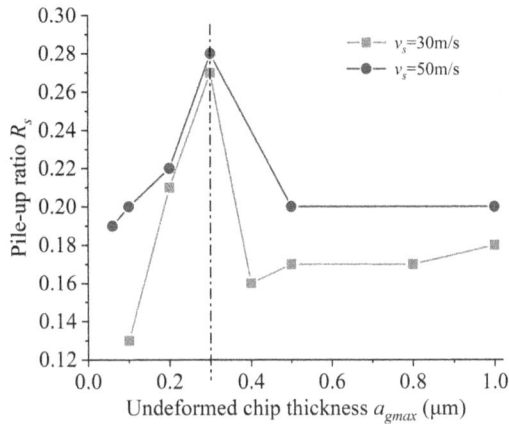

FIGURE 11.4   Pile-up ratio versus UCT.

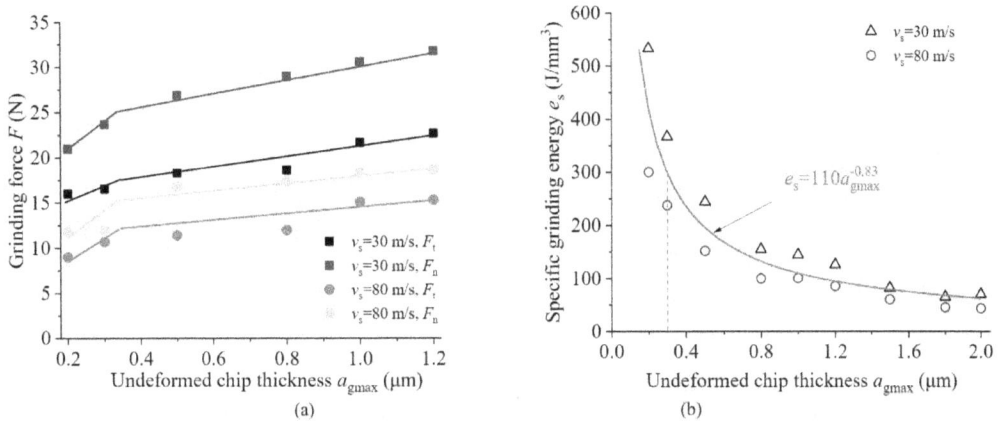

FIGURE 11.5   Size effect on grinding force and specific grinding energy: (a) Grinding force, (b) specific grinding energy.

the undeformed chip is in a wedge shape, which can be approximated as a triangle. The specific grinding energy e_s is defined as the energy consumed to remove one unit volume of material, which can be expressed as:

$$e_s = W / V = (F_t l_s) / (a_{g\max} l_s b / 2) = 2F_t / (a_{g\max} \cdot b) \tag{11.5}$$

where W is the energy consumption to remove the material in a single-grain workpiece contact zone, V is the volume of undeformed chip material to be removed, $l_s = (a_p \cdot d_s)^{1/2}$ is the length of the grain-workpiece contact zone, and $b = 0.3\,\text{mm}$ is the width of the undeformed chip.

It is reported that specific grinding energy is consumed using primary and secondary rubbing energy, ploughing energy, and chip formation energy. Among them, the ploughing energy and chip formation energy are constant for the determinate material, which is determined by the material's mechanical properties, such as dynamic yield shearing strength and workpiece hardness. However, the rubbing energy is inversely proportional to the UCT. According to Eq. (11.5), the variation of specific grinding energy with increasing UCT is plotted in Figure 11.5b. The specific grinding energy obeys the typical size effect rule in grinding. In the case of a small UCT (<0.3 μm, particularly), the specific grinding energy has larger values and decreases quickly with the increase of the UCT, which indicates that it is mainly composed of high rubbing energy and relatively low ploughing energy. However, as the UCT value gets larger, the specific grinding energy decreases at a slower rate, which shows the grinding process mainly consists of the cutting stage with small cutting energy. Based on the results shown in Figure 11.5b, the relationship between specific grinding energy and UCT could be calculated by Eq. (11.6):

$$e_s = 110 a_{g\max}^{-0.83} \qquad (11.6)$$

11.4 CLASSIFICATION OF GRAIN WEAR BEHAVIOUR IN SINGLE-GRAIN GRINDING

In general, during the single-grain grinding of Inconel 718, the diamond grain wear behaviour shows mainly four types: crescent depression on the rake face, abrasion on the flank face, grain microfracture, and grain macrofracture, as schematically displayed in Figures 11.6 and 11.7. The grain wear behaviour obtained at different UCT values in the current investigation is tabulated in Table 11.2. It shows the wear evolution of diamond grains could be divided into two stages: (i) the initial wear stage, acting as the formation of crescent depression on the rake face and the abrasion wear on the flank face; (ii) the steady wear stage, in which the wear behaviour on the rake and flank face is dramatically influenced by different UCT values.

According to the critical UCT value in Section 11.3, the rubbing, ploughing, and cutting stages at UCTs of 0.2, 0.5, and 1 μm are schematically shown in Figure 11.8. The percentage of these three stages in the grain-workpiece contact zone is significantly different. When

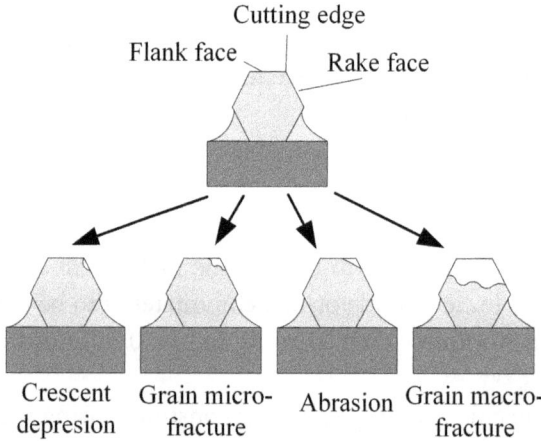

FIGURE 11.6 Classifications of grain wear behaviour.

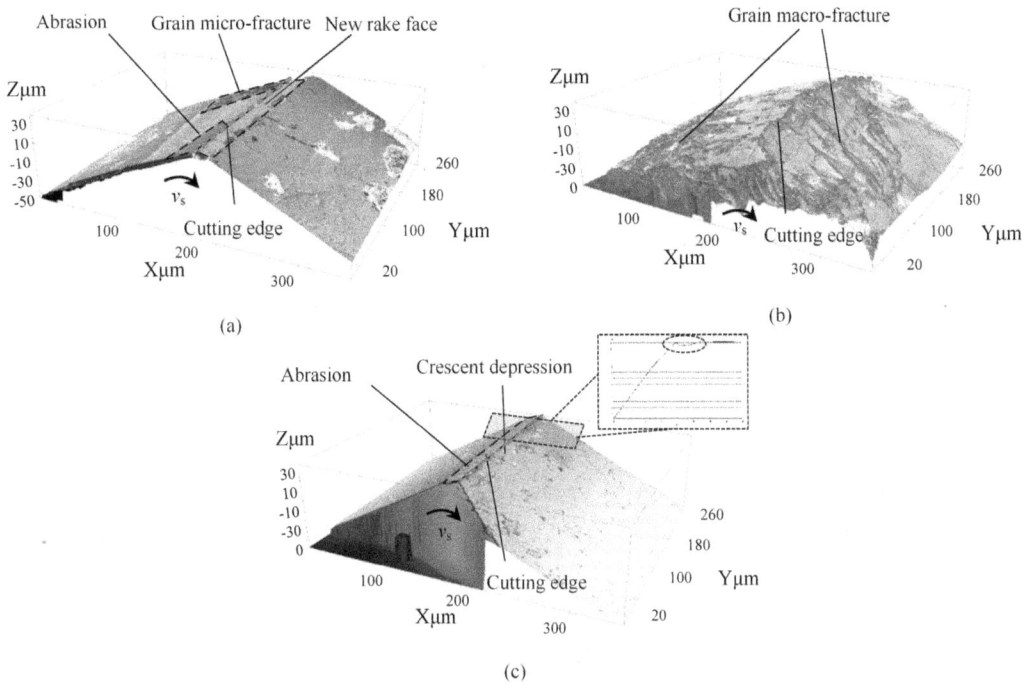

FIGURE 11.7 Morphology of grain wear in single-grain grinding: (a) $a_{gmax} = 0.2\,\mu m$, $\Delta V' = 5.44\,mm^3/mm$, (b) $a_{gmax} = 0.5\,\mu m$, $\Delta V' = 22.2\,mm^3/mm$, (c) $a_{gmax} = 1\,\mu m$, $\Delta V' = 2.88\,mm^3/mm$.

TABLE 11.2   Grain Wear Behaviour at Given UCTs

Classification	$a_{gmax}=0.2\,\mu m$	$a_{gmax}=0.5\,\mu m$	$a_{gmax}=1\,\mu m$
Crescent depression	No	Yes	Yes
Abrasion	Yes	Yes	Yes
Grain microfracture	Yes	Yes	No
Grain macrofracture	Yes	Yes	No

only the rubbing and ploughing stages take place in the grinding process, as shown in Figure 11.8a, the stress concentration on the cutting edge is so severe that abrasion and grain fracture appear due to the attrition wear on the flank face and the spring-back effect from the elastic deformation of the ground surface material, respectively. However, no crescent depression occurs because the chip formation of the material is rather difficult, which is necessary for the rake-face attritious wear. As seen in Figure 11.8b, when the dwell time of the cutting stage is approximately the same as that of the rubbing and ploughing stages, the removed chips flow on the rake face, which increases the crescent depression caused by the frictional interaction with a removed chips. When the cutting stage dominates, as displayed in Figure 11.8c, most of the load is shared by the rake face. Thus, the spring-back effect from the ground surface in the rubbing and ploughing stages is too weak to fracture the grains. Theoretically, according to the increasing SMRV, the grain wear behaviour could be interpreted in Figure 11.9 based on the experimental results.

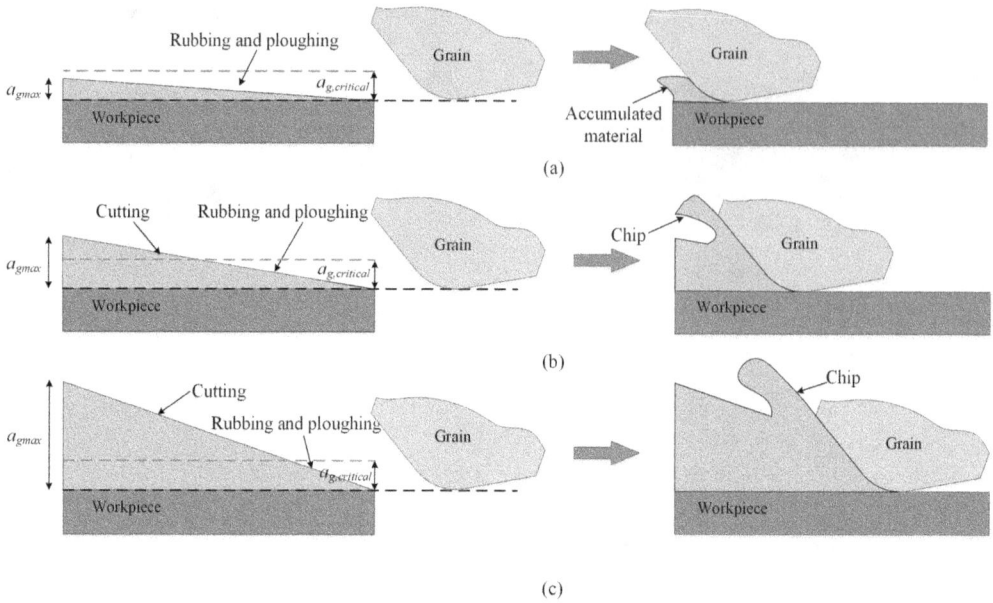

FIGURE 11.8 Schematic diagram of the rubbing, ploughing, and cutting stages: (a) $a_{gmax} = 0.2\,\mu m$, (b) $a_{gmax} = 0.5\,\mu m$, (c) $a_{gmax} = 1\,\mu m$.

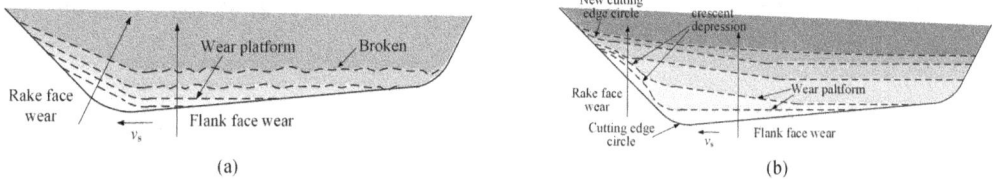

FIGURE 11.9 Schematic diagram of wear behaviour of diamond grain in the case of UCT of 0.2 and 1 μm: (a) $a_{gmax} = 0.2\,\mu m$, (b) $a_{gmax} = 1\,\mu m$.

11.5 ANALYSIS OF GRAIN WEAR BEHAVIOUR FROM THE VIEWPOINT OF THE GRINDING FORCE

Figure 11.10 shows the relationship between grinding force and SMRV at three given levels of UCT in the present investigation. With the increase in SMRV, the normal grinding force F_n and tangential grinding force F_t tend to increase first and then stay stable. However, the growth rate of the grinding force is different at each level of UCT values. When the UCT is 0.2 or 0.5 μm, the grinding force rises at a relatively higher growth rate in the initial wear stage. Moreover, the growth rate at the UCT of 0.2 μm is higher than that at 0.5 μm. This indicates stress concentration on the cutting edge is more prominent in the case of small UCT, which rapids the grain wear and increases the grinding force. When the UCT further increases to 1 μm, the load mostly acts on the rake face. Thus, the grain wear around the cutting edge is not remarkable. Under such conditions, the grinding force increases slowly in the initial wear stage. However, when the SMRV reaches 6.8 mm³/mm, due to the appearance of crescent depression on the rake face, the

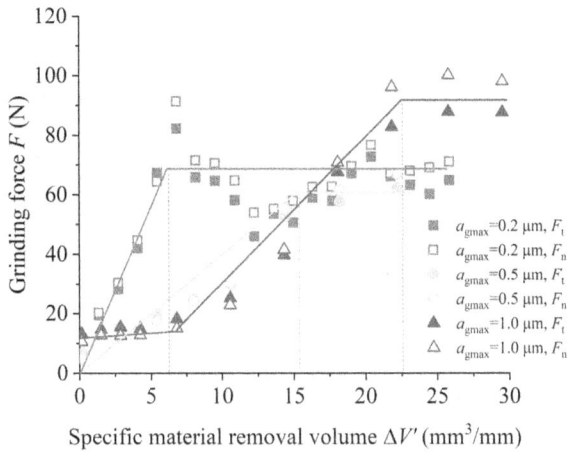

FIGURE 11.10   Grinding force versus SMRV.

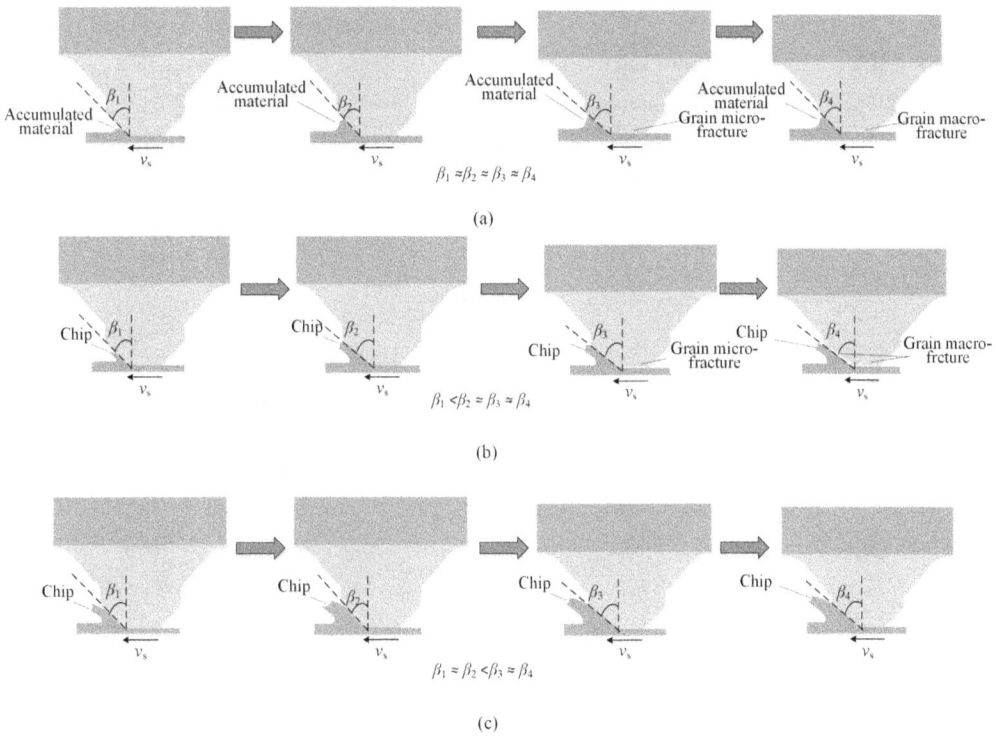

FIGURE 11.11   Schematic diagram of material removal behaviour and the variation of the negative rake angle in grinding: (a) $a_{gmax}=0.2\,\mu m$, (b) $a_{gmax}=0.5\,\mu m$, (c) $a_{gmax}=1\,\mu m$.

grinding force grows quickly with a growth rate approximate to that in the case of UCT of $0.5\,\mu m$.

The changes in grinding force with increasing SMRV could be attributed to grain wear behaviour. When the UCT is $0.2\,\mu m$, as seen in Figure 11.11a, since small chips are generated by removing material, more and more material accumulates on the rake face. Due to the stress concentration on the cutting edge, grain wear takes place quickly. As a result, the grinding force rises at a large growth rate. Once the new rake face is formed, the amount of accumulated material stays nearly constant, which stabilizes the grinding force. Because of the continuous formation of new rake faces in the grain wear procedure, the grinding force fluctuates between 50 and 70 N. When the UCT is 0.5 μm, as displayed in Figure 11.11b, the crescent depression on the rake face keeps expanding in the early stage, which results in an increase in the contact area between grain and flowing chips. Thus, the grinding force gradually increases first and then remains stable until grain fracture appears around the cutting edge. When the UCT reaches 1 μm, the crescent depression forms gradually in the initial wear stage, making the amount of flowing chip vary slightly, as demonstrated in Figure 11.11c. Therefore, the grinding force increases slowly. However, because the abrasion wear of a diamond grain on the flank face is getting serious, the load on the cutting edge is increased, which results in the suddenly expanded crescent depression. While the crescent depression disappears gradually with the increase in SMRV, the negative rake angle is enlarged. As a result, the grinding force increases. At last, once the crescent depression is completely transferred into the new sharp cutting edge, the grinding force remains stable.

It could be found by further analysis of grinding force that, when the grinding force remains stable, the SMRV values are different at three levels of UCT. The critical values are 8.16, 14.4, and 21.84 mm³/mm at the UCT of 0.2, 0.5, and 1 μm, respectively. It is inferred that, by increasing the UCT value, more material would be removed when the grinding force becomes stable in the wear process. That is to say, the grain wear rate would be effectively reduced if the UCT was increased to twice the radius of the cutting edge circle.

11.6 RELATIONSHIP BETWEEN GRINDING FORCE RATIO AND NEGATIVE RAKE ANGLE

Figure 11.12 provides the relationship between the grinding force ratio and the SMRV at different UCT values. When the UCT is 0.2 μm, the grinding force ratio always fluctuates between 0.9 and 1.2. When it is 0.5 μm, the grinding force ratio first increases gradually and then remains around 1.08 when the SMRV increases to 10 mm³/mm. When the UCT is 1 μm, the grinding force ratio first keeps a constant of 0.82 until the SMRV reaches 10 mm³/mm, then gradually rises to 1.12 as the SMRV increases to 20 mm³/mm, and finally, they

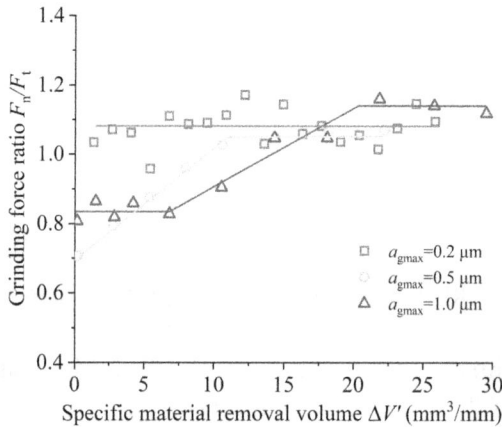

FIGURE 11.12 Grinding force ratio versus SMRV.

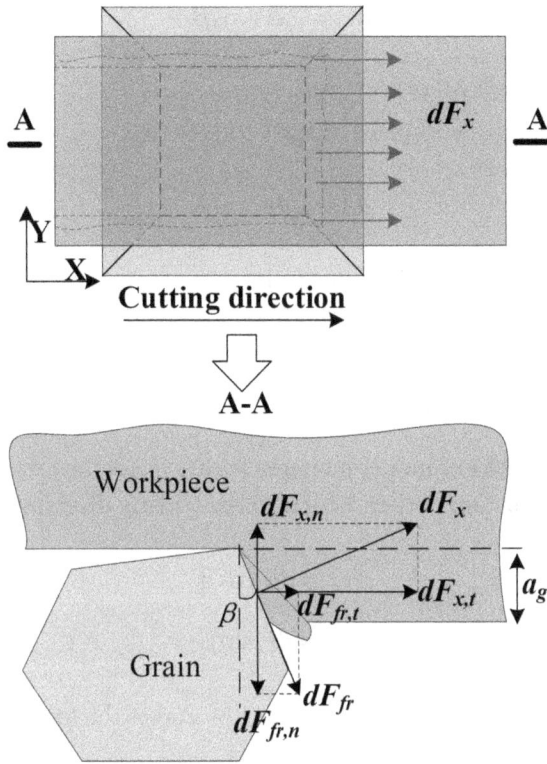

FIGURE 11.13   Grinding force model in single-grain grinding.

remain constants again at 1.12. It could be inferred from the grain wear behaviour and the variation of the grinding force ratio that the gradually disappearing crescent depression on the rake face would increase the actual negative rake angle, which is diagrammed in Figure 11.11. In the case of a UCT of 0.2 μm, since the workpiece material always presses the cutting edge, the rake and flank faces wear synchronously, as displayed in Figure 11.11a. Thus, a little change in the negative rake angle ($\beta_1 \approx \beta_2 \approx \beta_3 \approx \beta_4$) makes the grinding force ratio a constant. While in the case of UCT of 0.5 μm, seen from Figure 11.11b, the negative rake angle is enlarged ($\beta_1 < \beta_2$) due to the expanding of crescent depression in the initial wear stage. When the grain fracture happens around the cutting edge, the rake and flank face also wear synchronously, which causes the negative rake angle to change slightly ($\beta_2 \approx \beta_3 \approx \beta_4$). However, when the UCT is 1 μm, the generated crescent depression is insignificant in the beginning, so the negative rake angle doesn't alter very much ($\beta_1 \approx \beta_2$), as displayed in Figure 11.11c. Once the crescent depression disappears gradually, the negative rake angle is increased ($\beta_2 < \beta_3$). Finally, the negative rake angle remains constant ($\beta_3 \approx \beta_4$) again as the new cutting edge is generated.

The specific grinding force (F_p) is defined as the force per unit area in the grain moving direction. The forces on the two sides of the grain are balanced out by each other. Only the forces on the rake face are considered, as diagrammed in Figure 11.13.

In every cross-section X-X, the grinding force dF_x perpendicularly loads on the rake face, which is calculated by:

$$dF_x = F_p dA \cos\beta \tag{11.7}$$

where β is the negative rake angle.

The cell area in the grain-workpiece contact zone is:

$$dA = a_g dy / \cos\beta \tag{11.8}$$

where a_g is the actual cutting depth of the grain.

Thus, Eq. (11.7) can be rewritten as:

$$dF_x = F_p a_g dy \tag{11.9}$$

According to the grinding force model shown in Figure 11.13, the force dF_x could be decomposed into the force along and perpendicular to the cutting direction, which is written as:

$$dF_{x,t} = dF_x \cos\beta \tag{11.10}$$

$$dF_{x,n} = dF_x \sin\beta \tag{11.11}$$

The total friction force along the grain rake face dF_{fr} makes the formed chip flow out of the grain-workpiece contact zone, which is defined as:

$$dF_{fr} = \mu dF_x \tag{11.12}$$

where μ is the friction coefficient, and the value is 0.23.

As such, the friction force can also be decomposed into the force along and perpendicular to the cutting direction, which is expressed as follows:

$$dF_{fr,t} = dF_{fr} \sin\beta \tag{11.13}$$

$$dF_{fr,n} = dF_{fr} \cos\beta \tag{11.14}$$

Finally, by combining the component forces, the tangential and normal forces on the sampling unit of a single grain are calculated as:

$$dF_{tg} = dF_{x,t} + dF_{fr,t} = (\cos\beta + \mu\sin\beta)F_p a_g dy \tag{11.15}$$

$$dF_{ng} = dF_{x,n} - dF_{fr,n} = (\sin\beta - \mu\cos\beta)F_p a_g dy \tag{11.16}$$

Thus, considering that the actual load width of a single grain is a the total grinding force on the whole grain is integrated as:

$$F_{tg} = \int_{-\frac{a}{2}}^{\frac{a}{2}} dF_{tg} = F_p a a_g (\cos\beta + \mu\sin\beta) \tag{11.17}$$

$$F_{ng} = \int_{-\frac{a}{2}}^{\frac{a}{2}} dF_{ng} = F_p a a_g (\sin\beta - \mu\cos\beta) \qquad (11.18)$$

Commonly, the grinding force ratio ε is used to reflect whether the grain is blunt or not. From the given grinding force model in this article, it can be calculated as:

$$\varepsilon = F_{ng}/F_{tg} = \frac{\sin\beta - \mu\cos\beta}{\cos\beta + \mu\sin\beta} = \frac{\tan\beta - \mu}{1 + \mu\tan\beta} \qquad (11.19)$$

which can be further rewritten as:

$$\varepsilon = \frac{1}{\mu} - \frac{\mu^2 + 1}{\mu(1 + \mu\tan\beta)} \qquad (11.20)$$

According to Eq. (11.20), it is obvious that, with the increase of the negative rake angle β, the grinding force ratio ε is getting larger. In other words, a diamond grain is becoming blunt due to the grain wear behaviour on the rake face. The calculated grinding force ratio in the case of $a_{gmax} = 1\,\mu m$ under the condition of a negative rake angle ranging from 51.5° to 60.5° is provided in Figure 11.14. The calculated results match very well with the tested data, with an error of less than 5%. During the increase of SMRV from 6.8 to 21.84 mm³/mm, the measured negative rake angles are 51.8°, 54°, 57°, 59°, and 60.6°, while the corresponding calculated grinding force ratios are 0.8, 0.87, 0.97, 1.04, and 1.1, compared with tested values of 0.83, 0.9, 1.05, 1.05, and 1.16.

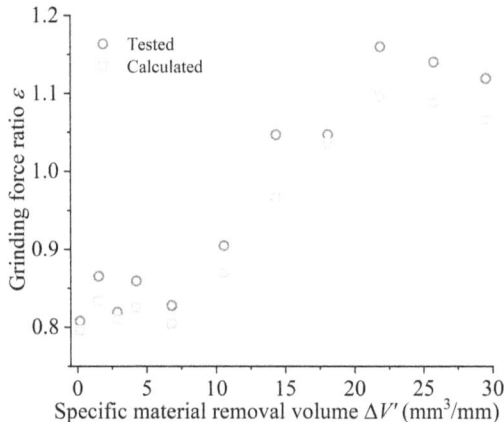

FIGURE 11.14 Comparison of the tested and calculated grinding force ratio in the case of $a_{gmax} = 1\,\mu m$.

11.7 INFLUENCE OF GRAIN WEAR BEHAVIOUR ON THE MATERIAL REMOVAL PROCESS

In the above sections, the chip formation characteristics are discussed in terms of grain wear-free status. However, the material removal ability has decreased due to attritious wear. To further investigate the influence of grain wear behaviour on the material removal process, the cross-section profiles of the scratch marks are obtained in different grain wear stages first, and then the pile-up ratio is discussed according to the increasing SMRV.

The cross-section profiles for various single-grain grinding conditions are shown in Figure 11.15. As expected, due to the bluntness of a single grain, there is a clear tendency for large material pile-up, manifested as pronounced higher and wider shoulders for all the UCT values, accompanied by a large groove width, thus constituting an extent of material rebuild rather than removal. Nevertheless, distinctly different material pile-up character-istics are produced at the given UCT values: a larger UCT value results in less amount of material pile-up in the sharp grain condition (Figure 11.15a, c, and e); that is likely due to a longer period of cutting, rather than rubbing and ploughing stages at UCT values of 0.5 and 1 μm to cut the material across the whole cross-section, resulting in minor material rebuild.

Furthermore, it is noticed from Figure 11.15b, d, and e that the shape of the scratch cross-section bottom changes from a line to an irregular curve or an arc due to the wear of the cutting edge. The reason can be interpreted from two aspects: one is the grain frac-ture caused by the spring-back effect of material elastic deformation in the rubbing and ploughing stages, e.g., at the UCT values of 0.2 and 0.5 μm; the other is the abrasion at the two corners of the cutting edge due to stress concentration, e.g., at the UCT values of 0.5 and 1 μm. Also, when a grain has been seriously worn, the pile-up area is bulky in cases of a higher value of UCT. That is because of the regenerated multi-cutting edges caused by grain fracture at a UCT of 0.2 μm, which facilities cutting the material more efficiently and results in narrow scratch shoulders. However, compared with multi-cutting edges, the cut-ting efficiency in the case of abrasion at the two corners of the initial cutting edge is so low that the pile-up shoulders exhibit a little flat-wide.

The pile-up ratio at various increasing SMRVs is calculated, as are the measured widths of new rake face and crescent depression. As shown in Figure 11.16a, when the grinding process mainly consists of rubbing and ploughing stages, the pile-up ratio varies nearly in the same direction as the width of the new rake face, and this trend results from the growing length of the grain-workpiece contact zone on the rake face. Thus, bulk mate-rial rebuilding and minor material removal occur. However, according to the decreased pile-up ratio, it is known that the regenerated multi-cutting edges due to grain fracture at SMRV of 8 mm³/mm can improve the material removal ability.

On the other hand, as shown in Figures 11.16b and c, the cutting stage dominates in grinding. The pile-up ratio is increased because of the increased crescent depression on the rake face. It is noted that the pile-up ratio increases rapidly when the width of the crescent depression is diminished (e.g., in the range of SMRV from 8 to 15 mm³/mm in the case of UCT of 0.5 μm and from 14 to 22 mm³/mm in the case of UCT of 1 μm). This is due to the increasing negative rake angle in the absence of crescent depression as a result of lower

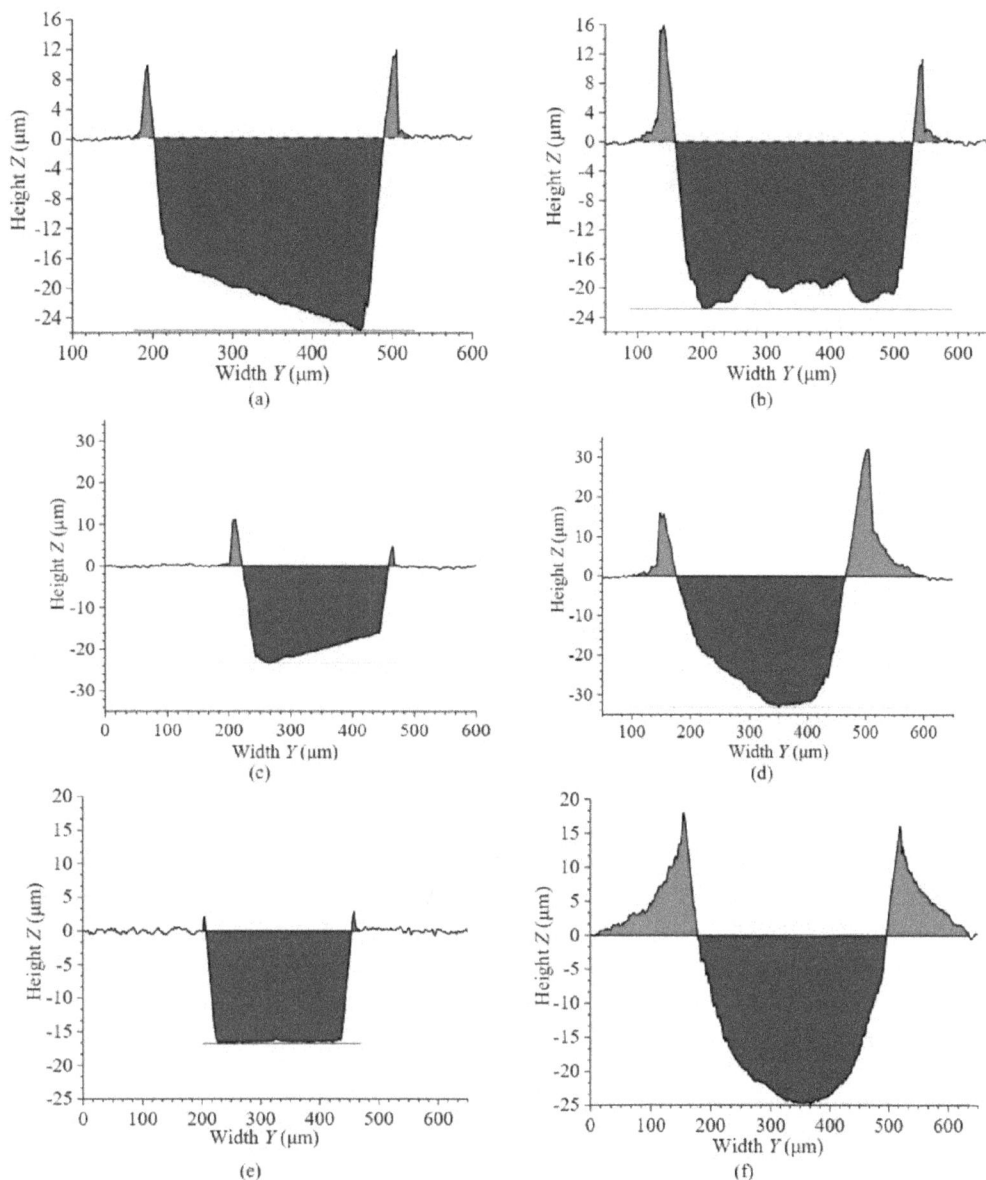

FIGURE 11.15 Influence of gain wear behaviour on cross-section profiles: (a) $a_{gmax}=0.2\,\mu m$, $\Delta V'=2.72\,mm^3/mm$, (b) $a_{gmax}=0.2\,\mu m$, $\Delta V'=25.84\,mm^3/mm$, (c) $a_{gmax}=0.5\,\mu m$, $\Delta V'=2.8\,mm^3/mm$, (d) $a_{gmax}=0.5\,\mu m$, $\Delta V'=22.2\,mm^3/mm$, (e) $a_{gmax}=1\,\mu m$, $\Delta V'=2.88\,mm^3/mm$, (f) $a_{gmax}=1\,\mu m$, $\Delta V'=29.52\,mm^3/mm$.

MMR and an enlarged grinding force ratio. Furthermore, when the pile-up ratio reaches about 0.33, the grain wear is expanded from the rake face to the two corners of the cutting edge because of the enlarged scratch shoulders. At this time, the material removal behaviour improves a bit due to the better material flowing ability around the corners.

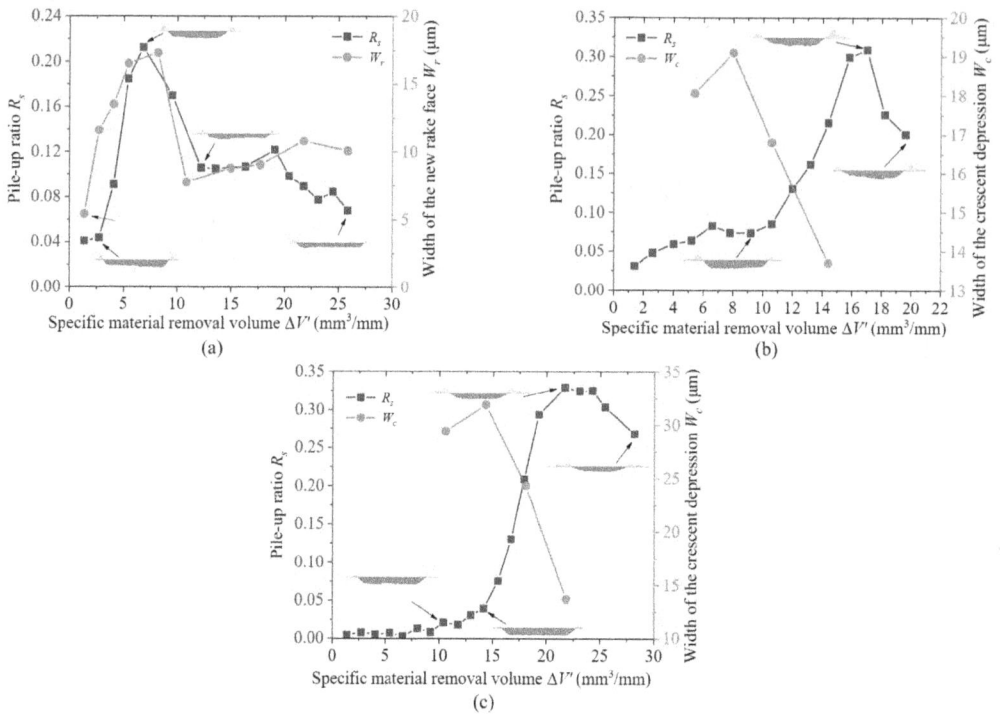

FIGURE 11.16 Relation between pile-up ratio and rake face wear: (a) $a_{gmax} = 0.2\,\mu m$, (b) $a_{gmax} = 0.5\,\mu m$, (c) $a_{gmax} = 1\,\mu m$.

Besides, it could also be found that, before the material removal efficiency decreases obviously, the value of SMRV for high material removal efficiency ($R_s < 0.1$) is increased with a bigger UCT (i.e., 5, 11, and 15 mm³/mm in cases of 0.2, 0.5, and 1 μm, respectively), which is attributed to a lower grain wear rate using a relatively larger UCT.

REFERENCES

Buhl, S., C. Leinenbach, R. Spolenak, et al. 2013. Failure mechanisms and cutting characteristics of brazed single diamond grains. *International Journal of Advanced Manufacturing Technology*, 66: 755–786.

Dai, C. W., W. F. Ding, J. H. Xu, et al. 2017. Influence of grain wear on material removal behavior during grinding nickel-based superalloy with a single diamond grain. *International Journal of Machine Tools and Manufacture*, 113: 49–58.

Kuram, E., and B. Ozcelik. 2016. Effects of tool paths and machining parameters on the performance in micro-milling of Ti6Al4V titanium with high-speed spindle attachment. *International Journal of Advanced Manufacturing Technology*, 84: 691–703.

Liu, H. T., Y. Z. Sun, Y. Q. Geng, et al. 2015. Experimental research of milling force and surface quality for TC4 titanium alloy of micro-milling. *International Journal of Advanced Manufacturing Technology*, 79: 705–716.

Pazmiño, T., I. Pombo, J. Girardot, et al. 2023. Multiscale simulation of volumetric wear of vitrified alumina grinding wheels. *Wear*, 530–531: 205020.

Shen, J. Y., J. Q. Wang, B. Jiang, et al. 2015. Study on wear of diamond wheel in ultrasonic vibration-assisted grinding ceramic. *Wear*, 332–333: 788–793.

Wu, H. Y., H. Huang, F. Jiang, et al. 2016. Mechanical wear of different crystallographic orientations for single abrasive diamond scratching on Ta12W. *International Journal of Refractory Metals and Hard Materials*, 54: 260–269.

Xiao, G., B. Zhao, W. Ding, et al. 2021. On the grinding performance of metal-bonded aggregated cBN grinding wheels based on open-pore structures. *Ceramics International*, 47(14): 19709–19715.

Zhan, Z. B., N. He, L. Li, et al. 2015. Precision milling of tungsten carbide with micro PCD milling tool. *International Journal of Advanced Manufacturing Technology*, 77: 2095–2103.

Zhang, X. W., K. F. Ehmann, T. B. Yu, et al. 2016. Cutting forces in micro-end-milling processes. *International Journal of Machine Tools and Manufacture*, 107: 21–40.

Zhao, B., W. Ding, G. Xiao, et al. 2021. Effects of open pores on grinding performance of porous metal-bonded aggregated cBN wheels during grinding Ti-6Al-4V alloys. *Ceramics International*, 47(22): 31311–31318.

Undeformed Chip Thickness Nonuniformity When Considering Abrasive Wheel Wear

12.1 INTRODUCTION

Grinding technology is often used as the final step due to its high precision, surface finishing, and stability. However, the stochastic behaviour of grinding makes it challenging for the prediction of ground surface roughness and topography (Zhang et al. 2023; Ding et al. 2017). Jiang et al. (2013a) established the 2D and 3D ground surface topography models based on the micro-interaction between grains and workpieces in the grinding contact zone. Denkena et al. (2013) predicted the resulting surface by taking into account the grinding layer topography and the grinding kinematics. Liu et al. (2013) have presented an efficient kinematic simulation to predict the workpiece surface finishing in grinding with the consideration of both the grain shape and a dressing model. Besides, the ground surface roughness was also obtained from the predicted workpiece photography with corresponding experimental validation. To optimize the performance of a textured wheel, Koshy et al. (2003) numerically generated the ground workpiece surface based on a formulation of a theoretical framework. In his work, the three-dimensional structure of a cubic volume of a diamond wheel was first simulated concerning the abrasive grit size and concentration, assuming the abrasives to be spherical. Then, a ground surface was generated concerning the wheel topology sample and the process kinematic parameters in a single-pass plunge surface grinding operation. It is noted that, however, the wheel topology applied to predict the ground workpiece surface is usually numerically generated without considering the measured wheel surface topology, especially the grain protrusion height nonuniformity of

DOI: 10.1201/9781032678047-14

the entire wheel. As a result, the influence of undeformed chip thickness nonuniformity on the predicted workpiece surface is absent during grinding.

In the past decades, the influence of the undeformed chip thickness of a single grain on the ground surface finishing has been widely investigated. Therefore, the undeformed chip thickness has been broadly utilized to better understand the grinding mechanism. According to the previous studies (Zhang et al. 2018; Dai et al. 2019), it has been known that an optimum value of undeformed chip thickness exists for a given combination of abrasive grains and workpiece material. Under such optimization, the workpiece material can be removed to form the desired ground surface. As a result, better workpiece surface finishing would be achieved if the undeformed chip thicknesses of all grains could be controlled within a small range. However, the undeformed chip thickness is usually considered a constant in the literature, which does not conform to the actual grinding condition where the grain protrusion height is nonuniform. Theoretically, it can be determined by the wheel topology and process parameters. Accordingly, different models of the undeformed chip thickness have been developed. For instance, according to the geometric models of the wheel and workpiece, Pahlitzsch and Helmerdig (1943) calculated the undeformed chip thickness with consideration of the spacing between active grains on the wheel surface. Shaw and Reichenbach (1956) developed the model by calculating the average volume of the cutting layer based on the number of active grains. Werner (1971) acquired a calculation formula from the aspect of the average area of the chip cross-section. Recently, Agarwal and Rao (2013) reconstructed the undeformed chip thickness model from randomly distributed cutting edges, elastic deformation in the contact zone, and the microinteraction of cutting edges. Based on their models, the ground surface roughness, grinding forces, and temperature could be predicted. However, all the above formulas are limited by only considering the mean value of undeformed chip thickness without considering its nonuniformity, which degrades the predicted accuracy. Therefore, it is necessary to investigate the distribution characteristics of the undeformed chip thickness according to the actual topology of the grinding wheels. Under such conditions, some models have been established to calculate the undeformed chip thickness, as listed in Table 12.1. For example, Malkin (2008) developed a model to obtain the undeformed chip thickness from grain height characteristics and the spacing between active grains. Additionally, Younis and Jiang (1984), Hecker et al. (2007), and Khare and Agarwal (2015) used the Rayleigh distribution to calculate the undeformed chip thickness, which is still in the absence of the nonuniformity prediction. According to these models, it can be interpreted that the spacing between active grains, the grain protrusion height, and the grinding process parameters are the key factors in determining the undeformed chip thickness, which is the scope of the current paper.

In the meantime, the textured single-layer cubic boron nitride (CBN) wheels have shown great potential for grinding difficult-to-cut metallic materials, e.g., titanium alloys and nickel-based superalloys. Aurich et al. (2003) have conducted numerical simulations on grinding forces and surface roughness to optimize the surface texture (also called grain distribution patterns) of single-layer CBN wheels. Compared to regular grinding wheels, the textured wheels have shown better and more controllable grinding performance

TABLE 12.1   Several Developed Models to Calculate the Undeformed Chip Thickness

Researchers	Formulas
Pahlitzsch and Helmerdig (1943)	$a_{\text{gmax}} = 2\lambda \dfrac{v_{\text{w}}}{v_{\text{s}}}\sqrt{\dfrac{a_{\text{p}}}{d_{\text{s}}}}$
Shaw and Reichenbach (1956)	$a_{\text{gmax}} = \left[\dfrac{4v_{\text{w}}}{v_{\text{s}}N_{\text{d}}c}\sqrt{a_{\text{p}}/d_{\text{s}}}\right]^{1/2}$
Werner et al. (1971)	$a_{\text{gmax}} = \dfrac{1}{A}\left(\dfrac{2}{N_{\text{st}}}\right)^{1/\alpha+1}\left(\dfrac{v_{\text{w}}}{v_{\text{s}}}\right)^{1/\alpha+1}\left(\dfrac{a_{\text{p}}}{d_{\text{s}}}\right)^{1/2(\alpha+1)}$
Agarwal et al. (2012, 2013)	$E(a_{\text{gmax}}) = \sqrt{\dfrac{a_{\text{e}}v_{\text{w}}}{Crv_{\text{s}}}\dfrac{1}{l_{\text{c}}}}$
Malkin (2008)	$a_{\text{gmax},j} = 2s_j\left(\dfrac{a_{j-1}}{d_{\text{s}}}\right)^{1/2} - \delta_j$
Younis and Jiang (1984), Hecker et al. (2007; 2003), and Khare and Agarwal (2015)	$f(a_{\text{gmax}}) = \begin{cases} \left(a_{\text{gmax}}/\sigma^2\right)e^{-a_{\text{gmax}}^2/2\sigma^2} & a_{\text{gmax}} \geq 0 \\ 0 & a_{\text{gmax}} < 0 \end{cases}$

(e.g., grinding forces, grinding temperature, workpiece integrity, and wheel wear). Recently, Denkena et al. (2015) further enhanced the grinding performance of textured wheels, which are machined with a patterning tool by using fly-cutting kinematics. In the current work, the textured single-layer CBN wheels applied are fabricated by using the brazing technique, in which the CBN grains, Cu-Sn-Ti bonding material, and the AISI 1045 steel substrate are joined at elevated temperatures. Under such conditions, firm joining could be achieved through chemical and metallurgical reactions to avoid premature grain pull-out in grinding. Furthermore, it is easy for the single-layer brazed CBN wheel to obtain high grain protrusion and a textured wheel surface. Additionally, similar to a single-layer electroplated CBN wheel, the dressing procedure is not usually used. However, it is noted that the undeformed chip thickness nonuniformity also exists in the application of this textured single-layer brazed CBN wheel since the grain size has a distribution. Therefore, a model to analyze the effect of undeformed chip thickness nonuniformity on the ground workpiece surface finishing is highly needed.

In this work, the abrasive wheel wear experiment has been conducted to investigate the evolution and influence of grain protrusion height nonuniformity. The surface topology of the textured single-layer CBN wheels has been measured using replica samples. A Johnson transformation and its inverse transformation method have been used for the wheel topology reconstruction. The reconstructed wheel surface information has been used for the undeformed chip thickness nonuniformity investigation, ground surface roughness model development, and ground surface topography reconstruction. Different wheel wear statuses have been considered.

12.2 EXPERIMENTAL MATERIALS AND PROCEDURE

12.2.1 Grinding Wheel Topology Measurement

For a textured single-layer brazed CBN abrasive wheel with designed grain patterns, the undeformed chip thickness nonuniformity of different grains is mainly determined by their actual depth of cut, which is controlled by the actual grain protrusion height and the radial run-out of the abrasive wheel. To obtain the grain protrusion height for further analysis, the wheel topology needs to be first measured and reconstructed. Recently, some methods have been developed to obtain the grinding wheel topology and the machined surface topography. For example, Yan et al. (2011) characterized the three-dimensional wheel topology directly using a white light interferometer; at the same time, Nilsson and Ohlsson (2001) used three kinds of replica material to copy different machined surfaces. These methods provide an effective way of measuring grinding wheel topology at different tool wear stages. In this work, the textured single-layer brazed CBN abrasive wheel has been installed on the grinding machine spindle, marked with numbers from 1 to 8, whose run-out is 0, 0, 0, −3, −2, −8, −8, and −8 μm, respectively (Figure 12.1). The tested textured wheel has a diameter of 400 mm and a width of 10 mm, with 80/100 US mesh grains and a patterned spacing of 1.2 mm. As shown in Figure 12.1, the replica samples made of polyaddition silicone rubber impression material have been utilized to copy the wheel topology of these eight marked zones. Then, 3D photography of a replica sample will be obtained by the 3D optical profiler (SENSOFAR S NEOX). The actual grain protrusion height h_g is the distance from the grain top to the wheel groove in a drawn cross-section. Thus, the equivalent grain protrusion height h_{eg} can be defined as the sum of the actual grain protrusion height and radial run-out.

12.2.2 Grinding Experiment Setup

A grinding experiment has been conducted to investigate the evolution of the textured single-layer CBN abrasive wheel topology. The experimental setup is shown in Figure 12.2. The experiment has been carried out on a computerized numerical control (CNC) surface

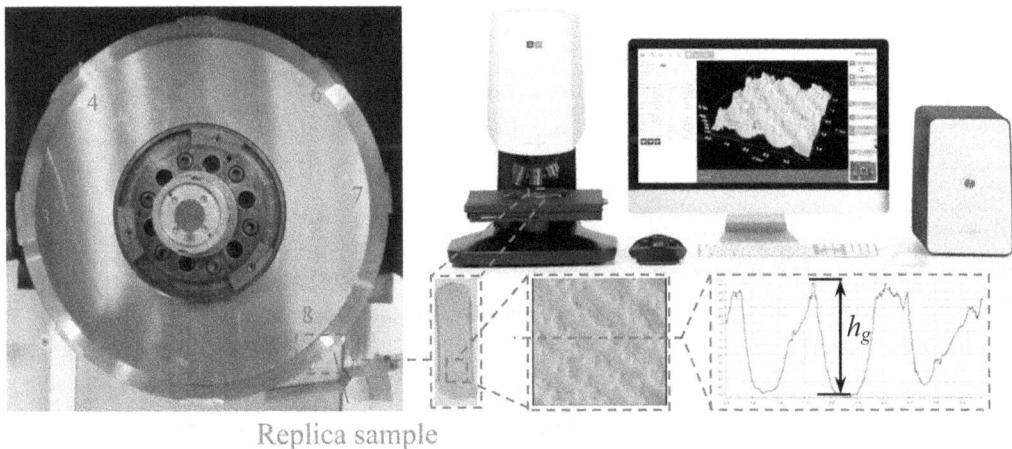

Replica sample

FIGURE 12.1   Experimental setup of wheel topology measurement.

FIGURE 12.2   Grinding experimental setup.

grinder (PROFIMAT MT 408). A fresh wheel with the same specification was utilized. The workpiece material is nickel-based superalloy Inconel 718. Two workpiece blocks have been prepared: one is for the use of conducting wear tests, in dimensions of $60 \times 80 \times 20$ mm; the other is for the measurement of ground surface roughness and grinding forces, in dimensions of $60 \times 5 \times 30$ mm.

A typical combination of grinding parameters has been applied as follows: wheel speed v_s of 30 m/s, workpiece infeed speed v_w of 3 m/min, and depth of cut a_p of 20 μm. During grinding, the collection of ground surface roughness and grinding forces are carried out in the absence of the axial infeed motion. This grinding operation is very common in the profiled grinding of jet engine blade root due to the specially designed surface shape, e.g., fir tree root. For each time a workpiece material volume of 1,920 mm^3 has been removed, the grinding wheel topology will be analyzed. The tangential and normal grinding forces have been measured in down grinding using a quartz piezo-electric-type dynamometer (Kistler 9272) attached to a multichannel charge amplifier (Kistler 5070A10100). During the grinding wheel wear process, the specific tangential and normal grinding forces are gradually increased to 3 and 10 N/mm, with a slowdown feed rate. Additionally, the ground surface roughness has been measured using a roughness tester (Mahr Perthometer M1).

12.3 WHEEL TOPOLOGY RECONSTRUCTION

12.3.1 Procedure for Wheel Topology Reconstruction

The applied replica sampling method can only measure a proportion of CBN grains on the abrasive wheel surface. To obtain the topology of the entire wheel, reconstruction would be investigated with several assumptions. In this aspect, Jiang et al. (2013b) have established a grain protrusion height reconstruction model based on the grain protrusion height normal distribution and the grain spatially random distribution. Furthermore, Nguyen and Butler (2005) reconstructed the wheel topology through Johnson transformation and wavelet transformation. In the real case, the wheel topology tends to be in the form of a non-normal distribution. Therefore, in the current study, the Johnson transformation has been applied to reconstruct the topology of the textured single-layer CBN wheels.

This section approaches the topology reconstruction process of the textured CBN abrasive wheels, which contains three steps: normality evaluation, Johnson transformation, and inverse Johnson transformation. First, the normality of the equivalent grain protrusion height values h_e has been evaluated. The measured h_{eg} data has been manipulated by using the Johnson transformation to reconstruct it into a normal distribution form $N(0,1)$, defined as below:

$$f(x) = \frac{1}{\sqrt{2\pi}} e^{\left(-\frac{x^2}{2}\right)} \tag{12.1}$$

Based on the expanded data from the function of $N(0,1)$, the equivalent grain protrusion height h_e could be extracted through an additional inverse Johnson transformation, which reconstructed the original distribution. The reconstruction process is schematically illustrated in Figure 12.3.

A total of 700 grain protrusion heights have been utilized in the current analysis. By adding the radial run-out of each marked zone, the equivalent grain protrusion height distribution is obtained. The distribution properties of measured and reconstructed data are listed comparatively in Table 12.2. It can be found that the equivalent grain protrusion height distribution has been successfully reconstructed with a mean value of 130.6 μm and a standard deviation of 13 μm. It can be inferred that the majority of the data are located around the mean value, and the left side has more weight.

12.3.2 Normality Evaluation

For the statistical analysis, the method of quantile linear fitting plot can be used to evaluate the normality degree. The equivalent grain protrusion height data $h_{eg,i}$ is sorted as

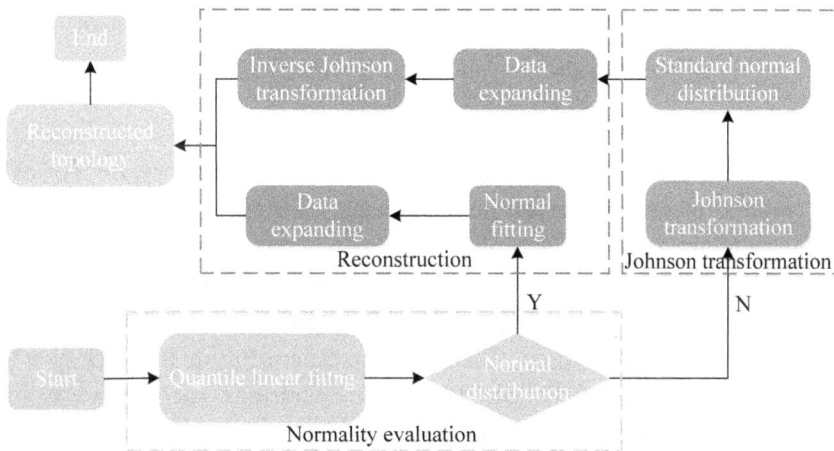

FIGURE 12.3 The process of the topology reconstruction of equivalent grain protrusion height.

TABLE 12.2   Comparison between the Measured and Reconstructed Data

Parameters (μm)	Measured	Reconstructed	Error (%)
Mean	130.64	130.68	0.03
Standard deviation	13.30	13.00	2.3
Maximum	161.62	163.35	1.1
Minimum	69.77	69.56	0.3
Range	91.85	93.79	2.1

$h_{eg,1} < h_{eg,2} < \ldots < h_{eg,n}$. The corresponding quantile x_q of a standard normal distribution is calculated from the inversed function of $N(0,1)$, shown as:

$$x_q = f^{-1}\left(\frac{i - 0.375}{n + 0.25}\right), \qquad 1 \le i \le n \tag{12.2}$$

where n is the total number of grains.

The distribution (x_q, h_{eg}) of each equivalent grain protrusion height has been plotted in Figure 12.4a. The non-normal distribution can be fitted as a linear relationship as follows:

$$h_{eg} = \sigma_{eg} x_q + \mu_{eg} \tag{12.3}$$

where μ_{eg} is the mean value of the sampling data, and σ_{eg} is the standard deviation.

12.3.3 Johnson Transformation

Principle and calculation proceeds of the Johnson transformation

Three distribution families can be transformed into the standard normal distribution. These families are represented as S_B (bounded), S_L (lognormal), and S_U (unbounded), in which the parameter constraints and variable ranges are shown in Table 12.3.

$$\text{where } \sinh x = \frac{e^x - e^{-x}}{2}, \sinh^{-1} x = \ln\left[x + \left(x^2 + 1\right)^{\frac{1}{2}} \right].$$

For a specific non-normal distribution, it is crucial to select an appropriate type of Johnson curve. The method of sample percentile has been widely applied for the Johnson curve selection. The steps are as follows:

Firstly, an appropriate quantile z of a standard normal distribution is chosen, and the distribution probability $\{p_{-3z}, p_{-z}, p_z, p_{3z}\}$ corresponding to $\{-3z, -z, z, 3z\}$ in the normal distribution table is located. Then the corresponding quantile $\{x_{-3z}, x_{-z}, x_z, x_{3z}\}$ in the sample can be proposed. Followed by some operations, such as $m = x_{3z} - x_z$, $n = x_{-z} - x_{-3z}$, $p = x_z - x_{-z}$, and the quantile ratio $QR = mn/p^2$. Finally, if $QR < 1$, the S_B family is selected; if $QR = 1$, the S_L family is selected; and if $QR > 1$, the S_U family is selected. The specific parameters of these Johnson families can be calculated as follows:

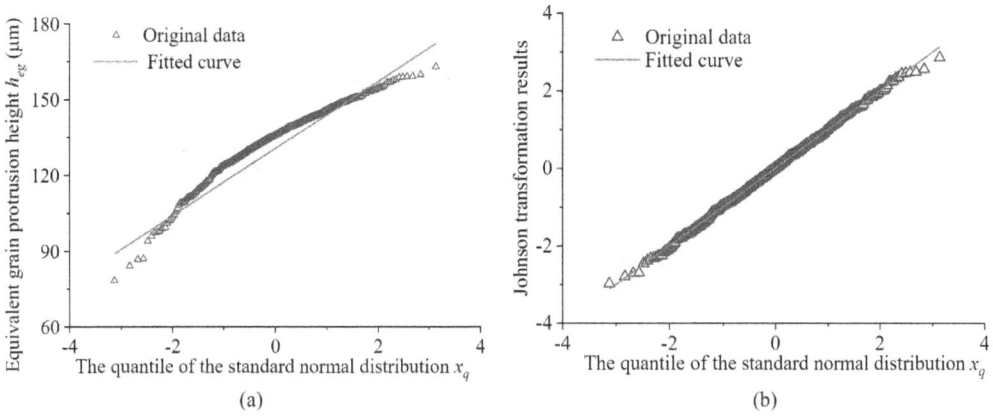

FIGURE 12.4 Quantile liner fitting plot: (a) sampling data, (b) Johnson transformation results.

TABLE 12.3 Johnson Distribution System

Johnson System	Johnson Curve	Normal Transformation	Parameter Constraints	X Constraints
S_B	$k_1 = \ln \dfrac{x-\varepsilon}{\zeta+\varepsilon-x}$	$z = \gamma + \eta \ln \dfrac{x-\varepsilon}{\zeta+\varepsilon-x}$	$\eta, \zeta > 0$ $-\infty < \gamma < \infty$ $-\infty < \varepsilon < \infty$	$\varepsilon < x < \varepsilon + \zeta$
S_L	$k_2 = \ln(x-\varepsilon)$	$z = \zeta + \eta \ln(x-\varepsilon)$	$\eta > 0$ $-\infty < \gamma < \infty$ $-\infty < \varepsilon < \infty$	$x > \varepsilon$
S_U	$k_3 = \sinh^{-1} \dfrac{x-\varepsilon}{\zeta}$	$z = \gamma + \eta \sinh^{-1} \dfrac{x-\varepsilon}{\zeta}$	$\eta, \zeta > 0$ $-\infty < \gamma < \infty$ $-\infty < \varepsilon < \infty$	$-\infty < x < \infty$

For the S_B family:

$$
\left\{
\begin{aligned}
\eta &= z \left\{ \cosh^{-1} \left[\frac{1}{2} \left[\left(1+\frac{p}{m}\right)\left(1+\frac{p}{n}\right) \right]^{\frac{1}{2}} \right] \right\}^{-1} \\[2mm]
\gamma &= \eta \sinh^{-1} \left\{ \left(\frac{p}{n}-\frac{p}{m}\right)\left[\left(1+\frac{p}{m}\right)\left(1+\frac{p}{n}\right)-4 \right]^{\frac{1}{2}} \left[2\left(\frac{p^2}{mn}-1\right) \right]^{-1} \right\} \\[2mm]
\zeta &= p \left\{ \left[\left(1+\frac{p}{n}\right)\left(1+\frac{p}{m}\right)-2 \right]^2 -4 \right\}^{\frac{1}{2}} \left(\frac{p^2}{mn}-1\right)^{-1} \\[2mm]
\varepsilon &= \frac{x_z + x_{-z}}{2} - \frac{\zeta}{2} + p\left(\frac{p}{n}-\frac{p}{m}\right)\left[2\left(\frac{p^2}{mn}-1\right) \right]^{-1}
\end{aligned}
\right.
\tag{12.4}
$$

For the S_L family:

$$
\left\{
\begin{aligned}
\eta &= \frac{2z}{\ln \dfrac{m}{p}} \\[2em]
\gamma &= \eta \ln \frac{\dfrac{m}{p}-1}{p\left(\dfrac{m}{p}\right)^{\frac{1}{2}}} \\[2em]
\varepsilon &= \frac{x_z - x_{-z}}{2} - \frac{p}{2}\left(\frac{\dfrac{m}{p}+1}{\dfrac{m}{p}-1}\right)
\end{aligned}
\right.
\tag{12.5}
$$

For the S_U family:

$$
\left\{
\begin{aligned}
\eta &= 2z\left\{\cosh^{-1}\left[\frac{1}{2}\left(\frac{m}{p}+\frac{n}{p}\right)\right]\right\}^{-1} \\[1.5em]
\gamma &= \eta \sinh^{-1}\left\{\left(\frac{n}{p}-\frac{m}{p}\right)\left[2\left(\frac{mn}{p^2}-1\right)^{\frac{1}{2}}\right]^{-1}\right\} \\[1.5em]
\zeta &= 2p\left(\frac{mn}{p}-1\right)^{\frac{1}{2}}\left[\left(\frac{m}{p}+\frac{n}{p}-2\right)\left(\frac{m}{p}+\frac{n}{p}+2\right)\right]^{\frac{1}{2}} \\[1.5em]
\varepsilon &= \frac{x_z + x_{-z}}{2} + p\left(\frac{n}{p}-\frac{m}{p}\right)\left[2\left(\frac{m}{p}+\frac{n}{p}-2\right)\right]^{-1}
\end{aligned}
\right.
\tag{12.6}
$$

The Johnson transformation in this work

The Johnson transformation Eq. (12.7) has been used to transform the data from non-normal distribution to normal distribution $N(0,1)$ as pre-processing of the topology reconstruction, which enables to extend the equivalent grain protrusion height distribution to the whole textured wheel.

$$
z = \gamma + \eta \cdot f\left(\frac{x - \varepsilon}{\zeta}\right)
\tag{12.7}
$$

where x and z represent the non-normal distribution and normal distribution data, respectively. γ and η determine the distribution shape, ε is the central offset, and ζ is a scale factor. Eq. (12.7) can be rewritten according to different wheel statuses. Take one of the Johnson transformation functions, for example, which corresponds to the measured data:

$$z = 1.487 + 2.238 \times \sinh^{-1} \frac{x - 147.502}{21.528} \tag{12.8}$$

The frequency distribution histogram and the quantile linear fitting plot are displayed in Figures 12.6b and 12.4b, respectively. It is found that the distribution of Johnson transformation results is consistent with the linear relationship in Eq. (12.3). The transformed data fits the standard normal distribution $N(0,1)$ very well, with a mean value of 0.011 and a standard deviation of 1.001, and the error is less than 5%. The data will be expanded based on the normal distribution in a region of (z_{min}, z_{max}).

According to the designed pattern of the textured single-layer CBN wheel shown in Figure 12.5, the grains could be divided into 1,047 columns, with 50 grains per column $(50 \times 1,047)$. Figure 12.6c shows that the corresponding expanded distribution fits the normal distribution very well.

12.3.4 Inverse Johnson Transformation and Wheel Topology Reconstruction

The inverse Johnson transformation is used to reconstruct the wheel topology. For example, the function corresponding to the data in Figure 12.6c is the inverse of Eq. (12.8):

$$x = 147.502 + 21.528 \times \sinh \frac{z - 1.487}{2.238} \tag{12.9}$$

Figure 12.6d shows the corresponding reconstructed data. The surface topology reconstruction of the textured single-layer CBN wheels is a combination of a special grain pattern and an equivalent grain protrusion height distribution. The reconstructed surface topology with ordered pattern grains can be represented by a $50 \times 1,047$ matrix, as expressed in Eq. (12.10). $h_{eg,(i,j)}$ is the equivalent grain protrusion height in Row i and Column j, as shown in Figure 12.5.

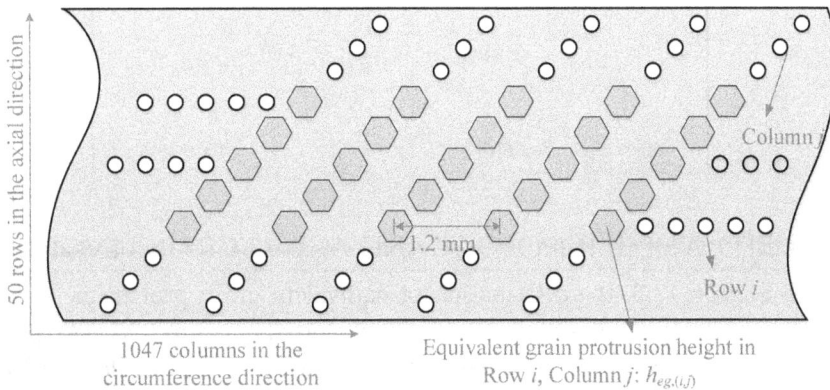

FIGURE 12.5   Schematic illustration of the textured single-layer CBN wheels.

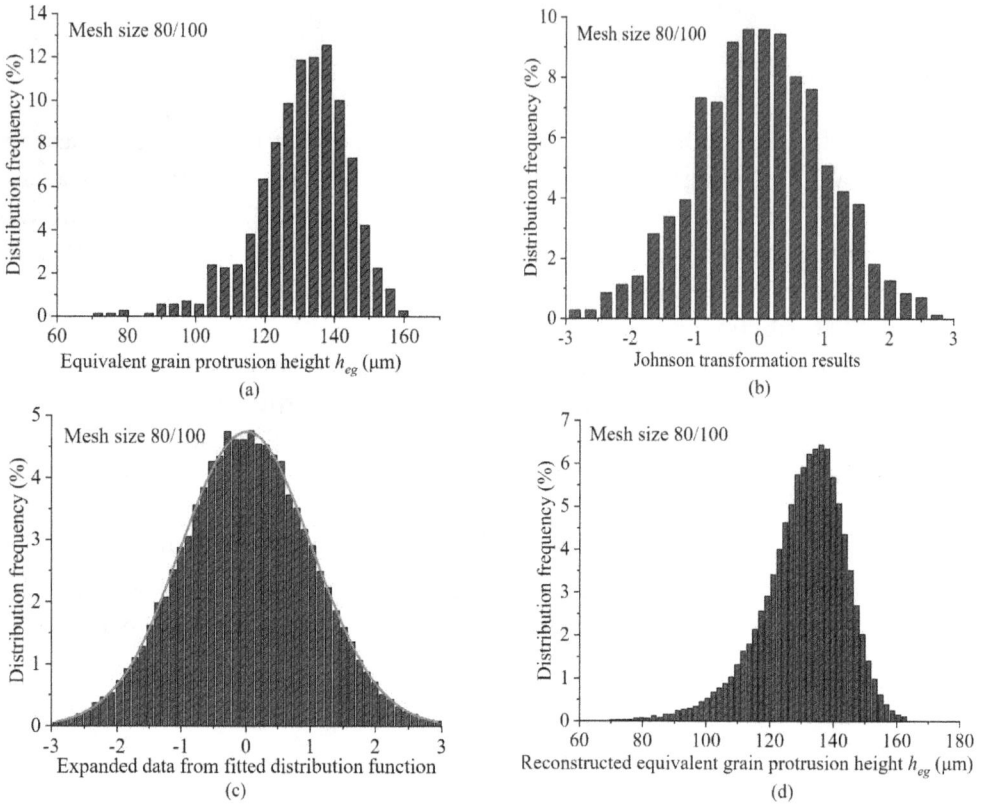

FIGURE 12.6 Frequency distribution histogram of equivalent grain protrusion height measured and reconstructed: (a) equivalent grain protrusion height measured, (b) Johnson transformation results, (c) expanded data from Johnson transformation, (d) equivalent grain protrusion height reconstructed.

$$
H = \begin{bmatrix}
h_{eg,(1,1)} & h_{eg,(1,2)} & \cdot & \cdot & \cdot & h_{eg,(1,1047)} \\
h_{eg,(2,1)} & h_{eg,(2,2)} & \cdot & \cdot & \cdot & h_{eg,(2,1047)} \\
\cdot & \cdot & \cdot & \cdot & \cdot & \cdot \\
\cdot & \cdot & \cdot & h_{eg,(i,j)} & \cdot & \cdot \\
\cdot & \cdot & \cdot & \cdot & \cdot & \cdot \\
h_{eg,(50,1)} & h_{eg,(50,2)} & \cdot & \cdot & \cdot & h_{eg,(50,1047)}
\end{bmatrix}
\tag{12.10}
$$

12.4 TEXTURED WHEEL TOPOLOGY AND ACTIVE GRAINS EVOLUTION

According to Section 12.3, the distribution of equivalent grain protrusion height h_{eg} at different states of wheel wear has been obtained and plotted in Figure 12.7. Two parameters are chosen to evaluate the grinding wheel status: the maximum equivalent grain protrusion height (h_{egmax}) and the percentage of active grains (α). The percentage of active grains α is defined as the ratio of the grains in the region ($h_{egmax}-a_p$, h_{egmax}) and the total number

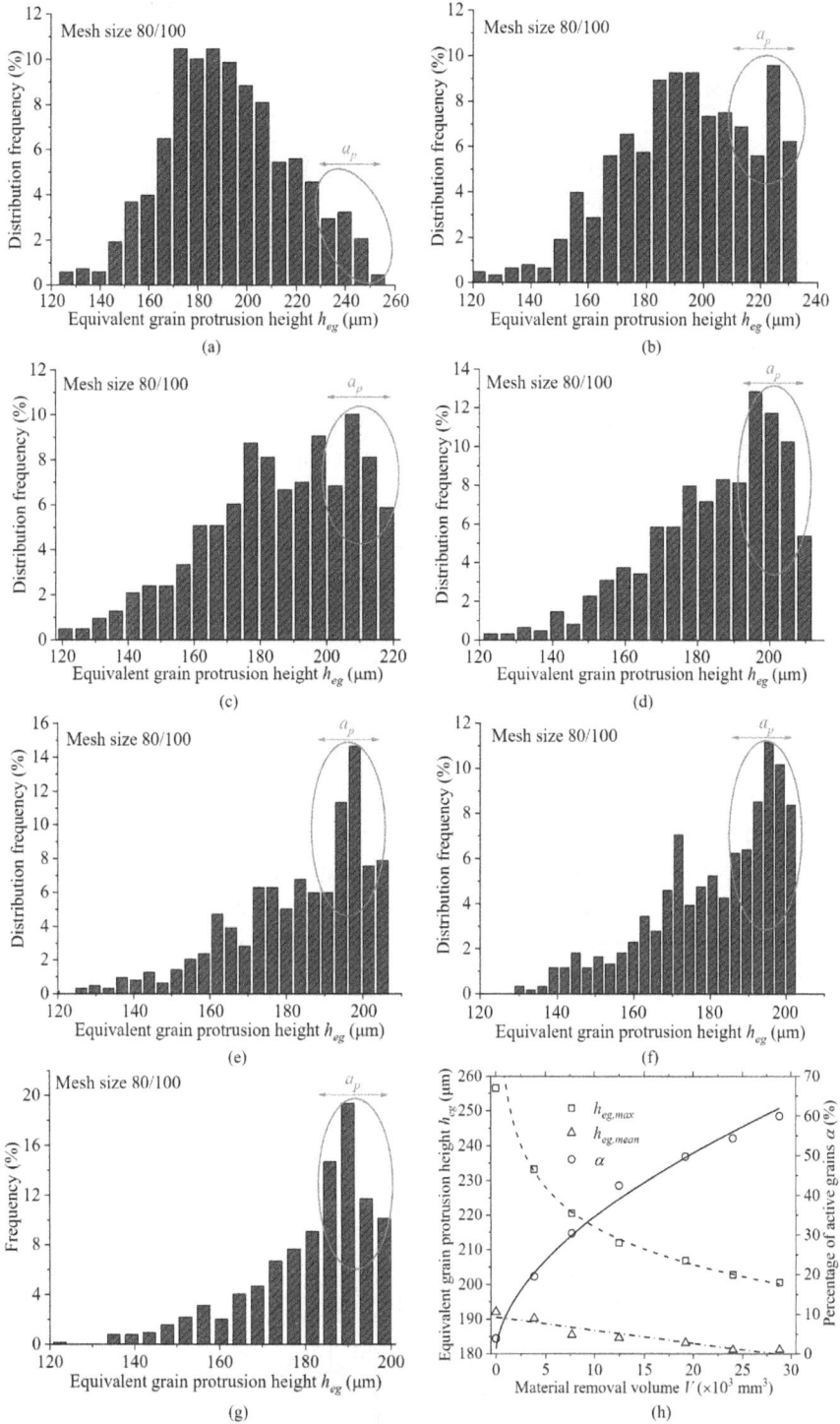

FIGURE 12.7 Equivalent grain protrusion height distribution at different *MRV*s: (a) initial status, (b) *MRV* = 3,840 mm³, (c) *MRV* = 7,680 mm³, (d) *MRV* = 12,480 mm³, (e) *MRV* = 19,200 mm³, (f) *MRV* = 24,000 mm³, (g) *MRV* = 28,800 mm³, (h) equivalent grain height versus the percentage of active grains.

of measured grains. By comparing the results at different statuses, the wear happens in the tall group ($h_{\text{egmax}}-a_{\text{p}}$, h_{egmax}), with a subtle change in the rest. With the increase in material removal volume (MRV), h_{egmax} decreases from 255 to 200 μm when α increases. The wear of the tallest grain group will facilitate the engagement of lower grains. From Figure 12.7h, it can be observed that with the increase of MRV, α increases due to the decrease of h_{egmax}. The increasing rate of α is high when h_{egmax} is large but decreases very rapidly. For instance, as h_{egmax} decreases from 255 to 205 μm, the increasing rate of α decreases by 2/3. This phenomenon indicates that only a few grains are involved in grinding at the fresh wheel state, which is the engaging stage. Entering the stable stage, the wear rate decreases due to more grains engaging. Finally, when a total of 28,800 mm³ of material is removed, the radial wear of the grinding wheel is 56 μm based on the variation of h_{egmax}. Also, during the grinding wear process, h_{eg} is becoming more uniform, and its mean value decreases linearly from 192 to 181 μm. All the experiment data will be used in the workpiece topography reconstruction in the next section.

12.5 MODEL DEVELOPMENT OF UNDEFORMED CHIP THICKNESS

The undeformed chip thickness can be calculated as follows, according to Malkin (2008):

$$a_{\text{gmax},j} = 2\lambda_j \frac{v_w}{v_s} \left(\frac{a_{j-1}}{d_s} \right)^{1/2} - \delta_j \tag{12.11}$$

$$\delta_j = a_{j-1} - a_j \tag{12.12}$$

where $a_{\text{gmax},j}$ is the undeformed chip thickness of the jth grain, λ_j is the spacing between two consecutive active grains, d_s is the diameter of the grinding wheel, δ_j represents the difference in undeformed chip thickness between any nearby active grains, and a_{j-1} and a_j are the actual depth of cut of the $(j-1)^{\text{th}}$ and j^{th} grains, which are defined as:

$$a_j = a_{\text{p}} - \left(h_{\text{egmax}} - h_{\text{eg},j} \right) \tag{12.13}$$

$$a_{j-1} = a_{\text{p}} - \left(h_{\text{egmax}} - h_{\text{eg},j-1} \right) \tag{12.14}$$

where $h_{\text{eg},j}$ and $h_{\text{eg},j-1}$ are the equivalent grain protrusion heights of the jth and $(j-1)^{\text{th}}$ grains, respectively.

For further analysis of Eq. (12.11), it has been found that, in the Malkin's model, δ_j is considered under the condition of a stationary wheel. It describes the difference between grain protrusion heights without considering the influence of kinetics. However, a_{gmax} is dependent on each grain moving at the same speed. The grinding kinematics effect on the undeformed chip thickness nonuniformity δ_j can't be ignored. Therefore, a more accurate model of undeformed chip thickness is developed with a corrected δ_j by considering grinding kinematics.

As shown in Figure 12.8, the protrusion heights of the jth and $(j-1)^{\text{th}}$ grains are different. The moving distances are l_j and l_{j-1}, respectively, which change the undeformed chip thickness of the jth grain by $\delta_{j,a}$, as follows:

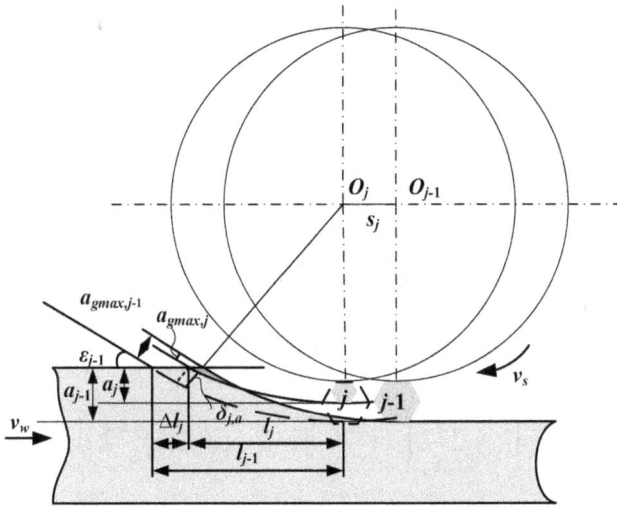

FIGURE 12.8   Schematic illustration of the undeformed chip thickness of a single grain.

$$\delta_{j,a} = \Delta l_j \sin \varepsilon_{j-1} \qquad (12.15)$$

where ε_{j-1} is the maximum cutting angle of the $(j-1)^{\text{th}}$ grain, for a small value of ε_{j-1}, $\sin\varepsilon_{j-1} \approx \tan\varepsilon_{j-1}$:

$$\delta_{j,a} = \Delta l_j \tan \varepsilon_{j-1} \qquad (12.16)$$

where

$$\Delta l_j = l_{j-1} - l_j \qquad (12.17)$$

Based on the motion of a single grain on the wheel, the dwell time of the jth grain during cutting can be calculated as:

$$t_j = \frac{1}{v_s}\sqrt{a_j d_j} \qquad (12.18)$$

The moving distance of the grain is composed of two parts: the distance of grain moving along the infeed direction and the moving distance of the workpiece, as shown in Eqs. (12.19) and (12.20).

$$l_{j-1} = \frac{v_w}{v_s}\left(a_{j-1}d_{j-1}\right)^{1/2} + \left[a_{j-1}\left(d_{j-1} - a_{j-1}\right)\right]^{1/2} \qquad (12.19)$$

$$l_j = \frac{v_w}{v_s}\left(a_j d_j\right)^{1/2} + \left[a_j\left(d_j - a_j\right)\right]^{1/2} \qquad (12.20)$$

Substitute Eqs. (12.19) and (12.20) into Eq. (12.17). Since d_j and d_{j-1} are much larger than a_j and a_{j-1}, Δl_j can be further described as:

$$\Delta l_j = d_j^{1/2} \left(a_{j-1}^{1/2} - a_j^{1/2} \right) \left(1 + \frac{v_w}{v_s} \right) \tag{12.21}$$

Substitute Eq. (12.21) into Eq. (12.16), where $\tan \varepsilon_{j-1} = 2\frac{v_w}{v_s} \left(\frac{a_{j-1}}{d_{j-1}} \right)^{1/2}$ (Malkin, 2008). Thus, $\delta_{j,a}$ can be expressed as:

$$\delta_{j,a} = 2\frac{v_w}{v_s} \left(1 + \frac{v_w}{v_s} \right) \left[a_{j-1} - \left(a_j a_{j-1} \right)^{1/2} \right] \tag{12.22}$$

Therefore, the improved model of the undeformed chip thickness is obtained as follows:

$$a_{gmax,j} = 2\lambda_j \frac{v_w}{v_s} \left(\frac{a_{j-1}}{d_s} \right)^{1/2} - 2\frac{v_w}{v_s} \left(1 + \frac{v_w}{v_s} \right) \left[a_{j-1} - \left(a_j a_{j-1} \right)^{1/2} \right], 0 \le a_{j-1}, a_j \le a_p \tag{12.23}$$

According to Eq. (12.23), the undeformed chip thickness of each grain on the textured single-layer CBN wheels can be calculated with the given wheel surface topology reconstruction information from Section 12.4.

To compare these two models, for example, when a small depth of cut $a_p = 0.05$ mm is taken, the wheel-workpiece contact length l_s is 4.5 mm. Since λ is larger than l_s and the nonuniformity of grain protrusion height (about 0.03 mm) is less than a_p, it ensures only one grain is in cutting. Thus, each grain can be considered an active grain, which indicates the percentage of active grain is 100%. It is obvious that each grain of different protrusion height will remove the workpiece material with a different undeformed chip thickness. In the absence of nonuniformity δ_j, the undeformed chip thickness a_{gmax} is around 0.12 µm. The results of the two models are shown in Figure 12.9. According to Malkin's model, a_{gmax} ranges from 0 to 12 µm, and α is only about 30%. However, based on the developed model, a_{gmax} varies from 0.09 to 0.12 µm and α reaches 100%.

Malkin (2008) developed the undeformed chip thickness model for non-uniform wheel topology with a stationary wheel assumption. The Malkin's undeformed chip thickness model shows a Rayleigh distribution with a larger value than the real case (Figure 12.9a). Hecker et al. (2007) assumed the undeformed chip thickness follows a Rayleigh probability for an aluminum oxide grinding wheel, considering dynamic cutting edge density and wheel microstructure. Using the same Rayleigh distribution as the undeformed chip thickness model, Hecker et al. have also predicted the arithmetic mean roughness, which has been found to be linear with the undeformed chip thickness. One drawback of the Rayleigh distribution is that it is a single-parameter defined probability density function. The developed model combines material hardness, wheel topology, and grinding kinematic conditions all into this single parameter, which degrades the prediction's robustness. The developed model in this article combines grinding kinematics and non-uniform active grains, and it has been found that the textured single-layer CBN wheel renders a normal distributed undeformed chip thickness (Figure 12.9b) rather than a Rayleigh distribution found for aluminium oxide grinding wheel. Therefore, the developed model shows better prediction on textured single-layer CBN wheels.

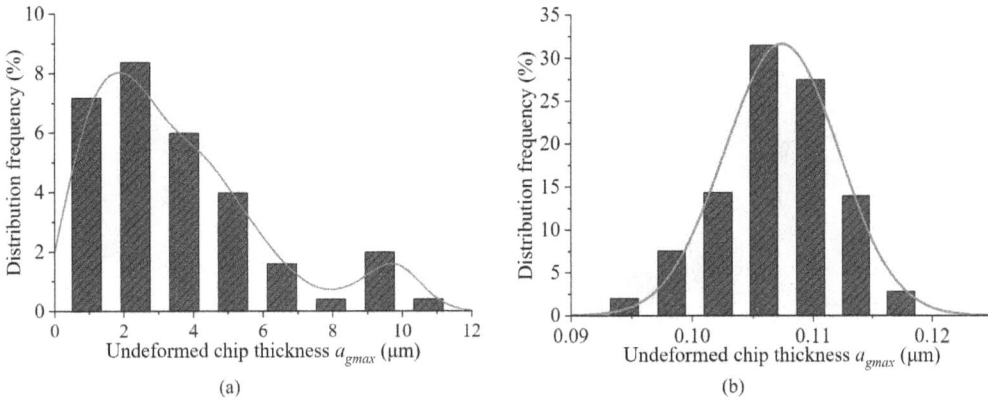

FIGURE 12.9   Results of undeformed chip thickness from the two models: (a) from Malkin's model, (b) from the improved model in this article.

12.6 RELATIONSHIP BETWEEN THE WHEEL WEAR STATUS AND THE NONUNIFORMITY OF UNDEFORMED CHIP THICKNESS

The undeformed chip thickness nonuniformity can be calculated from Eq. (12.23). A typical distribution histogram at RMV of 7680 mm³ has been shown in Figure 12.10a, which almost follows a normal distribution. The distribution of undeformed chip thickness a_{gmax} can be expressed as:

$$f\left(a_{\mathrm{gmax}}\right) = \frac{1}{\sqrt{2\pi}\sigma} e^{-\frac{\left(a_{\mathrm{gmax}} - \mu\right)^2}{2\sigma^2}} \tag{12.24}$$

where μ and σ are the mean value and standard deviation of a_{gmax}. The only difference among all the histograms is that μ and σ decrease simultaneously with the increase of MRV, as shown in Figure 12.10b. This phenomenon results from the increase of α. Since more grains are involved in grinding with wheel wear, the spacing between consecutive active grains has been reduced. According to Eq. (12.23), μ decreases with a smaller λ_j. Moreover, σ is also reduced with more uniform h_{eg}. It can also be observed that the decreasing rate of μ and σ is proportional to the increasing rate of α. For instance, when MRV accumulates to 7,680 mm³, μ and σ rapidly decrease, respectively, from 1.82 to 0.22 μm (by 88%) and from 1.32 to 0.08 μm (by 94%), with α increasing from 4% to 30% (by 6.5 times). However, when MRV is greater than 7,680 mm³, μ and σ change slightly as α grows up to 60% (by one time). Furthermore, as seen from Figure 12.10c, μ and σ are decreasing linearly with an increasing α, at a high rate first when α is less than 20% and then slowing down. This can be explained by the fact that the wheel wears fast in a fresh state, resulting in a quick drop of h_{eg}. While getting into the stable stage, the decreasing rate of h_{eg} decreases.

12.7 MODELLING OF GROUND SURFACE ROUGHNESS

The prediction of ground surface roughness is challenging in the stochastic grinding process; however, it is a crucial parameter to evaluate the grinding performance. The current work investigates the undeformed chip thickness non-uniformity-induced surface

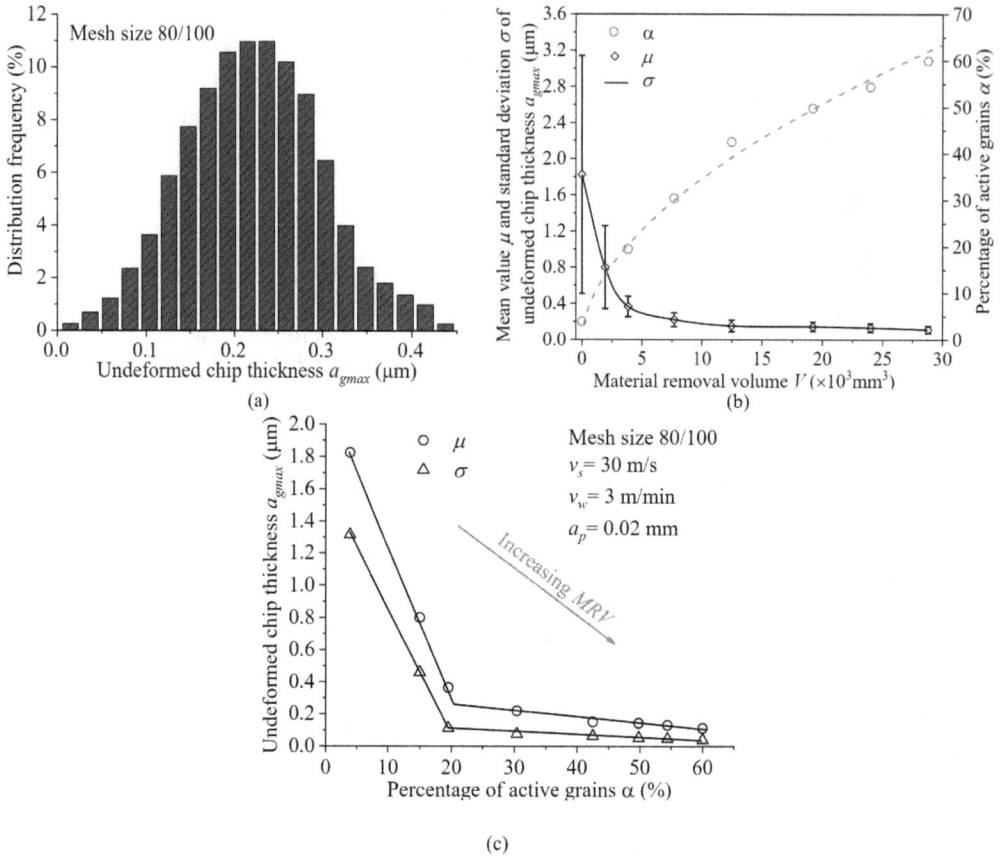

FIGURE 12.10 Distribution of undeformed chip thickness at different *MRVs*: (a) *MRV* = 7,680 mm³, (b) evolution of undeformed chip thickness, (c) relationship between the undeformed chip thickness and the percentage of active grains.

topography evolution at different wear stages. The measured ground workpiece surface roughness is shown in Figure 12.11 as triangles. The surface roughness decreases quickly from 1.55 to 0.85 μm and drops slowly below 0.8 μm. This is due to more active grains; the material removal process tends to be more uniform. Additionally, based on Figures 12.10b and 12.11, it can be found that the ground surface roughness has improved with a decrease of μ. Thus, the workpiece material removal becomes more uniform.

On the other hand, the expected surface roughness $E(Ra)$ has been modelled by Agarwal and Rao (2010):

$$E(Ra)=0.396(1-\phi)E(a_{gmax}) \tag{12.25}$$

Since the undeformed chip thickness distributes symmetrically, the mean value μ equals the expected value $E(a_{gmax})$. Meanwhile, ϕ is the overlap factor, which is the ratio of the removed workpiece material cross-section area with overlapping of consecutive grains to the cross-section area without overlapping (Agarwal and Rao 2010). It increases with a

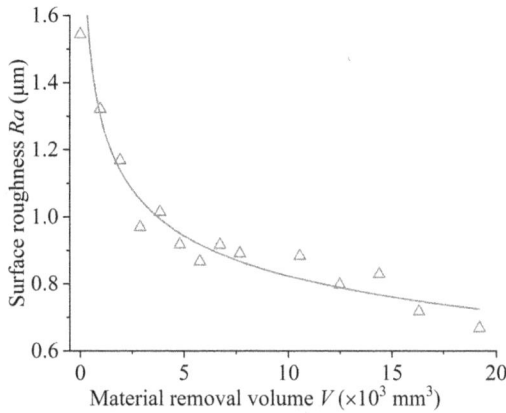

FIGURE 12.11   Variation of surface roughness versus material removal volume.

larger α when more grains are in cutting. Meanwhile, ϕ increases when the standard deviation σ of undeformed chip thickness decreases because the undeformed chip thickness is more uniform. Thus, ϕ it can be replaced by $K_w \alpha^w \sigma^r$, where K_w is a proportionality factor and w and r are the exponential parameters. Besides, the value of 0.396 in Eq. (12.25) is a process-dependent parameter that can be replaced by a proportionality factor K_r. Then, Eq. (12.25) can be further written as:

$$Ra = K_r\left(1 - K_w \alpha^w \sigma^r\right)\mu \tag{12.26}$$

Based on the experimental analysis of surface roughness, the values of K_r, K_w, r, and w in the current work are obtained. Then, Eq. (12.26) can be rewritten as follows, with $K_r = 3.6$ and $K_w = 0.02$:

$$Ra = K_r\left(1 - K_w \alpha^{3/2} \sigma^{-1/5}\right)\mu \tag{12.27}$$

According to Eq. (12.27), the modelling results of surface roughness and their errors against the experimental data are shown in Figure 12.12, which shows great agreement. For example, when MRV is increased from 3,840 to 19,200 mm³, the calculated surface roughness is decreased from 1.3 to 0.51 µm, compared with the tested data from 1.01 to 0.67 µm. It can be observed that with grinding wheel wear, both the mean undeformed chip thickness μ and the standard deviation σ will decrease. As shown in Eq. (12.27), both of these effects will reduce the surface roughness Ra, and the mean value μ has a larger effect on this.

12.8 NUMERICAL PREDICTION FOR THE WORKPIECE TOPOGRAPHY EVOLUTION

To better understand the formation of the ground surface at different wheel wear statuses, the workpiece surface topography has been predicted based on the grinding wheel topology reconstruction data generated. From an initial analysis of the ductile material for the

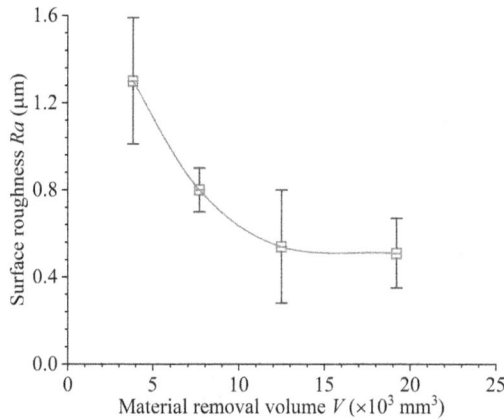

FIGURE 12.12   Modeling results of surface roughness.

specific grain shapes, it has been observed that the single grain produces a trace resembling the cross-sectional profile in combination with displaced material on the shoulders of the trace. According to the previous research on grinding nickel-based superalloys, the metallic workpiece used in this investigation is machined in a ductile mode, so the material is also removed as displacement/pileup. The schematic of the granular profile removing the workpiece material is shown in Figure 12.13a, where the grains are square and arranged in the same column. The displacement/pileup rebuild by the grains is marked with A, C, E, and G, accompanied by grooves with B, D, and F. Besides, it has been proven that the removal process of ductile materials in grinding is significantly influenced by the wheel topology, i.e., influenced by the actions of previous/trailing grains. Therefore, due to the overlapping among the grains in lots of columns, the material piled up by the leading column grains is removed by the trailing column grains on the wheel surface. As shown in Figure 12.13b, the displacement/pileup in areas A, C, E, and G is removed and gets into new grooves. Thus, the machined workpiece surface can be obtained by combining the reference points of each groove.

To better illustrate the grain-workpiece interaction, a schematic representation has been implemented, as shown in Figure 12.14a. The moving path for a single grain during grinding can be defined as (Malkin, 2008):

$$\begin{bmatrix} X \\ Y \end{bmatrix} = \begin{bmatrix} \dfrac{d_s}{2}\sin\psi - \dfrac{d_s v_w}{2}\psi \\ \dfrac{d_s}{2}(1-\cos\psi) \end{bmatrix}$$

(12.28)

where X and Y are the coordinates of the grain and ψ is the wheel rotation angle.

According to the above-mentioned section, the equivalent grain protrusion height h_{eg} is nonuniform, which makes the moving path of each grain also nonuniform. If the grain in the Row i and Column j is numbered as $g_{i,j}$, the coordination of an arbitrary grain on the textured single-layer CBN wheel can be expressed as:

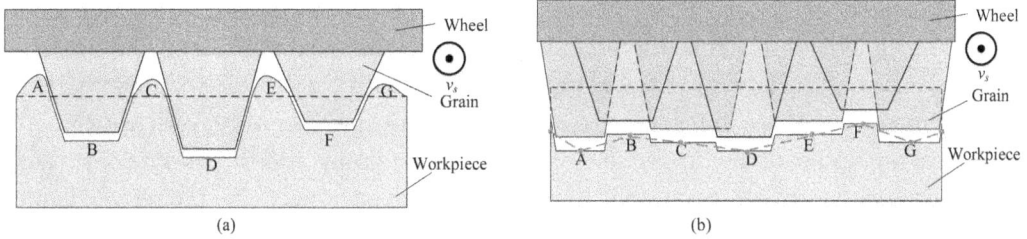

FIGURE 12.13 Schematics of the workpiece material removal in grinding: (a) with grains in one column, (b) with grains in overlapping columns.

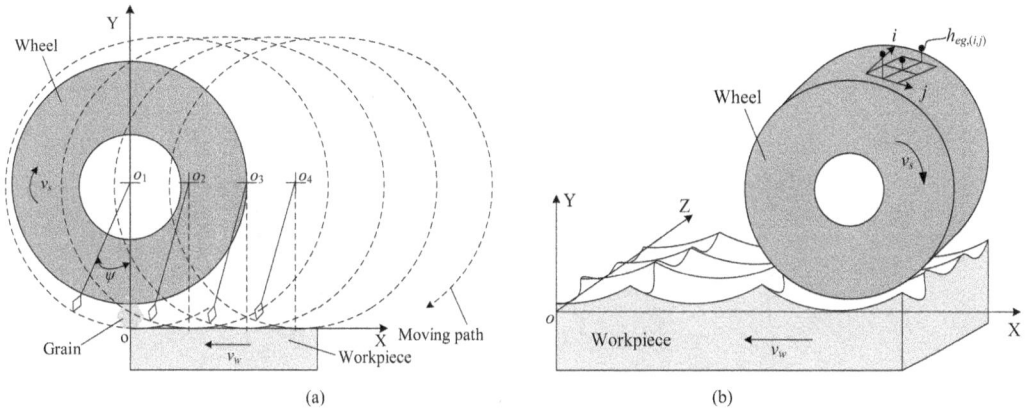

FIGURE 12.14 Schematics of grain-workpiece interaction: (a) 2D, (b) 3D.

$$
\begin{bmatrix} X_{i,j} \\ Y_{i,j} \\ Z_{i,j} \end{bmatrix} = \begin{bmatrix} \left(r_{\mathrm{b}} + h_{\mathrm{eg},(i,j)}\right)\left(\sin\psi - \dfrac{v_{\mathrm{w}}}{v_{\mathrm{s}}}\psi\right) + \lambda j \dfrac{v_{\mathrm{w}}}{v_{\mathrm{s}}} \\ \left(r_{\mathrm{b}} + h_{\mathrm{eg},(i,j)}\right)(1 - \cos\psi) - \left(h_{\mathrm{eg},(i,j)} - h_{\mathrm{egmax}}\right) \\ j \cdot d_{\mathrm{gave}} \end{bmatrix} \qquad (12.29)
$$

where $\lambda = 1.2\,\mathrm{mm}$ is the patterned grain spacing, $r_{\mathrm{b}} = 200\,\mathrm{mm}$ is the radius of the grinding wheel, $h_{\mathrm{eg},(i,j)}$ is the equivalent grain protrusion height of the grain $g_{i,j}$, and $d_{\mathrm{gave}} = 160\,\mu\mathrm{m}$ is the average dimension of the CBN grains. The coordinate reference point is set to the highest active grain (as shown in Figure 12.14b). Thus, according to the reconstructed wheel topology model, the moving path of each active grain can be calculated with an iteration program. The path of each grain can be mapped into the spatial coordinates of the workpiece surface by numerically solving Eq. (12.29) for the wheel rotation angle ψ when the grain is at a given Z-coordinate. The assumption here is that even multiple grains can be cut at the same X-coordinate on the workpiece surface; however, only the minimum Y-coordinate can contribute to the finished workpiece surface.

Several predicted workpiece surfaces at different wheel statuses are drawn in Figure 12.15. Obviously, from the predicted surfaces, the maximum depth of scratches left on the ground surface decreases from 12 to 1.2 μm with the decrease of the maximum equivalent grain protrusion height, h_{egmax}. Meanwhile, the number of peaks also increased, which are the results of the interaction between moving grains and the workpiece surface. By comparing the surface topography in region A, it can be noticed that the scratch length between two neighbouring peaks is shortened along the workpiece infeed pass (direction X) with the increase of active grains.

Moreover, the mean value μ and standard deviation σ of the undeformed chip thickness a_{gmax} are also diminished during the evolution of wheel surface topography. The workpiece surface material is removed uniformly in much smaller chips, which generates shallow scratches. As a result, the ground surface roughness has improved. Figure 12.16 shows the

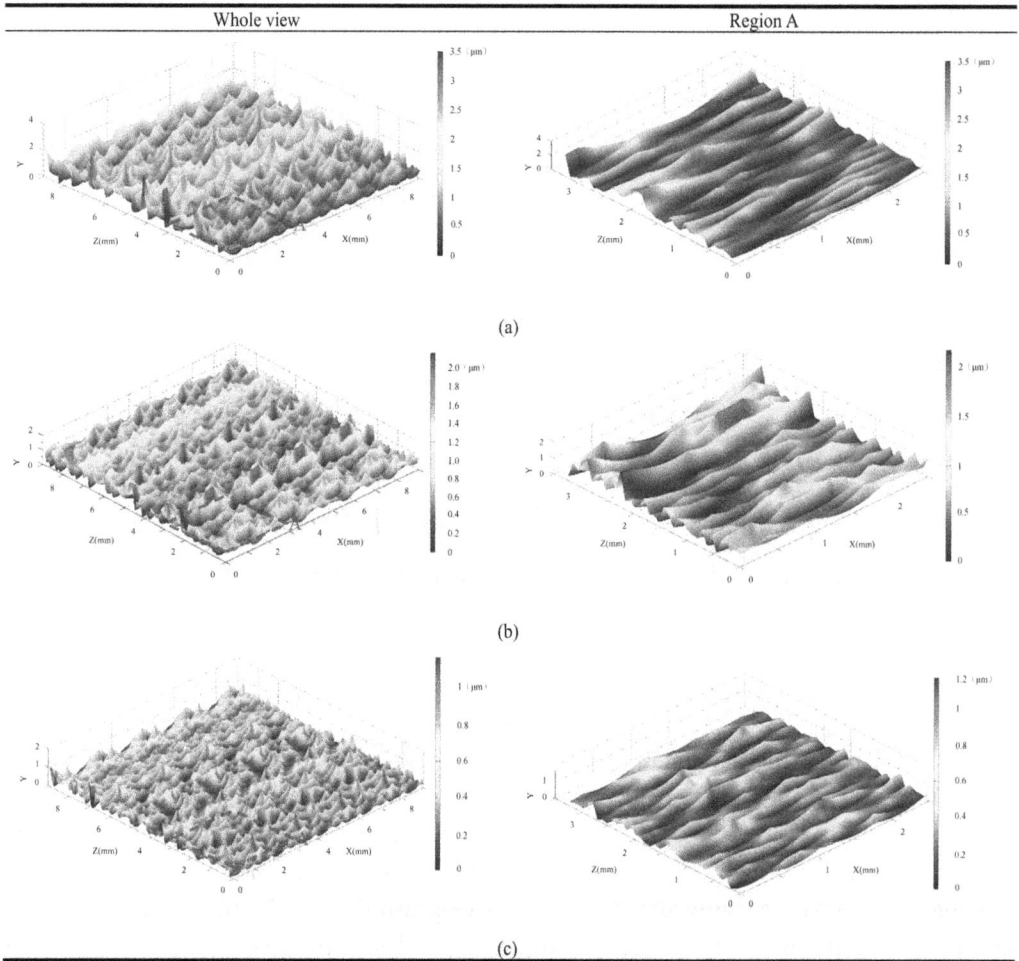

FIGURE 12.15 Simulated ground surface results: (a) $MRV = 7,680\,\text{mm}^3$, (b) $MRV = 19,200\,\text{mm}^3$, (c) $MRV = 28,800\,\text{mm}^3$.

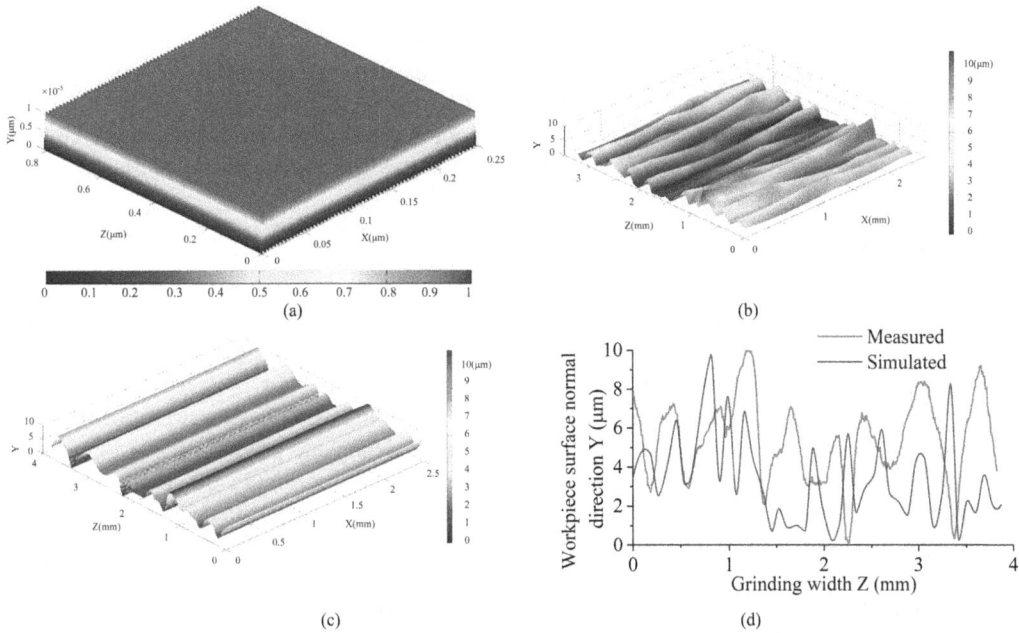

FIGURE 12.16 Comparison of simulated and measured workpiece surfaces at: (a) $MRV=0$: Simulated (nonuniformity absent), (b) Simulated (nonuniformity considered), (c) Measured, (d) 2D cross-section profile at X=0.5 mm.

predicted and measured workpiece surface at the fresh wheel state. When the nonuniformity of wheel surface topography or undeformed chip thickness is absent (namely, an ideal wheel), the workpiece surface is generated by regular short waves in a much smaller dimension, i.e., $10^{-5}\mu m$, due to the uniform undeformed chip thickness. In addition, no scratches along direction X have been observed. However, as shown in Figure 12.16b and c, it is obvious that both the predicted workpiece surface with consideration of nonuniformity and the measured surface are formed by a few narrow scratches along the workpiece infeed direction X, which reflect the interaction results of grain moving paths and the workpiece. This is due to the fact that the workpiece material is removed at a nonuniform, undeformed chip thickness. It can be seen from Figure 12.16d that the predicted and measured ground profiles have discrepancies, but the valley-peak values are matched well, i.e., 0–10 μm. The grain height distribution is reproduced in this study with a randomized spatial distribution; the ground profile discrepancies are results of the grain height randomized spatial distribution, which means the grain in an exact location may have a different size than the real case. However, Figure 12.15d shows the modelling results are accurate and can help to understand the material removal mechanism during grinding. The developed model can help better design the textured grinding wheels as well as the grinding process with a predictable workpiece surface topography.

REFERENCES

Agarwal, S., and P. V. Rao. 2010. Modeling and prediction of surface roughness in ceramic grinding. *International Journal of Machine Tools and Manufacture*, 50(12): 1065–1076.

Agarwal, S., and P. V. Rao. 2013. Predictive modeling of force and power based on a new analytical undeformed chip thickness model in ceramic grinding. *International Journal of Machine Tools and Manufacture*, 65: 68–78.

Aurich, J. C., O. Braun, and G. Warnecke. 2003. Development of a superabrasive grinding wheel with defined grain structure using kinematic simulation. *CIRP Annals-Manufacturing Technology*, 52: 275–280.

Dai, C., Z. Yin, W. Ding, et al. 2019. Grinding force and energy modeling of textured monolayer CBN wheels considering undeformed chip thickness nonuniformity. *International Journal of Mechanical Sciences*, 157–158: 221–230.

Ding, W. F., C. W. Dai, T. Y. Yu, et al. 2017. Grinding performance of textured monolayer CBN wheels: Undeformed chip thickness nonuniformity modeling and ground surface topography prediction. *International Journal of Machine Tools and Manufacture*, 122: 66–80.

Denkena, B., T. Grove, T. Gottsching, et al. 2015. Enhanced grinding performance by means of patterned grinding wheels. *International Journal of Advanced Manufacturing Technology*, 77(9): 1935–1941.

Denkena, B., J. Köhler, and M. Van der Meer. 2013. A roughness model for the machining of biomedical ceramics by toric grinding pins. *CIRP Journal of Manufacturing Science and Technology*, 6: 22–33.

Hecker, R. L., S. Y. Liang, X. J. Wu, et al. 2007. Grinding force and power modeling based on chip thickness analysis. *International Journal of Advanced Manufacturing Technology*, 33: 449–459.

Jiang, J. L., P. Q. Ge, W. B. Bi, et al. 2013a. 2D/3D ground surface topography modeling considering dressing and wear effects in grinding process. *International Journal of Machine Tools and Manufacture*, 74: 29–40.

Jiang, J. L., P. Q. Ge, and J. Hong. 2013b. Study on micro-interacting mechanism modeling in grinding process and ground surface roughness prediction. *International Journal of Advanced Manufacturing Technology*, 67: 1035–1052.

Khare, S. K., and S. Agarwal. 2015. Predictive modeling of surface roughness in grinding. *Procedia CIRP*, 31: 375–380.

Koshy, P., A. Iwasald, and M. A. Elbest. 2003. Surface generation with engineered diamond grinding wheels: Insights from simulation. *CIRP Annals-Manufacturing Technology*, 52: 271–274.

Liu, Y., A. Warkentin, R. Bauer, et al. 2013. Investigation of different grain shapes and dressing to predict surface roughness in grinding using kinematic simulations. *Precision Engineering*, 37: 758–764.

Malkin, S. 2008. *Grinding technology- theory and applications of machining with abrasives_2th Edition*. Industrial Press, New York.

Nguyen, T. A., and D. L. Butler. 2005. Simulation of precision grinding process, part I: generation of the grinding wheel surface. *International Journal of Machine Tools and Manufacture*, 45: 1321–1328.

Nilsson, L., and R. Ohlsson. 2001. Accuracy of replica materials when measuring engineering surfaces. *International Journal of Machine Tools and Manufacture*, 41: 2139–2145.

Pahlitzsch, G., and H. Helmerdig. 1943. Determination and significance of chip thickness in grinding. *Workshop Technology*, 12: 397–401.

Shaw, M. C., and G. S. Reichenbach. 1956. The role of chip thickness in grinding. *Transactions of the ASME*, 18: 847–850.

Werner, G. 1971. *Kinematic and mechanic during grinding processes*. TH Aachen.

Yan, L., Y. M. Rong, F. Jiang, et al. 2011. Three-dimension surface characterization of grinding wheel using white light interferometer. *International Journal of Advanced Manufacturing Technology*, 55: 133–141.

Younis, M. A., and H. Jiang. 1984. Probabilistic analysis of the surface grinding process. *Transactions of the CSME*, 8(4): 208–213.

Zhang, Y., C. Fang, G. Huang, et al. 2018. Modeling and simulation of the distribution of undeformed chip thicknesses in surface grinding. *International Journal of Machine Tools and Manufacture*, 127: 14–27.

Zhang, B., S. Lu, M. Rabiey, et al. 2023. Grinding of composite materials. *CIRP Annals*, 72(2): 645–671. https://doi.org/10.1016/j.cirp.2023.05.001.

PART III

Grinding Performance and Mechanism

CHAPTER 13

Grinding Behaviour and Surface Integrity of Titanium Alloy

13.1 INTRODUCTION

Titanium alloys are widely employed in such fields as aerospace, automobiles, and the defence industry because of their excellent mechanical strength and resistance to surface degradation. However, the favourable physical and chemical properties that make these alloys suitable for many applications also result in the difficulty with which they are machined, especially the critical surface damage of ground specimens (Xiao et al. 2023; Ronoh et al. 2022; Liu et al. 2023). cubic boron nitride (CBN) abrasive wheels have been one of the most popular tools to machine titanium alloys. Being regarded as a substitute for electroplated wheels, the single-layer brazed CBN abrasive wheels provide higher crystal exposure, bond uniformity, and better grit retention. Moreover, the abrasive grains are hard to pull out, and the inter-grit chip space is larger and can thereby contain more coolant and take more heat away. At the same time, the combination of grain material, brazing material, and a core material is characterized by high thermal conductivity to ensure heat dissipation from the grinding tool tip of the grains, which would benefit the finished surface integrity, especially in creep-feed grinding for its long contact zone.

At the same time, titanium alloys tend to be complex in profile, whose characteristics such as dimension deviation, roughness, type and magnitude of residual stresses, hardness alterations, and plastic deformation affect loading capacity, wear behaviour, and fatigue resistance. The present investigation was undertaken on Ti-6Al-4V alloy to evaluate the ground surface integrity of a complex profile while using a monolayer brazed CBN grinding wheel, which will expand the application of brazed CBN grinding wheels, and a vitrified CBN wheel was used for comparison.

DOI: 10.1201/9781032678047-16

13.2 EXPERIMENTAL DETAILS

The blade tenon of an aeroengine would be machined in this experiment. The profile was shown in Figure 13.1. Creep-feed profile grinding experiments were carried out on a MKL7150×16/2 CNC Profile Grinder. Two types of CBN-shaped grinding wheels, single-layer brazed and vitrified, were used. Further experimental conditions are listed in Table 13.1.

The single-layer brazed CBN abrasive wheel was developed in our laboratory. The grain distribution is shown in Figure 13.2. It needs to be mentioned that, because of the difficulty and high cost of dressing, the profile of the vitrified CBN-shaped wheel was not identical to the designed profile.

In the experiment, the dimension accuracy was assessed by a measurement taken on six points using a coordinate measuring machine and a KH-7700 3D viewer. Surface roughness measurements were carried out with a surface roughness tester on six points with a cut-off length of 1.75 mm, and measurement traces were perpendicular to the grinding direction. In doing so, each of the six points was measured three times before an average was obtained. The metallographic sample was etched for about 2–5 s with a mixture of HF and HNO_3, and the volume ratio was 5:95. The microstructure was analysed with optic microscopy. Microhardness was measured by a HXS-1000A microhardness tester. The load adopted was 100 g with a holding time of 15 s. Residual stress measurements were made on the surface after grinding using X-rays by the $sin^2\psi$ technique for biaxial stresses, where Ψ is the angle of incidence. $Ti(\alpha+\beta)$ {4 1 2} reflection and CuKα radiation were chosen. The penetration of CuKα was 5 μm, giving a peak at $2\theta \approx 140.86°$, where θ is the angle of diffraction. Reflections were recorded in a 2θ measuring range of 130°–170°.

13.3 DIMENSIONAL ACCURACY OF GROUND SPECIMENS

The photographs of ground specimens are shown in Figure 13.3, in which the profiles of both specimens are smooth and clear. The specimen 1# was ground with brazed CBN wheels, and the specimen 2# was ground with vitrified CBN wheels. The dimensions of

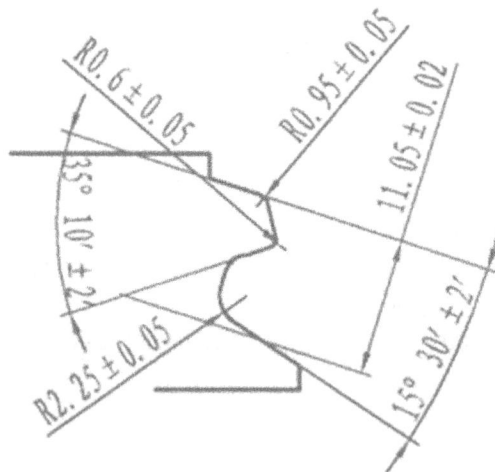

FIGURE 13.1   Designed profile of blade tenon.

TABLE 13.1   Grinding Experimental Conditions

Item	Description
Abrasive wheel	Single-layer brazed CBN wheel
	Vitrified CBN wheel (150% concentration)
Grain size	80/100 US mesh size
Wheel diameter d_s (mm)	175
Cutting velocity v_s (m·s^{-1})	18.5
Table speed v_w (mm·min^{-1})	100
Depth of cut a_p (mm)	1.0
Coolant	5% solution of an oil-in-water emulsion
	90 L/min
	0.8 MPa

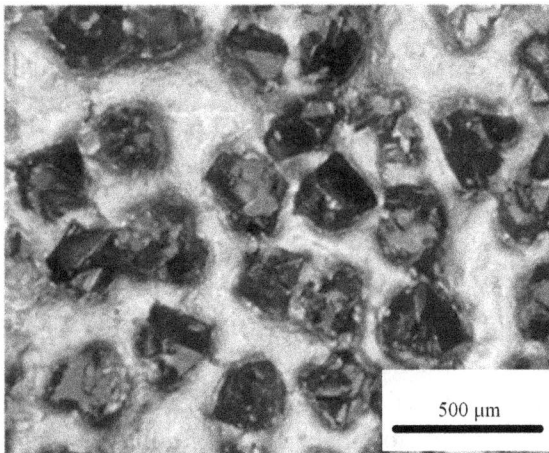

FIGURE 13.2   Grain distribution of the brazed CBN wheel.

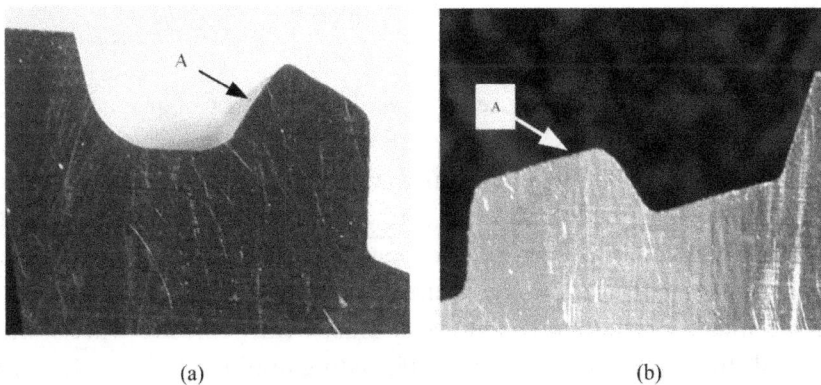

(a)   (b)

FIGURE 13.3   Photograph of ground specimen: (a) 1# specimen, (b) 2# specimen.

TABLE 13.2   Main Dimensions of Specimens Ground by Brazed CBN Wheels

No. of Workpiece		No. of Specimen			
		1	**3**	**5**	**10**
Distance/mm	Measured value	11.07	11.06	11.05	11.05
	Designed value	11.05±0.02			
	Deviation	+0.02−0.00			
Corner radius/mm	Measured value	2.23	2.24	2.26	2.26
	Designed value	2.25±0.05			
	Deviation	+0.01−0.02			
Angle	Measured value	35°8′	35°8′	35°10′	35°10′
	Designed value	35°10′±2′			
	Deviation	−2′ −0			

specimens ground with single-layer brazed CBN wheels were measured, and the results are listed in Table 13.2, which indicates that the accuracy of the profile completely meets the requirements of the design. Other dimensions were also measured, and all the deviations were within tolerance.

Profile wear is manifested as changes in the shape of the profile formed on the grinding wheel as a result of wear during the grinding process. In creep-feed profile grinding, profile wear as well as radial wear affect the shape and accuracy of ground profiles. Thus, the profile wear of the wheel can be reflected in the changes in the dimensions of the ground specimen. As can be seen from Table 13.2, the main dimensions of ground blade tenons remain nearly constant since the fifth one, i.e., the brazed CBN abrasive wheel gives less profile wear and hence more excellent form retention. Therefore, the single-layer brazed CBN abrasive wheels are suitable for applications with high dimensional accuracy in the creep-feed profile grinding of titanium alloy.

The dimensions of the specimens ground by vitrified CBN wheels were not measured for the reason that the profile of the vitrified ones was not identical to that of the brazed CBN wheels.

13.4 SURFACE INTEGRITY OF GROUND SPECIMENS

13.4.1 Surface Roughness of the Ground Specimen

Surface roughness is a variable often used to describe the quality of ground surfaces as well as to evaluate the competitiveness of the overall grinding system, which is represented by the arithmetic mean value R_a, the root mean average square R_q, and the maximum roughness height R_t. Generally, the longitudinal surface roughness has a lower value than the traverse surface roughness; the latter is more frequently used in the industry. Therefore, the R_a of all specimens after grinding was measured, and the results are shown in Figure 13.4.

For the specimens ground with single-layer brazed CBN abrasive wheels, the surface roughness tends to decrease with continuous grinding until the fifth specimen, at which point the surface roughness remains nearly constant at about 0.75 μm. On the other hand, the surface roughness of the specimens ground with vitrified CBN wheels stabilizes at

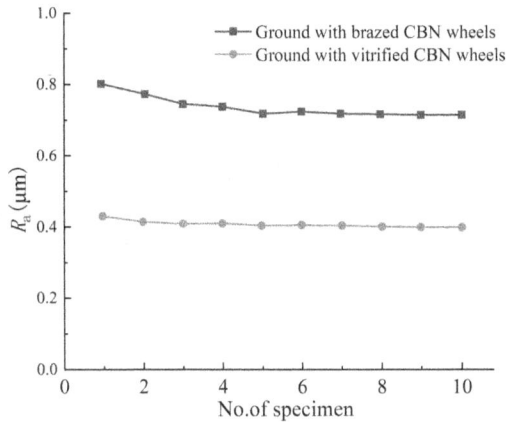

FIGURE 13.4   Surface roughness of specimens after grinding.

about 0.40 µm since the second one, and the R_a of the tenth specimen only decreases slightly in comparison with the second one. Therefore, the wear resistance of CBN wheels is excellent in the creep-feed grinding of titanium alloy.

The distance of the abrasive grain tips from the newly brazed CBN wheel substrate surface is unequal without truing and dressing, though some new technologies were used to control the grain protrusion in this work. In the initial stage of grinding, these high grains prevented the shorter ones from participating. This results in fewer overlapping cuts of grain, leading to transverse surface roughness that is sometimes substantially higher than the acceptable value. The absence of the cross feed during profile grinding worsens the situation. So the surface roughness of specimens ground with a newly brazed CBN wheel, about R_a 0.80 µm, is quite higher than that ground with the dressed vitrified CBN wheel, about R_a 0.43 µm. Since most grains of the brazed wheel even have protrusion height with continuous grinding, almost all grains gradually become active and participated in grinding; hence, the surface roughness R_a decreases little by little until it becomes constant at about 0.75 µm.

The values of R_a obtained are below 0.80 µm for all specimens, which meet the requirement of blade tenon used in aero-engines and are appropriate for surfaces having to slide over each other, such as high-performance gears, sliding plates, guides, etc.

13.4.2 Morphology and Microstructure of the Ground Specimen

At surface A, the morphology of the ground surface and the microstructure of the cross-section, which is perpendicular to the grinding direction, were studied using a KH-7700 3D viewer. In Figure 13.5, the typical morphology of ground surface A is shown. Comparing the morphologies of the ground with two types of CBN wheels, surface micro-defections are found on the surface of the 2# specimen (see Figure 13.5b), which do not seem to occur on the surface of the 1# specimen (see Figure 13.5a). As is well known, a single grain plays three roles in grinding: rubbing, ploughing, and cutting. With rubbing, the elastic/plastic deformation occurs without noticeable material removal. In ploughing, the material

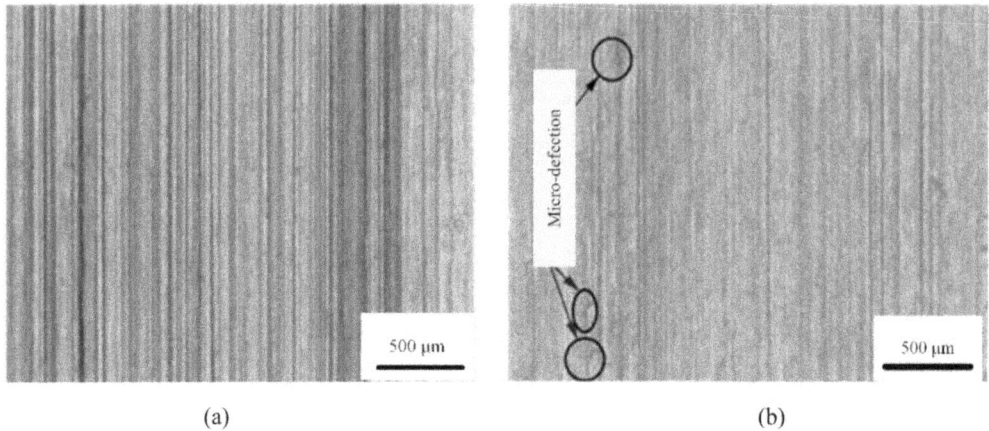

FIGURE 13.5   Morphology of ground surface: (a) 1# specimen, (b) 2# specimen.

is plastically extruded in both the grinding direction and the direction perpendicular to it. The workpiece material is mainly removed by shearing via the cutting action, which depends on the sharpness of the abrasive cutting edges. The morphology generation of a ground surface is influenced by several factors, including the characteristics of the used wheel, the properties of the workpiece material, and the grinding conditions. In this work, the same material was ground on the same machine tools with nominally identical grinding conditions; thus, the causes of the above morphology differences can be reasonably attributed to the wheel performance during the material removal process.

The above result also attributes to the effective coolant flow of the brazed wheel. It is well known that the coolant plays three major roles in grinding: heat removal, lubrication, and chip cleaning (prevention of the grinding wheel surface from chip loading). While creep-feed grinding with vitrified wheels, the chips were hardly completely cleaned, so some chips peeled off from the wheel surface might reenter the grinding zone. Under high pressure, these chips could plough the material and result in defection on the finished surface. Due to the larger inter-grit chip space, while grinding with brazed wheels, more coolant enters the grinding zone, which leads to more effective cooling and better lubrication, promotes chip cleaning, and thus prevents the re-entrance of chips into the grinding zone. Consequently, the sharpness of the active grains is maintained. Ultimately, the brazed CBN wheel is found to allow a higher depth of cut and minimize the frequency of wheel dressing and truing.

The microstructure of the cross-section perpendicular to the grinding direction is illustrated in Figure 13.6, which shows that there is no microcrack in the ground surface/subsurface layer. In contrast to the substrate, the grains of the ground surface layer is slightly finer, which results mainly from the sliding, ploughing, and cutting during the grinding process. The distortion of the crystal lattice tends to strengthen the resistance of transmutation and thus results in the work-hardening of the surface layer. If the degree of work-hardening is excessive, it would undoubtedly lead to the degradation of the performance of substrate materials.

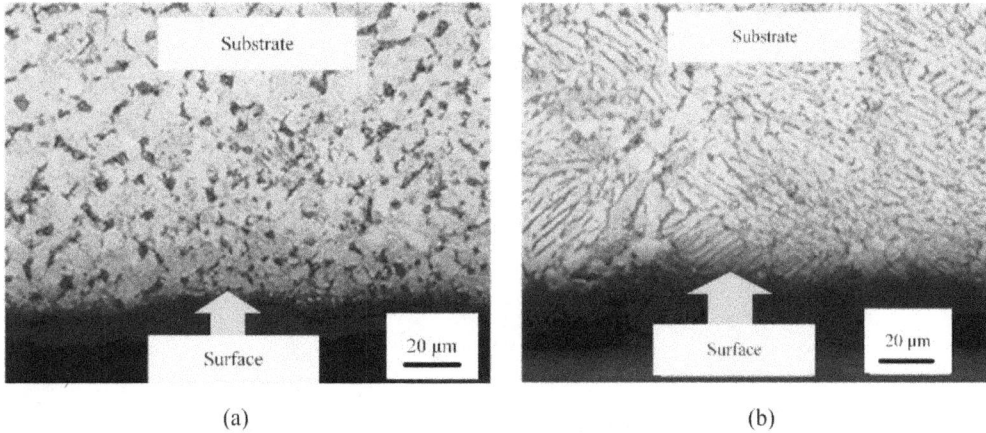

FIGURE 13.6 Microstructure of cross section of different ground specimens: (a) 1# specimen, (b) 2# specimen.

13.4.3 Microhardness of Ground Specimen

In order to ascertain the grinding affected zone, the microhardness on the section perpendicular to the surface layer and grinding direction was measured, as shown in Figure 13.7a, in which the angle is about 10° between the line of measure points and the ground surface A, and the perpendicular distance of each two points is 5 μm. The results are shown in Figure 13.7b and c, where HK is the Knoop hardness. In general, depending on the temperature of the grinding process, annealing of the workpiece may occur during grinding, causing softening close to the finished surface. The values of microhardness are both maximum on the very surface and tend to decrease in the sub-surface layer, which indicates that softening does not occur, i.e., the temperature of the grinding zone is lower than the temperature of phase transformation in the grinding process of titanium alloy with CBN wheels. The microhardness of the ground surface of 1# specimen (about HK 2,800 MPa) rises by 30.2% in comparison with that of the substrate (about HK 2,150 MPa), while that of 2# specimen (about HK 2,900 MPa) rises by 36.1%, higher than the specimen ground by brazed CBN wheels. In both cases, the depth of work-hardening is less than 40 μm.

As the single-layer brazed CBN abrasive wheel provides higher grain exposure than vitrified ones, its inter-grit chip space is larger, hence there is more coolant to remove more heat from the contact zone. Meanwhile, the higher thermal conductivity of the brazing material compared with that of the vitrified bond is characterized by ensuring heat dissipation from the grinding tool tip of the grains. Then the surface temperature of the ground with braze CBN wheels is lower than that of the ground with vitrified wheels, which results in a slighter thermal effect and less thermal damage.

13.4.4 Surface Residual Stress of Ground Specimen

Figure 13.8 shows the mean value of measured surface residual stresses along the grinding direction. It can be seen that a high compressive surface stress (about −245 MPa) is

(a)

(b)

(c)

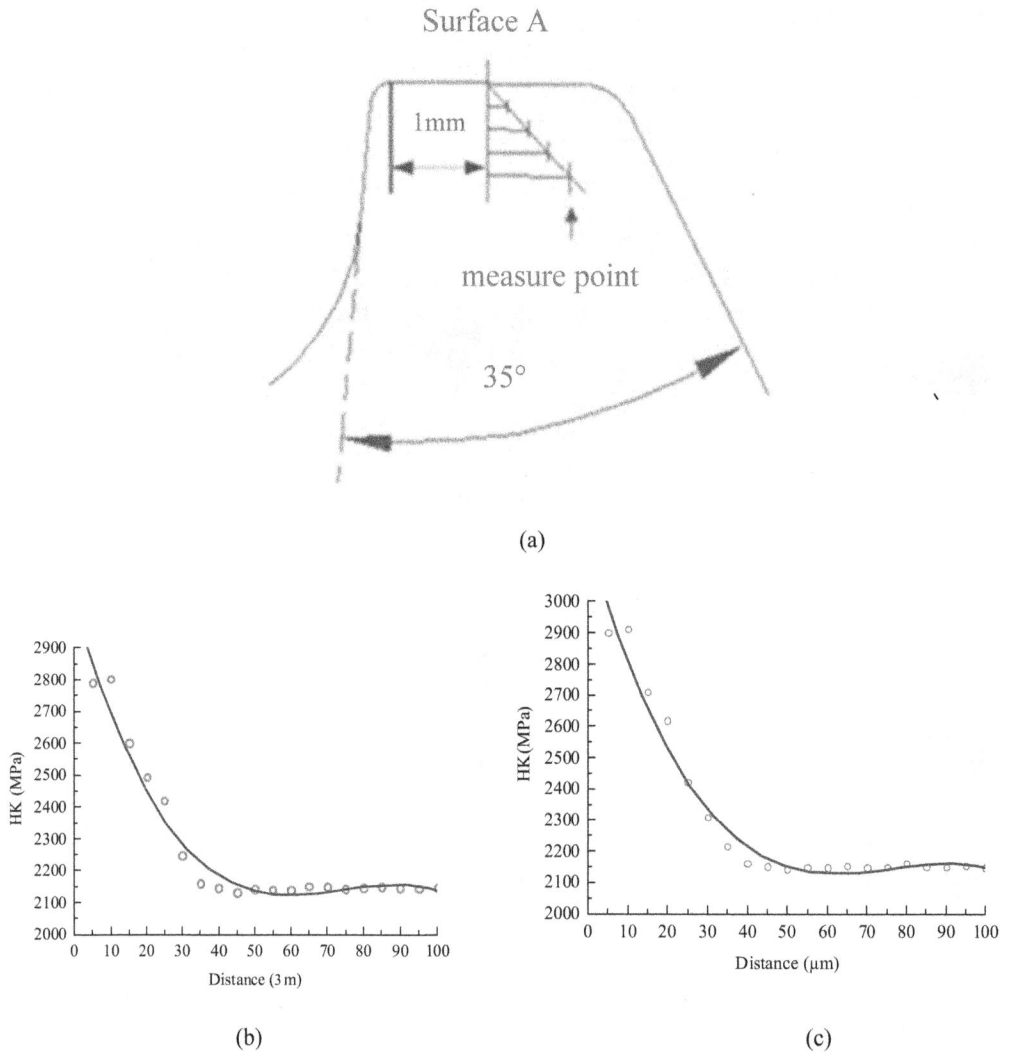

FIGURE 13.7 Microhardness of the ground surface: (a) Schematic of measure points, (b) 1# specimen, (c) 2# specimen.

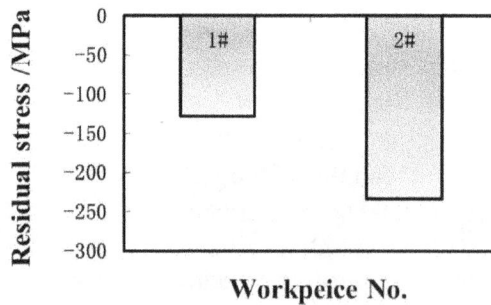

FIGURE 13.8 Surface residual stress of ground specimens.

induced by the vitrified wheel. When using the brazed wheel, the residual surface stress is less compressive (about −128 MPa).

Residual stress generation in grinding is caused by three principal sources: mechanically induced plastic deformation, thermal deformation, and phase transformation. In this work, no phase transformation is found on the surface or subsurface of both specimens. Also, the microhardness of the ground specimens is higher than the ones before grinding, which may be due to the cold hardening caused by grinding. The observations above show that mechanical and thermal deformations are the sources of residual stress formation here.

In the grinding processes, the workpiece-wheel contact zone can reach very high-temperature gradients, which produces tensile residual stresses during the rapid cooling process upon contraction of the heated area (thermal effect). At the same time, surface and sub-surface layers get plastically deformed during grinding, which leads to compressive stresses.

The fact that residual grinding stresses are all compressive highlights the preponderance of mechanical over thermal effects, which suggests that the temperature gradients are very low while using CBN wheels due to their high thermal conductivities.

13.5 COMPARATIVE PERFORMANCE OF BRAZED AND ELECTROPLATED CBN ABRASIVE WHEELS IN GRINDING TITANIUM ALLOY

For evaluating the comprehensive performance of the brazed CBN abrasive wheels, comparative machining tests on Ti-6Al-4V alloy have been carried out between two groups of CBN wheels: one is a three-brazed tool with uniform grain distribution, and the other is an electroplated tool with stochastic grain distribution. It should be noted that the brazed tools utilized were fabricated under quite identical conditions; moreover, the electroplated tools, of which the CBN grains are the same model as those of the brazed tools, were manufactured in a famous factory. Similar to brazed diamond tools, the performances of brazed CBN grinding wheels could also be estimated by the following factors: machining characteristics, fabrication, and use cost.

Figure 13.9 shows the comprehensive machining characteristics of the brazed and electroplated abrasive wheels. All the tests were executed at wheel peripheral speed $v_s = 25$ m/s, workpiece infeed speed $v_w = 0.3$ m/s, and depth of cut per pass $a_p = 0.08$ mm. The synthetic water-based coolant fluids were utilized as refrigerant throughout the machining process. As seen in Figure 13.9a, the brazed CBN abrasive wheels exhibit tool life at least three times as long as conventional electroplated tools. To further assess the suitability of the brazed tools, the surface quality of the ground workpieces has been measured, as illustrated in Figure 13.9b. Obviously, the surface roughness at the initial stage obtained with the brazed wheels was a little above that with their electroplated counterparts, which was induced by several CBN grains with a higher exposing height than most grains on the surface of the brazed tools. After about 1,500 passes have been machined, however, in excess of the working capacity of the electroplated tools, roughness Ra dropped below 0.50 μm. Here, the electroplated tools have failed because of the pullout and abrasion

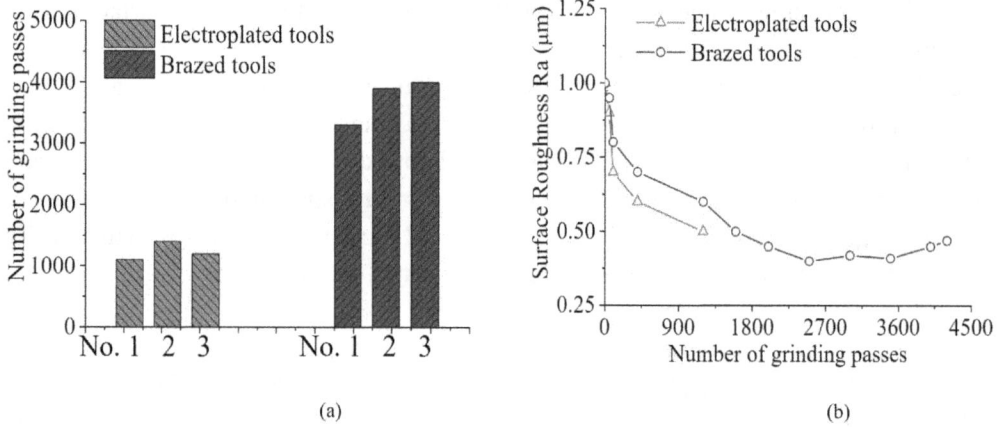

(a)

(b)

FIGURE 13.9 Tool life of the electroplated and brazed CBN grinding wheels and the corresponding surface roughness of the workpiece: (a) tool life, (b) surface roughness ($v_s = 25\,m/s$, $v_w = 0.3\,m/s$, $a_p = 0.08\,mm$).

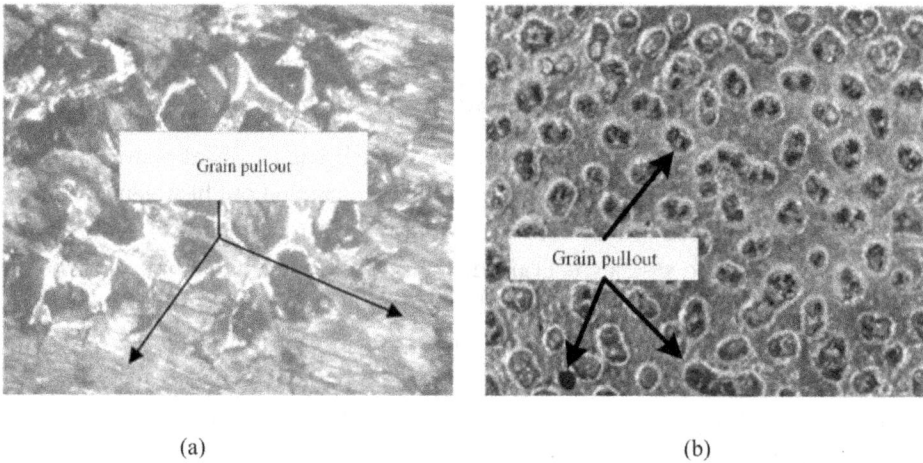

(a)

(b)

FIGURE 13.10 Different wear patterns of electroplated and brazed CBN abrasive wheels in grinding Ti-6Al-4V titanium alloy: (a) electroplated tool, (b) brazed tool.

behaviour of the grains, as shown in Figure 13.10a. A fairly steady value of about 0.4 μm for the surface finish of the workpiece machined by the brazed tools, however, established itself and remained almost constant up to over 4,000 passes. Clearly, after an early transient stage, relying on the abrasion-wearing behaviour instead of the pullout effect of the CBN grains (Figure 13.10b), the brazed tools display an unvarying working state in spite of their single-layer structure, which strongly supports the claimed value and quality of uniform grain distribution.

REFERENCES

Liu, M., C. Li, M. Yang, et al. 2023. Mechanism and enhanced grindability of cryogenic air combined with biolubricant grinding titanium alloy. *Tribology International*, 187: 108704.

Ronoh, K., F. Mwema, S. Dabees, et al. 2022. Advances in sustainable grinding of different types of the titanium biomaterials for medical applications: A review. *Biomedical Engineering Advances*, 4: 100047.

Xiao, G., Y. Zhang, B. Zhu, et al. 2023. Wear behavior of alumina abrasive belt and its effect on surface integrity of titanium alloy during conventional and creep-feed grinding. *Wear*, 514–515: 204581.

CHAPTER 14

Grinding Behaviour and Surface Integrity of Nickel-Based Superalloy

14.1 INTRODUCTION

Nickel-based superalloys are a diverse group of materials commonly used for elevated temperature applications in which high strength, excellent corrosion resistance, and good fatigue resistance are required. Their major applications are in gas and steam turbine components and aircraft engine component construction. However, the high strengths at elevated temperatures, high work hardening, and low thermal diffusivity are generally associated with the poor grindability of nickel-based superalloys. It always leads to high temperatures at the grinding zone and possible thermal damage to the workpiece during grinding with abrasive wheels (Liang et al. 2022; Xu et al. 2022; Xiao et al. 2022; Ding et al. 2010).

The technique of creep-feed grinding is a promising machining process because it combines the advantage of high shape accuracy, which is the character of conventional grinding, with the merit of a high material removal rate, which is the quality of tool cuttings. The large depth of cut, the long arc of cut, and the very slow workpiece speed are the predominant features of the creep-feed grinding process. In theory, it could offer high productivity and a smooth surface finish for components with complex profiles. At present, the creep-feed grinding technique has been applied to machine nickel-based superalloys, titanium alloys, SiC, and Si_3N_4 ceramics, and cemented tungsten carbide materials using electroplated cubic boron nitride (CBN) abrasive wheels, resin-bonded diamond wheels, and vitrified-bonded SiC abrasive tools. Whereas, the thermal damage of the machined materials and the drastic wear of the abrasive wheels have constituted the major problems in the creep-feed grinding process. In particular, premature grain pullout behaviour from the tool substrate under heavy loads has been one of the important wear patterns.

DOI: 10.1201/9781032678047-17

On the other hand, single-layer brazed CBN abrasive wheels have outperformed their widely used electroplated and vitrified counterparts. The brazed abrasive tools have a strong potential to realize high-efficiency grinding of difficult-to-cut materials owing to the high bonding strength of the abrasive grains and the sufficient storage space for chips. In this work, single-layer brazed CBN abrasive wheels are utilized to creep-feed profiled grind cast nickel-based superalloy K424 for fabricating straight grooves, which is projected to have a large application in high-efficiency profiled machining of aero-engine blade roots in the near future. It is known from the theory of metal cutting that the investigation of surface integrity is the most effective way of understanding the machining characteristics of a material. Thus, measurement of the machining characteristics of nickel-based superalloys represents one of the most important aspects of the analysis of machining processes. Therefore, the present paper reports on the investigation of grindability (e.g., grinding force and temperature, specific grinding energy) and surface integrity (e.g., size stability, surface topography, microhardness, microstructure, and residual stress as well) of K424 superalloy in creep-feed grinding with brazed CBN abrasive wheels.

14.2 EXPERIMENTAL DETAILS

A cast nickel-based superalloy coded K424 was used as the workpiece material in this work. This kind of nickel-based superalloy is a vital material for turbine blades and parts of aeroengines. The chemical composition and mechanical properties of K424 superalloy are given in Tables 14.1 and 14.2, respectively. The content of Ti and Al in the K424 superalloy is higher than that of other nickel-based superalloys, such as GH4169 (also called Inconel 718 in some cases). Moreover, K424 superalloy has high strength and good ductility. It is regarded as one of the most difficult-to-cut materials with the smallest relative cutting coefficient.

The brazed CBN abrasive wheels were manufactured at a brazing temperature of 920° for a dwell time of 5 min, as displayed in Figure 14.1. It had a segmented structure with a slot ratio of 0.75. The wheel size was 265 mm in diameter and 4.5 mm in width in the working zone. AISI 1045 steel was utilized as the tool substrate. $(Ag_{72}Cu_{28})_{95}Ti_5$ (wt.%) alloy was

TABLE 14.1 Chemical Composition of Cast Nickel-Based Superalloy K424

Elements	Minimum (wt.%)	Maximum (wt.%)
C	0.14	0.20
Cr	8.50	10.50
Co	12	15
W	1.0	1.8
Mo	2.7	3.4
Al	5.0	5.7
Ti	4.2	4.7
Nb	0.5	1.0
V	0.5	1.0
Fe	-	2.0
Ni	Balance	

TABLE 14.2 Mechanical Properties of Cast Nickel-Based Superalloy K424

Properties	@20°C	@800°C
Tensile strength (MPa)	1,010	935
Yield strength (MPa)	755	770
Hardness (HRC)	37.2	-
Thermal conductivity (W/m·K)	10.85	21.95
Density (kg/m³)	8,200	-
Linear thermal expansion (10^{-6}/°C)	14.1	-

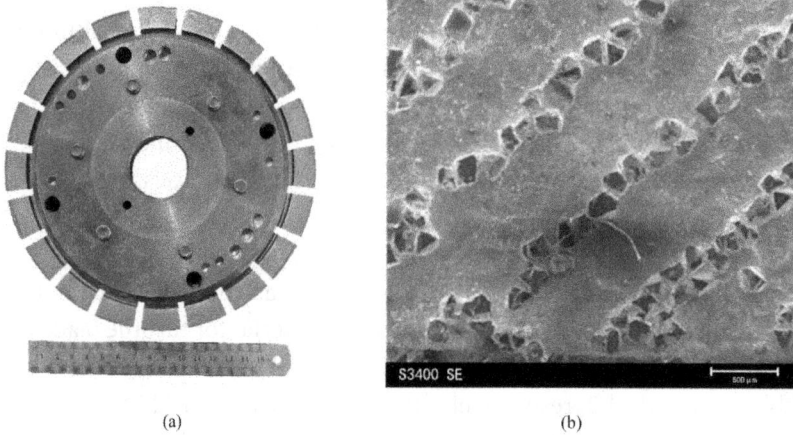

(a)   (b)

FIGURE 14.1 Morphology of single-layer brazed CBN abrasive wheel: (a) whole morphology, (b) grain distribution.

applied as the bonding material for CBN grains and tool substrate. The wheel was made of 80/100 US mesh CBN abrasive grains. Or, equivalently, the average grain size was 150 μm. In particular, the CBN grains were distributed linearly in the working zones of the abrasive wheels (Figure 14.1b).

A MMD7125 precision surface grinding machine with a minimum workpiece speed of 0.1 m/min was employed. Experimental conditions are listed in Table 14.3. The experimental workpiece of K424 superalloy was a pair of small blocks measuring 60 mm in length, 10 mm in width, and 10 mm in height. Figure 14.2 shows the schematic illustration of the cross-section of the wheel-workpiece couple during creep-feed profiled grinding. Thus, a straight groove may be fabricated. In the current experiments, the brazed CBN abrasive wheels were dressed in a single-point diamond dresser to ensure a sharp, clean tool surface. A set of initial grinding passes were carried out before the real grinding and data logging to stabilize the grinding performance.

To fully explore the creep-feed grinding process of K424 superalloy, the vertical and horizontal forces, F_V and F_H, were measured using a piezoelectric transducer-based type dynamometer (type Kistler 9265B), coupled to charge amplifiers and a PC running Dynoware software. For each set of process parameters, three grinding passes were undertaken to ensure reliable machining results. At the same time, a constantan wire-workpiece

TABLE 14.3   Conditions for the Grinding Tests

Machine Tools	MMD7125 Precision Surface Grinding Machine
Grinding mode	Plunge surface down-grinding
Abrasive wheels	Brazed CBN wheels
Average grain size (μm)	150
Wheel speed v_s (m/s)	17.5~25.0
Workpiece speed v_w (m/s)	0.1~0.4
Depth of cut a_p (mm)	0.08~0.02
Grinding width b (mm)	4.5
Cooling mode	Emulsified liquid; 5% dilution; 90 L/min; Pressure at 0.4 MPa

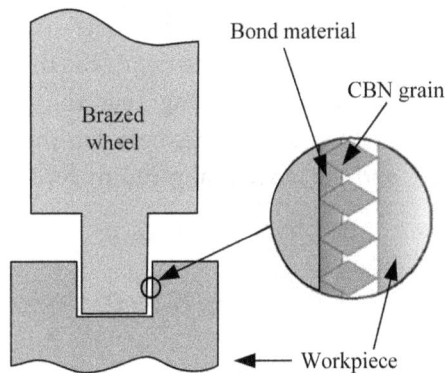

FIGURE 14.2   Schematic illustration of the creep-feed profiled grinding.

semi-natural thermocouple was employed to measure grinding zone temperature. The grinding temperature signal was captured and recorded directly by the HP3562 dynamic signal analyzer. The Constantan wire was very thin with a diameter of 0.02 mm, and its response was rapid enough to record the temperatures caused by individual abrasive grains. Because the method to measure the grinding force and temperature in this work is similar to those applied in the corresponding literature, it is not described or illustrated more in detail.

Firstly, the cross-section profile of the ground samples was observed using a KH-7700 optical microscope. The surface roughness of the samples was measured using an MAHR perthometer M2 surface roughness tester. The topography of the ground surface was detected with a QUANTA 200 scanning electron microscope (SEM) in secondary electron (SE) mode. The residual stresses were evaluated using the ψ tilt X-ray method using an XD-3A diffractometer with Cu Kα radiation. The measured zone of the ground surface was $6 \times 20 \, \text{mm}^2$.

Secondly, the ground samples were sectioned perpendicular to the grinding direction. Then the sectioned surfaces were polished. The hardness of the sub-surface was measured using a HXS-1000A microhardness tester at a load of 100 g for a dwell time of 15 s.

Finally, the polished samples were etched using a solution containing 6 vol.% HF, 50 vol.% HNO_3, and balance water. The metallurgical images of the sub-surface and the bulk material of the ground components were recorded with a KH-7700 optical microscope.

14.3 GRINDABILITY OF K424 NICKEL-BASED SUPERALLOY

14.3.1 Grinding Force

Generally, grinding force comprises four elements, including cutting, ploughing, sliding, and coolant dynamics. The grinding force, to a vast extent, affects the machined surface roughness, the work hardening, the power consumption, the heat flux at the contact zone between the abrasive wheel and the workpiece, the residual gradient stresses, the surface defects, and the wear of the abrasive grains. On this basis, for K424 superalloy, it is necessary to express the relationships between the grinding forces and the process parameters, i.e., wheel speed, workpiece speed, and depth of cut.

In a creep-feed grinding process, because the depth of cut is large, the grinding force components are always calculated by assuming that the average resultant grinding force acts at the middle point of the contact arc between the wheel and the workpiece. The normal grinding force F_n and the tangential grinding force F_t are described as:

$$F_n = F_V \cos\theta - F_H \sin\theta \qquad (14.1)$$

$$F_t = F_V \sin\theta + F_H \cos\theta \qquad (14.2)$$

where F_v is the measured vertical force; F_H is the measured horizontal force; θ is the included angle between the vertical force perpendicular to the workpiece and the normal force vertor acting on the grinding zone, which may be written as

$$\theta = \sqrt{a_p / d_s} \qquad (14.3)$$

where a_p is the depth of cut and d_s is the wheel diameter.

Figure 14.3 illustrates the representative grinding force curves measured under the conditions of a wheel speed of 22.5 m/s, a workpiece speed of 0.1 m/min, and a depth of cut of 0.2 mm. Serials of grinding tests with different process parameters were undertaken in this investigation. Figure 14.4 demonstrates the measured grinding force as a function of wheel speed, workpiece speed, and depth of cut, respectively. It was found that the increase in wheel speed led to a slight decrease in both normal and tangential forces, as shown in Figure 14.4a. However, a faster workpiece speed resulted in a larger grinding force, and the increase in depth of cut also induced a greater grinding force (Figure 14.4b and c). This is attributed to the fact that the increase in workpiece speed and depth of cut increased the undeformed chip thickness. At the same time, the magnitude of the normal force is significantly greater than that of the tangential force. For example, as seen in Figure 14.4c, when the depth of cut is increased from 0.08 to 0.20 mm, the normal force has been raised rapidly from 12 to 28 N, while the tangential grinding force is merely increased from 9 to 13 N. Under such conditions, the force ratio of F_n versus F_t increases remarkably from 1.3 to 2.2.

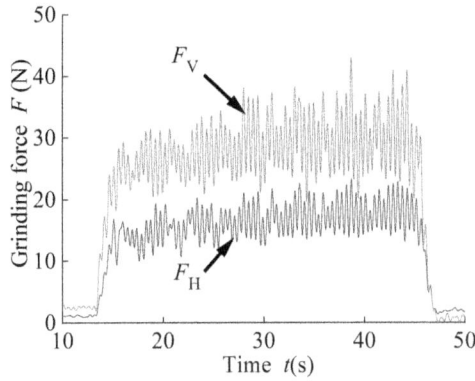

FIGURE 14.3   Typical grinding force signal of brazed CBN wheels.

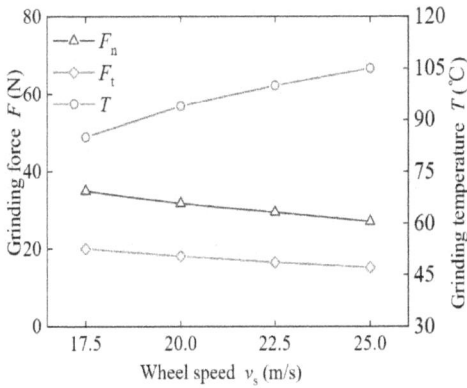

(a)

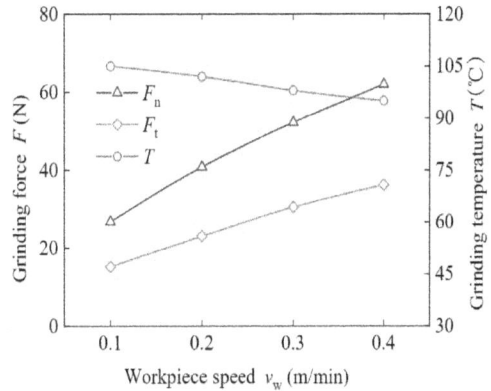

(b)

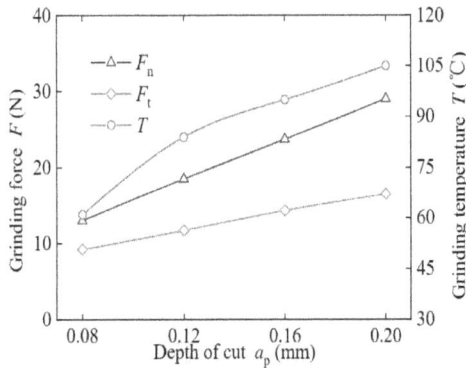

(c)

FIGURE 14.4   Influence of the process parameters on the grinding force and temperature: (a) v_w=0.1 m/min, a_p=0.2 mm, (b) v_s=22.5m/s, a_p=0.2 mm, (c) v_s=22.5 m/s, v_w=0.1 m/min.

14.3.2 Specific Grinding Energy

During grinding, specific energy is one of the most important grindability indices. It can be calculated with the use of the process parameters and the tangential grinding force measured in the experiment, as follows:

$$u = \frac{F_t v_w}{v_s a_p b} \tag{14.4}$$

where b is the width of the working zone of the grinding wheels, i.e., 4.5 mm in this investigation.

On the other hand, based on the geometrical characteristics of the chip formation, the maximum undeformed chip thickness of the segmented grinding wheels is written as:

$$a_{g\max} = [\frac{4v_w}{\eta v_s N_d C} \sqrt{a_p / d}]^{1/2} \tag{14.5}$$

where N_d is the active cutting point density (6/mm^2 in this work), η is the slot ratio, i.e., 0.75 in this investigation, and C is a constant correlated with the angle of the grain tip. For the grain with a size of 150 μm, it is taken to be 6.928.

According to the grinding force and the process parameters shown in Figure 14.4, the experimental specific energy for creep-feed grinding K424 superalloy was obtained and plotted against the undeformed chip thickness, as displayed in Figure 14.5. With the application of the least squares curve-fitting technique to the experimental data, the specific energy in this investigation is represented by

$$u = 128.99 a_{g\max}^{-1.027} \tag{14.6}$$

As seen in Figure 14.5, the specific grinding energy of K424 superalloy is high, up to 200–300 J/mm^3. Compared with other nickel-based superalloys also used for turbine blades,

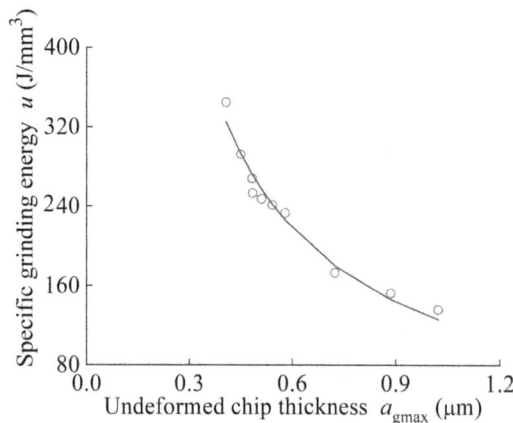

FIGURE 14.5   Relationship between specific grinding energy and undeformed chip thickness.

such as Udimet 520 and Inconel 713C, the specific grinding energy is only 130–160 and 80–120 J/mm³, respectively, in the case of the nearly identical undeformed chip thickness [Esmaeili et al. 2022; Suto et al. 1990]. This implies that the grindability of K424 superalloy is much worse than that of Udimet 520 and Inconel 713C, though the latter is also called the most difficult-to-cut materials.

On the other hand, the specific grinding energy of cast nickel-based superalloy K424 has been gradually decreased with an increase in a_{gmax}. This could be explained from the viewpoint of the abrasive-workpiece interaction. The major phenomena for chip formation during grinding are sliding, ploughing, and cutting. In the case of small undeformed chip thicknesses, the high specific grinding energy of K424 superalloy is mainly attributed to high ploughing and sliding energies. This is expended in grinding in excess of the energy of chip formation by cutting. However, when the undeformed chip thickness is increased, the percentage contributions from sliding and ploughing diminish, and thus specific grinding energy decreases with the undeformed chip thickness as well.

14.3.3 Grinding Temperature

Grinding nickel-based superalloys is a mechanical process that involves a great deal of energy per unit volume of removed material. This energy is almost all converted into heat, perhaps causing a significant rise in temperature above 700°C in the traditional plunge surface grinding process of DZ4 and GH4169 superalloy. Therefore, the ground surface of the workpiece always experiences rapid thermal cycles, while the rest of the part remains at lower temperatures. The increased temperature can have harmful effects on the workpiece material, such as thermal deflection, burnout, surface cracking, and residual stress.

A representative grinding temperature signal measured using the grindable thermocouple is displayed in Figure 14.6. The temperature response recorded by the foil/workpiece thermocouple is composed of two parts: a background temperature and periodic flash temperature impulses at the wheel rotational frequency. It has been suggested that flash temperature may be associated with the cutting action at individual grain cutting points. In this work, the grinding temperature means the temperature increase at the workpiece

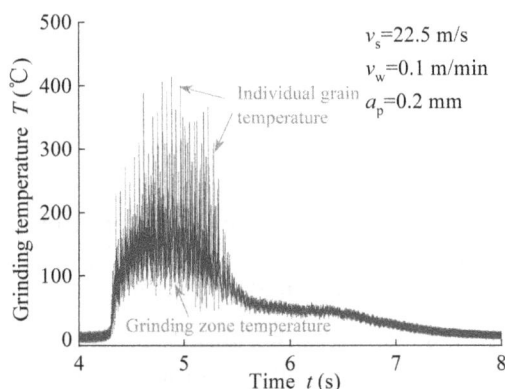

FIGURE 14.6 Typical temperature signal during creep-feed grinding.

surface, that is, the grinding zone temperature during creep-feed grinding. The highest temperature at the given condition was approximately 105°C.

The values of grinding temperature for K424 superalloy are plotted against the process parameters in Figure 14.4. It is seen that the grinding temperature is increased with the increase in both the wheel speed and the depth of cut in the current investigation. However, the grinding temperature decreases with an increase in the workpiece speed. This phenomenon is due to the change in the undeformed chip thickness of K424 superalloy during grinding.

The increase in wheel speed leads to a reduction in the undeformed chip thickness. Thus, the chips become smaller. The corresponding material deformation energy is increased. Moreover, the quantity of grains ploughing and sliding on the workpiece surface becomes larger. The friction behaviour between the wheel and the workpiece gets severe. Therefore, the grinding temperature is increased. On the other hand, when the workpiece speed is heightened, the grinding heat is released conveniently, though the maximum undeformed chip thickness is still increased. That is, high workpiece speed can substantially reduce the proportion of energy conducted into the workpiece.

It has to be noticed that, though the specific grinding energy of K424 superalloy is high up to 200–300 J/mm³, the highest grinding temperature of K424 superalloy is merely about 100°C in the current investigation. This favourable phenomenon is not only determined by the fluid film burn-out effect of the coolant in creep-feed grinding but also contributed to the following factors: excellent cutting ability of CBN abrasive grains, efficient chip storage space of the tools, good thermal conductivity property among the CBN grains, the bonding layer of the Ag-Cu-Ti alloy, and the tool substrate of AISI 1045 steel.

14.4 DIMENSIONAL ACCURACY OF THE GROUND GROOVES

Based on the experimental results of grinding force and temperature mentioned above, creep-feed profiled grinding of K424 superalloy was finally carried out with a combination of the process parameters, i.e., $v_s = 22.5$ m/s, $v_w = 0.1$ m/min, and $a_p = 0.2$ mm. Grinding tests were stopped at a stock removal of 4,050 mm³. Accordingly, three grooves with a length of 60 mm and a depth of 5 mm were fabricated. At this time, the brazed CBN wheels still have good machining performance.

Figure 14.7 presents the typical morphology of the ground grooves. For evaluating the dimensional accuracy, the groove width was measured at the entrance and the departure, respectively. The results are listed in Table 14.4. Obviously, the requirement for expected dimensional accuracy has been fulfilled.

The comparison in the cross-sectional profile of the ground grooves is displayed in Table 14.5. It is found that the corner profile formed in creep-feed grinding with a brazed CBN abrasive wheel remains nearly in the initial state up to the stock removal of 4,050 mm³. More importantly, the corner radius at the bottom of the grooves is constantly maintained at 0.50 mm, both at the entrance and at the departure. As reported in the previous literature, the profile wear and the radial wear of the applied abrasive wheels significantly affect the shape and accuracy of the ground grooves during creep-feed profiled grinding. Profile wear is the change in shape of the profile formed on the grinding

(a)

(b)

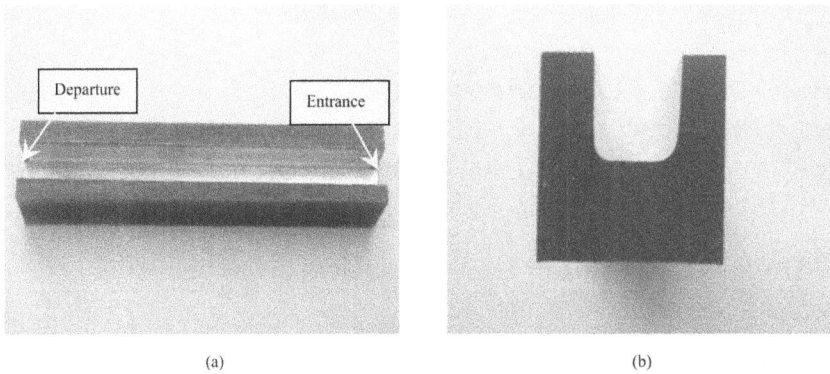

FIGURE 14.7 Typical morphology of the ground grooves: (a) whole morphology, (b) entrance cross-section.

TABLE 14.4 Width Variance in the Cross-Sectional Profile of the Ground Grooves

Samples No.	Entrance Width (mm)	Departure Width (mm)	Designed Width (mm)	Entrance Tolerance (mm)	Departure Tolerance (mm)	Groove Tolerance (mm)
1	4.54	4.55	$4.5^{+0.05}_{-0.05}$	$4.5^{+0.04}_{+0.02}$	$4.5^{+0.05}_{+0.03}$	$4.5^{+0.05}_{+0.02}$
2	4.53	4.53				
3	4.52	4.54				

TABLE 14.5 Morphology Variance in the Cross-Sectional Profile of the Ground Grooves

Stock Removal (mm³)	1,350	2,700	4,050
Corner at the entrance			
Corner at the departure			

wheel as a consequence of wheel wear. According to the experimental results in Table 14.5, it is known that the brazed CBN abrasive tools give less profile wear and therefore better form retention. This is due to the fact that the very high hardness of CBN abrasive grains provides good size-holding ability for CBN grinding. As a result, brazed CBN abrasive wheels are suitable for applications with the requirement of high size stability in creep-feed profiled grinding of K424 superalloy.

14.5 SURFACE INTEGRITY OF THE GROUND GROOVES

14.5.1 Surface Topography

The quality of the surface generated by grinding determines many workpiece characteristics, such as the minimum tolerances, the lubrication effectiveness, and the component life, among others. When the workpiece is machined using the conventional milling and grinding method, it is not easy to meet the surface quality requirements.

Figure 14.8a and b display the typical surface morphology of the workpiece ground under the condition $v_s=22.5$ m/s, $v_w=0.1$ m/min, and $a_p=0.2$ mm. Here, the surface roughness R_a is 1.0 μm, which is lower than the requirement provided in the corresponding literature (R_a below 2 μm). Lower surface roughness on a component tends to induce higher fatigue strength as compared to a coarser surface.

Additionally, a ground surface may be characterized by clean cutting paths and ploughed materials to the sideway. The feed mark, which is a natural defect because of infeeding, is observed on all the ground surfaces. Figure 14.8b shows the representative image of the feed marks. Particularly, the severity of the grinding condition and the progressive wear of the CBN grains produce a more significant feed mark on the machined surface. If the workpiece speed is increased further to 0.3 m/min, the tearing surface is randomly found on the ground surface, as demonstrated in Figure 14.8c.

As is well known, in the creep-feed grinding process, three bodies are contacted in a sliding direction, i.e., abrasive grains, a machined surface, and a small part of the bonding layer of the brazed CBN wheels. Considering that the CBN grains are much harder than the workpiece material, they slid between the wheel and the machined surface. At last, the grains scratch and tear away the ground surface (Figure 14.8c). Consequently, it is thought that the combination of the beneficial effects of the brazed CBN abrasive wheels and the optimum grinding parameters led to the generation of a defect-free machined surface.

14.5.2 Microhardness and Microstructure Alteration of the Sub-surface

In the grinding process, once the workpiece material is subjected to high grinding temperatures and large grinding pressure, it undergoes a competing process between thermal softening and work hardening. For instance, when nickel-based superalloys and titanium alloys are machined with vitrified or electroplated superabrasive wheels, the rate of thermal softening behaviour is always much greater than that of the work hardening effects in the process of the plastic deformation and the microstructure alteration of the sub-surface. As a result, a soft sub-surface is formed, and the mechanical properties of the workpiece material may deteriorate.

(a)

(b)

(c)

FIGURE 14.8   Typical topography of the ground surface: (a) whole, (b) regional, (c) tearing surface.

To analyse the sub-surface alteration of the ground grooves, the microhardness was measured using the Vickers method with $HV_{0.1}$ at the side surface and bottom surface of the grooves. Especially, the measurement of the hardness was conducted five times at every $5\,\mu m$ up to $80\,\mu m$ below the ground surface, and the average hardness value for each depth was recorded. The results in terms of hardness versus the depth below the side surface and the bottom surface of the ground grooves are plotted in Figure 14.9. No soft region but a hard one exists in the sub-surface of the ground components. That is, workpiece burning has been avoided. The maximum hardness value, about $540\ HV_{0.1}$, is reached at a depth of $10\,\mu m$. It is approximately 46% higher than the bulk material hardness of $380\ HV_{0.1}$. After the peak point has been obtained, the hardness value decreases gradually until it reaches the bulk material hardness at about $30\,\mu m$ below the ground surface. Then, minor

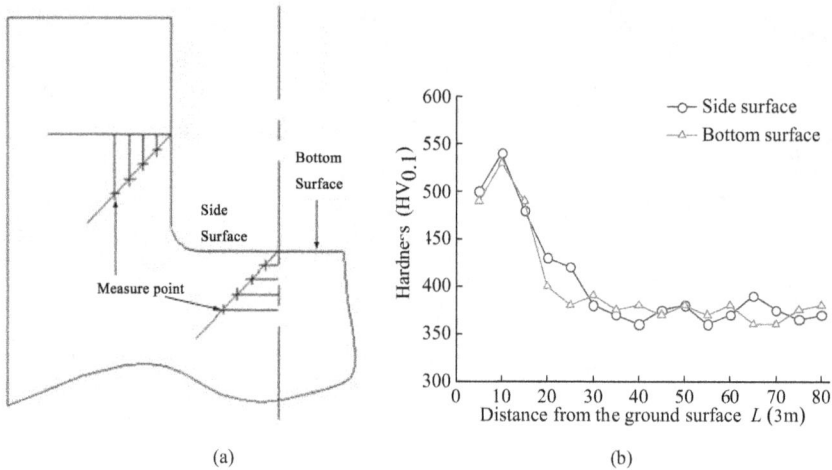

FIGURE 14.9   Hardness distribution at the sub-surface of the ground grooves: (a) schematic of the measuring point, (b) hardness distribution.

fluctuations have been observed up to 80 μm below the surface. The same trends in the hardness distribution in the corner zone have been found in the current investigation.

The phenomenon mentioned above is attributed to the fact that, during creep-feed grinding, the highest hardness value of the sub-surface zone is always formed at the position where the most drastic plastic deformation takes place due to the removal of the material under high grinding pressure. Moreover, this special position is dependent on the maximum ratio of the normal grinding force F_n to the tangential grinding force F_t. The ratio of F_n/F_t is larger because the position of the most drastic plastic deformation is farther from the ground surface.

The representative optical metallurgical image of the ground K424 superalloy is displayed in Figure 14.10. In particular, the metallurgy samples were carefully prepared to hinder smearing on the machined region. According to Figure 14.10, no microcrack is formed. Compared with the typical microstructure of the bulk material of K424 superalloy (Figure 14.10b), that is, γ phase, γ', and γ'' ones, the microstructure at the sub-surface region of the ground samples tends to exhibit plastic deformation and crystal lattice distortion to a small extent. This is the reason for the hardening of the sub-surface. The drastic plastic deformation does occur from a single droplet that has been pressed through the gap between the cutting edge of the CBN grains and the ground material and smeared over the ground surface. Therefore, the plastic deformation and induced work hardening of the sub-surface may be controlled by adjusting the grinding condition and alleviating wheel wear. To increase the material removal rate and further improve the machining quality of the ground components, it is necessary to discuss the wheel wear mechanisms during creep-feed grinding with brazed CBN abrasive wheels in the subsequent work.

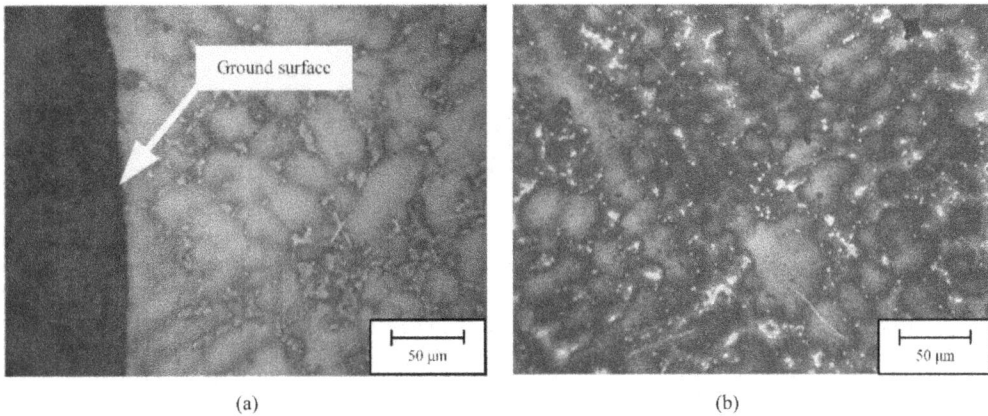

(a)   (b)

FIGURE 14.10   Microstructure alteration of the sub-surface and the bulk material: (a) sub-surface microstructure, (b) bulk material microstructure.

14.5.3 Residual Stresses on the Ground Surface

Low residual stress after grinding is an important requirement for the surface integrity of stress-sensitive components, i.e., the turbine blade roots. When the grooves are machined using the traditional grinding and milling technique, it is possible to form tensile residual stress up to 40 MPa in the sub-surface. The subsequent service life is therefore reduced under stress, corrosion, or fatigue conditions.

Among the three grinding parameters, the depth of cut has the most important influence on the residual stresses during creep-feed grinding. For a reason, the stresses in the direction parallel to the grinding direction, σ_x, and that perpendicular to the grinding direction, σ_y, are measured against the depth of cut. The result is plotted in Figure 14.11.

As seen in Figure 14.11, the compressive residual stresses remain in both directions on the ground surface. The highest compressive stress value is obtained for a depth of cut of 0.08 mm. In this case, the residual stress measured in the grinding direction is $\sigma_x = -300$ MPa, and that in the perpendicular direction is $\sigma_y = -340$ MPa. With the increase in depth of cut, the values of the compressive stresses decrease.

Moreover, the compressive stress in the direction perpendicular to the grinding direction is always greater than that in the direction parallel to the grinding direction. When the depth of cut is 0.2 mm, the compressive stress values of σ_x and σ_y are −40 and −120 MPa, respectively. This corresponds to a dominance of the mechanical effects compared to the thermal effects during chip formation at the grinding zone.

It is known that the residual stresses on a ground surface are primarily generated due to the following effects: thermal expansion and contraction during grinding, phase transformations due to high grinding temperatures, and plastic deformation caused by the cutting edges of the abrasive grains. Because the grinding temperature of K424 superalloy with brazed CBN abrasive wheels is as low as about 100°C in this work, the thermal effect may be neglected. Therefore, the effects generating residual stresses are governed by the wheel-workpiece mechanical interactions. In other words, the variance in the physical properties

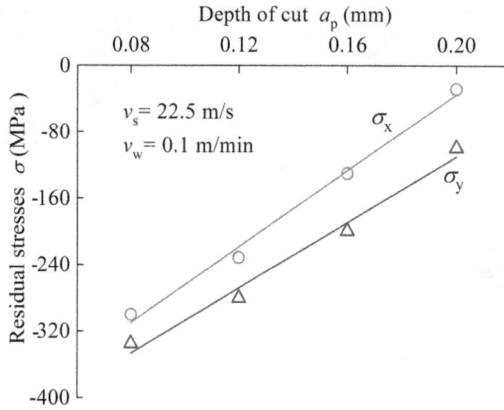

FIGURE 14.11   Residual stresses of the ground surface against the depth of cut.

of the machined surface is attributed mainly to the plastic deformation resulting from the grinding pressure. In the area around the tip of the cutting edge of the CBN grain, the compressive level has to be very high since this is the only way to ensure that the ground material plastifies to a suffiecient degree to allow chip formation. The high level of mechanical stress being exerted on the surface of the workpiece tends to induce compressive residual stresses. However, with the increase in depth of cut, the burnishing and scraping effects between the CBN abrasive grains and the K424 superalloy workpiece are weakened, and the magnitude of the compressive residual stresses on the ground surface is therefore decreased gradually.

14.6 COMPARATIVE PERFORMANCE OF DIFFERENT CBN ABRASIVES IN GRINDING NICKEL-BASED SUPERALLOY

The single-layer brazed abrasive wheels with 260 mm in diameter and 12 mm in width were fabricated based on the brazing procedures of polycrystalline CBN grains. The grinding performance of the brazed wheels with polycrystalline CBN (PCBN) grains was also evaluated through surface grinding experiments. The machine (Blohm Profimat MT 408) spindle could run up to 8,000 rpm for the maximum wheel diameter of 400 mm. The ground material was nickel-based superalloy (GH4169). Grinding was conducted on the 50×5 mm surface in a down-grinding mode. Various wheel speeds were utilized to investigate the impacts on the grinding forces and force ratio. Particularly, the grinding process was repeated three times under each set of parameters to obtain the average values of the grinding force. As a comparison, single-layer brazed abrasive wheels with monocrystalline CBN (McBN) were also fabricated and applied in the present grinding practice.

According to some reports, the grinding process carried out at a wheel speed above 80 m/s could be thought of as high-speed grinding for the difficult-to-cut materials. Figure 14.12 displays the grinding forces and force ratio according to the wheel speed of a single-layer brazed polycrystalline CBN wheel and a single-layer brazed monocrystalline CBN wheel, respectively. It is evident from Figure 14.12a and b that the grinding forces show a generally decreasing tendency when the wheel speed is increased from 60 to 90 m/s. Moreover, as

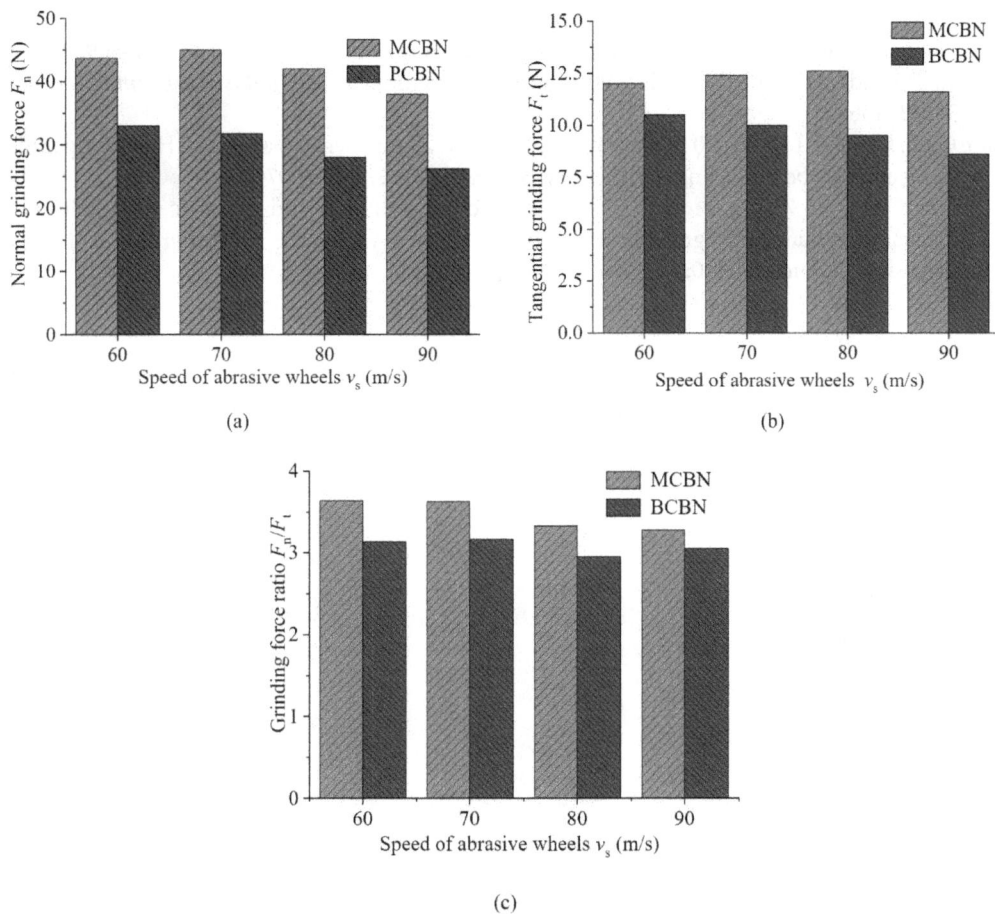

FIGURE 14.12 Grinding force and force ratio variation according to abrasive wheel speed: (a) normal grinding force, (b) tangential grinding force, (c) grinding force ratio of F_n/F_t.

seen in Figure 14.12c, the force ratio of a monocrystalline CBN wheel, i.e., 3.3–3.6, is always larger than that of a polycrystalline CBN wheel, i.e., 2.9–3.2. The grinding force ratio is tightly correlated with the sharpness of the grinding wheels. A low force ratio corresponds to a high sharpness degree. Therefore, the brazed polycrystalline CBN wheel shows better grinding performance compared to the brazed monocrystalline CBN wheel, though the latter also has excellent grinding performance.

REFERENCES

Ding, W. F., J. H. Xu, Z. Z. Chen, et al. 2010. Grindability and surface integrity of cast nickel-based superalloy in creep feed grinding with brazed CBN abrasive wheels. *Chinese Journal of Aeronautics*, 23(4): 501–510.

Esmaeili, H., H. Adibi, R. Rizvi, et al. 2022. Coupled thermo-mechanical analysis and optimization of the grinding process for Inconel 718 superalloy using single grit approach. *Tribology International*, 171: 107530.

Liang, C., Y. Gong, S. Qu, et al. 2022. Performance of grinding nickel-based single crystal superalloy: Effect of crystallographic orientations and cooling-lubrication modes. *Wear*, 508–509: 204453.

Suto, T., T. Waida, H. Noguchi, et al. 1990. Wheel designs for grinding. *Industrial Diamond Review*, 3: 133–136.

Xiao, G., B. Chen, S. Li, et al. 2022. Surface integrity and fatigue performance of GH4169 superalloy using abrasive belt grinding. *Engineering Failure Analysis*, 142: 106764.

Xu, Y., Y. Gong, W. Zhang, et al. 2022. Microstructure evolution and dynamic recrystallization mechanism induced by grinding of Ni-based single crystal superalloy. *Journal of Materials Processing Technology*, 310: 117784.

CHAPTER 15

Grinding Behaviour and Surface Integrity of Titanium Matrix Composites

15.1 INTRODUCTION

TiCp/Ti-6Al-4V particulate-reinforced titanium matrix composites (PTMCs for short in this work) have high specific strength, high wear resistance, and high-temperature durability, which show great application potential as important structural materials in the aerospace and military industries (Dong et al. 2022; Li et al. 2016; Klocke et al. 2015; Choi et al. 2014). However, similar to SiCp/Al aluminium matrix composites, it is also very difficult to machine PTMCs due to the co-existence of hard TiC reinforcements and high-strength Ti-6Al-4V matrix. Bejjani et al. (Qian et al. 2023; Bejjani et al. 2011) have studied the effects of cutting parameters with PCD on tool life, chip morphology, and surface integrity in laser-assisted turning of titanium metal matrix composite. They discovered that tool life could be enhanced for laser-assisted machining of titanium metal matrix composite; however, surface roughness increased moderately by up to 15%. Aramesh et al. (2014) tested the tool wear of turning titanium metal matrix composites by using a PCD cutting tool. A statistical model was developed to estimate the average residual life of the cutting inserts during turning Ti-MMCs. Initial wear, steady wear, and rapid wear regions in the tool wear curve were regarded as the different states in the statistical model. Besides the severe tool wear, poor surface quality is another machining problem in the cutting and grinding processes. For example, Blau and Jolly (2009) discovered that the reinforcing particles were easily fractured and pulled out in the grinding practice of titanium matrix composites with vitrified alumina abrasive wheels. At the same time, some clusters of fragments also remained on the surface or were trapped in microcracks. Under such conditions, the ground surface quality of PTMCs would be degraded even though

DOI: 10.1201/9781032678047-18

the maximum material removal rate applied is only 0.5 mm³/(mm·s). For the above reasons, how to realize high-efficiency and precision machining of PTMCs has become an important research topic in the present days.

High-speed grinding with cubic boron nitride (CBN) abrasive wheels has been regarded as a desirable method to improve the machined surface quality in high-efficiency grinding. The reason is mainly due to the decreasing undeformed chip thickness and the possible ductile removal mode of the brittle materials during the high-speed grinding process. In other words, the difficulty in machining difficult-to-cut materials is perhaps weakened to a certain extent in high-speed grinding. For example, Tian and Fu found that the grinding temperature could be controlled effectively in high-speed grinding of Ti-6Al-4V titanium alloy; accordingly, the grinding-induced surface quality was improved.

To provide an experimental and theoretical basis for high-efficiency and precision machining of PTMCs, high-speed grinding experiments are carried out with the single-layer electroplated CBN wheel and brazed CBN counterpart, respectively, in the current work. The comparative grinding performance of PTMCs with the two types of single-layer CBN wheels are analysed comprehensively in terms of grinding force, grinding temperature, and grinding-induced surface features and defects.

15.2 EXPERIMENTAL DETAILS

The PTMCs contain Ti-6Al-4V metal matrix and 10 vol.% TiC reinforcing particles, which is produced by the state key laboratory of metal matrix composites in China. The TiC reinforcing particles with a size of 1.2–8.4 μm are synthesized using the powder metallurgy technique in the fabrication process of PTMCs. The average quantity of TiC particles is about 1,500/mm², as displayed in Figure 15.1. Meanwhile, a small quantity of TiB whiskers is also contained in PTMCs. The composition and mechanical properties are listed in Table 15.1.

Grinding experiments are performed on a high-speed surface grinder, the BLOHM Profimat MT-408. The maximum rotational speed is 8000 r/min, and the output power is 45 kW. Figure 15.2 schematically demonstrates the experimental setup. Up-grinding mode is applied. Figure 15.3a displays the whole morphology of single-layer CBN wheels. The

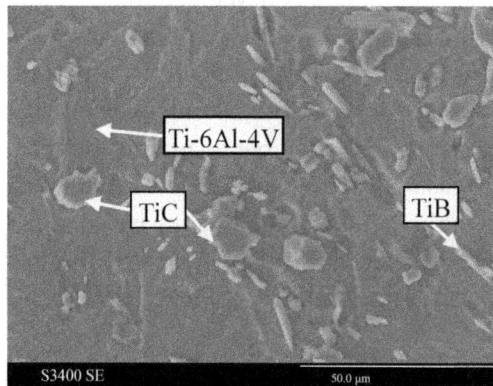

FIGURE 15.1 Microstructure of TiCp/Ti-6Al-4V composites.

TABLE 15.1 Composition and Mechanical Properties of PTMCs

Contents	Values
Matrix	Ti-6Al-4V
Reinforcements	10 vol.% TiC particles
Tensile strength	1102 MPa
Yield strength	972 MPa
Elongation rate	0.55%
Elasticity modulus	133 GPa
Poisson's ratio	0.34

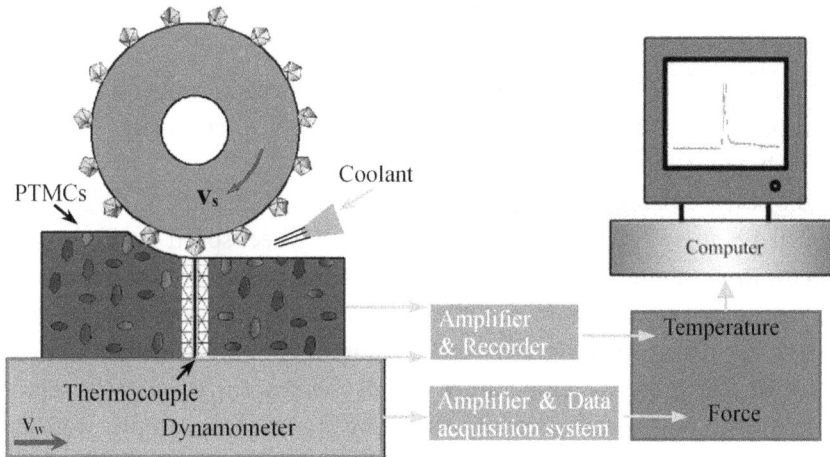

FIGURE 15.2 Schematics of the grinding experimental set-up with a single-layer CBN wheel.

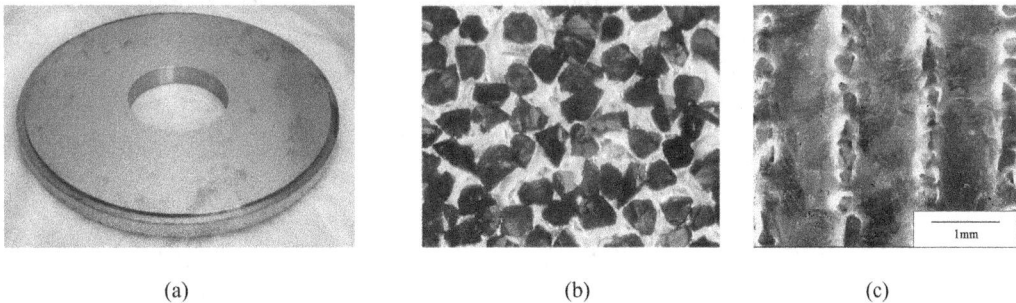

(a)   (b)   (c)

FIGURE 15.3 Single-layer CBN wheels: (a) whole profile, (b) electroplated CBN wheel, (c) brazed CBN wheel.

outer diameter of the wheel is 400 mm, and the grain size is 80/100 mesh. In particular, the grain particles are distributed randomly on the working surface of the electroplated CBN wheel (Figure 15.3b), while they are distributed linearly with a row interval of 1.2 mm on the working surface of the brazed wheel (Figure 15.3c). The electroplated CBN wheel is fabricated by Suzhou Wente Super-abrasive Tools Co. Ltd., China. The brazed CBN wheel is fabricated by the authors in Jiangsu Precision and Micro-manufacturing Lab, China. The

TABLE 15.2   Grinding experimental conditions

Contents	Values
Machine tool	Surface grinding machine BLOHM Profimat MT-408
Grinding mode	Up-grinding
Wheel speed v_s	120 m/s
Workpiece infeed speed v_w	3–12 m/min
Depth of cut a_p	0.005–0.020 mm
Grinding width b	5 mm
Cooling fluid	Emulsified liquid; 5% dilution; 90 L/min; Pressure at 15 MPa

fabrication technology of the single-layer brazed CBN wheels was introduced in the previous publication. The bonding material of the brazed CBN wheel is $(Ag_{72}Cu_{28})_{95}Ti_5$ (wt.%) alloy, and the brazing temperature is 920°C.

The details of the current grinding conditions are listed in Table 15.2. The wheel speed v_s was fixed at 120 m/s, which was a broadly accepted wheel speed for high-speed grinding of difficult-to-cut materials in the present days. The workpiece infeed speeds v_w were varied among 3, 6, 9, and 12 m/min, and the depth of cut a_p was among 0.005, 0.010, 0.015, and 0.020 mm. Grinding force signals were measured using the piezoelectric dynamometer Kistler 9272. The grinding temperature signals were captured using the semi-artificial thermocouple technique. Before grinding, the PTMCs workpiece samples were cut equally into two blocks with a size of 15 mm (length) × 5 mm (width) × 25 mm (height). A constantan foil of 0.020 mm thickness was sandwiched between two pieces of the workpiece and was insulated by two mica sheets. A hot junction for a constantan-workpiece semi-artificial thermocouple was made in the grinding process. The thermal electromotive signal was therefore formed, and the grinding temperature could be measured using a Labview data acquisition card connected to a computer, as displayed in Figure 15.2.

After grinding, the PTMC surface was observed using Hirox KH-7700 optical microscopy and Quanta-200 scanning electron microscopy (SEM). The corresponding defects were quantitatively and statistically analysed.

15.3 COMPARISON OF GRINDING FORCE IN HIGH-SPEED GRINDING

Figure 15.4 shows the measured signal of the normal force F_n and tangential force F_t, with different single-layer CBN wheels at identical grinding parameters, i.e., the wheel speed of 120 m/s, the workpiece infeed speed of 6 m/min, and the depth of cut of 0.010 mm during high-speed grinding of PTMCs. Figure 15.5 displays the effects of the grinding parameters on the grinding forces. As seen in Figure 15.5a, an increase in workpiece infeed speed leads to an increase in both the normal and tangential grinding forces. When the depth of cut is kept at 0.010 mm and the workpiece infeed speed is increased from 3 to 12 m/min, the normal force F_n is increased from 26.2 to 43.8 N (by 67%) with the electroplated CBN wheel and from 19.7 to 37.4 N (by 90%) with the

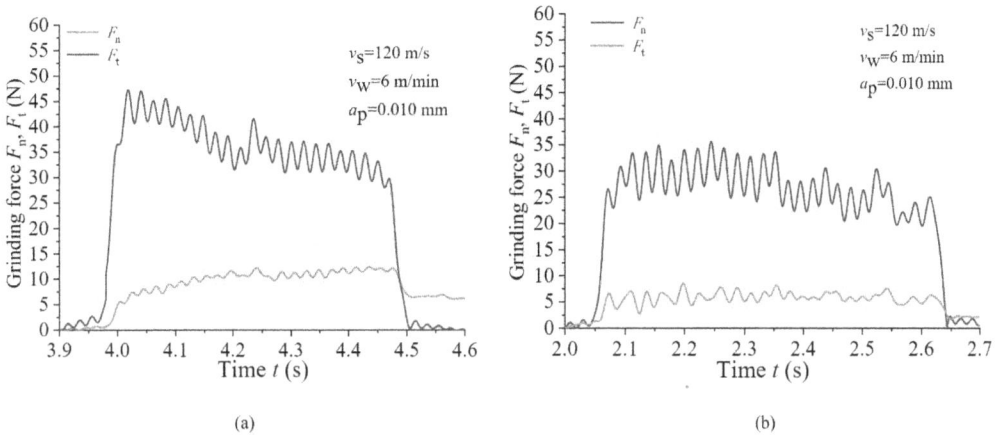

FIGURE 15.4   Grinding forces signal: (a) electroplated CBN wheel, (b) brazed CBN wheel.

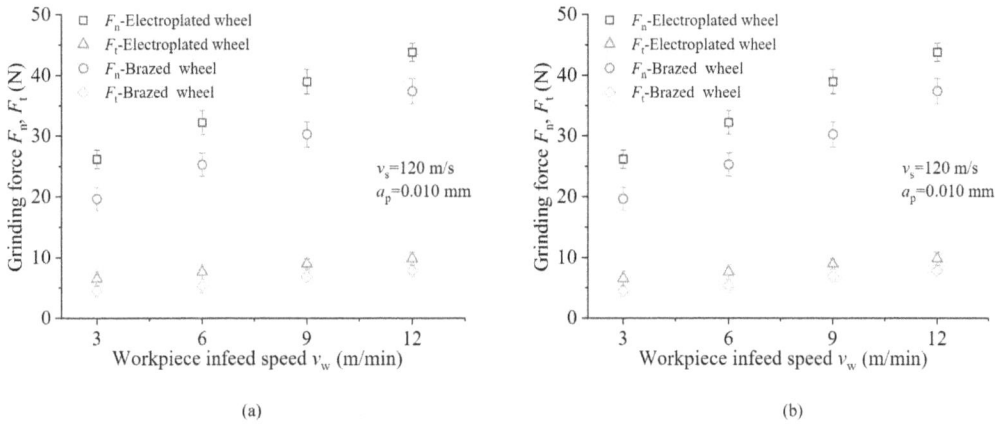

FIGURE 15.5   Effects of grinding parameters on grinding forces: (a) workpiece infeed speed, (b) depth of cut.

brazed counterpart, respectively. Meanwhile, the tangential force F_t is raised from 6.5 to 9.8 N (by 51%) with the electroplated CBN wheel and from 4.5 to 7.8 N (by 73%) with the brazed one, respectively.

When the depth of cut is increased from 0.005 to 0.020 mm and the workpiece infeed speed is fixed at 6 m/min, the normal force is raised rapidly from 26.5 to 48.7 N (by 84%), and the tangential force is also increased from 6.0 to 10.5 N (by 75%) with the electroplated CBN wheel (Figure 15.5b). For the brazed CBN wheel, the recorded normal force is raised from 20.0 to 43.0 N (by 115%), and the tangential force is increased approximately from 4.5 to 7.6 N (by 67%). In general, the lower grinding force obtained with the brazed CBN wheel is mainly due to the higher grain protrusion on the tool surface, which increases the sharpness degree and enlarges the chip storage space of the CBN super-abrasive wheel during grinding.

15.4 COMPARISON OF GRINDING TEMPERATURE IN HIGH-SPEED GRINDING

During grinding difficult-to-cut materials, a great amount of energy is converted to grinding heat within the tool-workpiece contact zone, most of which is eventually transferred into the workpiece and causes a grinding temperature rise. Figure 15.6 displays two typical grinding temperature curves measured using the thermocouples, which correspond to different single-layer CBN wheels at identical grinding parameters, i.e., the wheel speed of 120 m/s, the workpiece infeed speed of 6 m/min, and the depth of cut of 0.010 mm. The grinding temperature of 703°C with the electroplated CBN wheel is much higher than that of 569°C with the brazed counterpart. At this time, the burnout phenomenon could take place on the surface ground with the electroplated CBN wheel, as displayed in Figure 15.7a; on the contrary, the good ground surface is produced with the brazed wheel (Figure 15.7b).

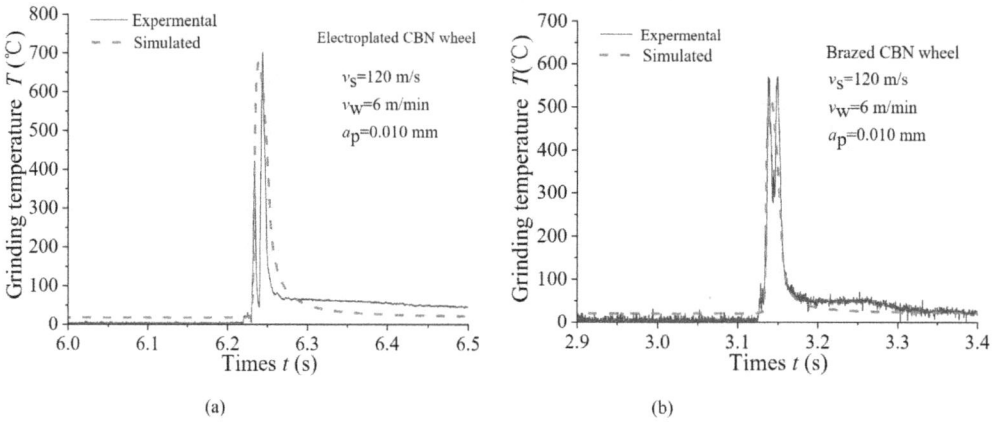

FIGURE 15.6 Typical grinding temperature curves: (a) electroplated CBN wheel, (b) brazed CBN wheel.

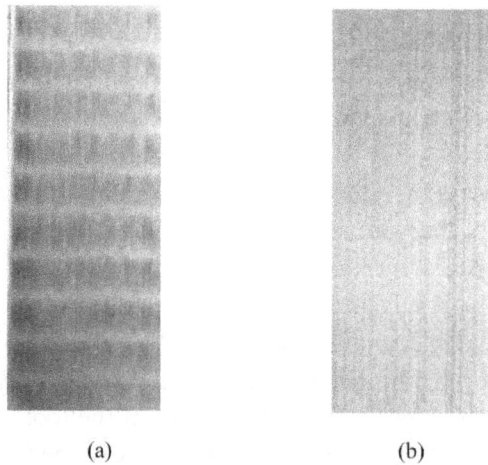

FIGURE 15.7 PTMCs surface produced in high-speed grinding: (a) electroplated CBN wheel, (b) brazed CBN wheel.

Because the grinding temperature beneath the workpiece surface could not be measured directly by using the semi-artificial thermocouple method in the current investigation, a three-dimensional (3D) finite element model (FEM) based on a moving triangular heat source was established to simulate the temperature distribution. The initial temperature considered for the workpiece is the room temperature, i.e., $T = 20°C$. In wet grinding, convective cooling is applied to the top surface to represent coolant application. A heat transfer coefficient of 82,000 W/m² is used to simulate the cooling effects of the grinding coolant fluid passing through the workpiece in a grinding pass. The bottom surface of the workpiece is regarded as thermally insulated from the environment. According to Newton's cooling law, this means that there is no heat loss from the workpiece to the environment. The characteristics of the moving triangular heat source and convective cooling during grinding are illustrated in Figure 15.8. The mesh method was applied to deal with the sharp temperature gradient in the workpiece top region while keeping a reasonable element size to reduce the calculating time. Therefore, the fine elements were designed in the top workpiece region, and the coarse ones were in the bottom region of the ground workpiece. The component was meshed using 20-node hex elements. The heat flux applied for the finite element simulation was computed from the average values of the tangential grinding forces, which varied from 3.49×10^7 W/m² to 6.48×10^7 W/m².

Figure 15.9 displays the results of the simulated temperature distribution contours within the workpiece ground with the electroplated and brazed CBN wheels, respectively. The grinding parameters are as follows: the wheel speed of 120 m/s, the workpiece infeed speed of 6 m/min, and the depth of cut of 0.010 mm. As for the grinding temperature, the 3D finite element model results and those measured using the semi-artificial thermocouple method show good

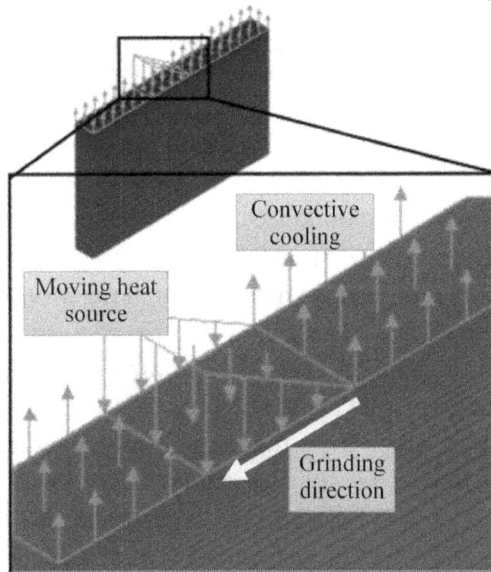

FIGURE 15.8 Finite element model based on the moving triangular heat source and convective cooling.

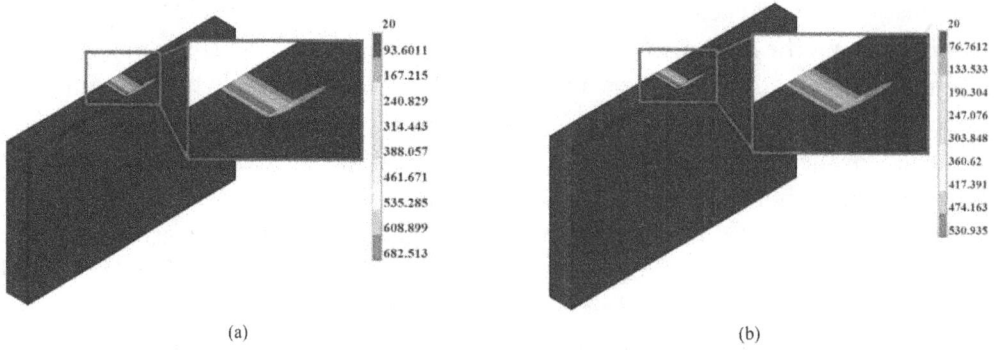

(a)   (b)

FIGURE 15.9 Simulated grinding temperature distribution: (a) electroplated CBN wheel, (b) brazed CBN wheel.

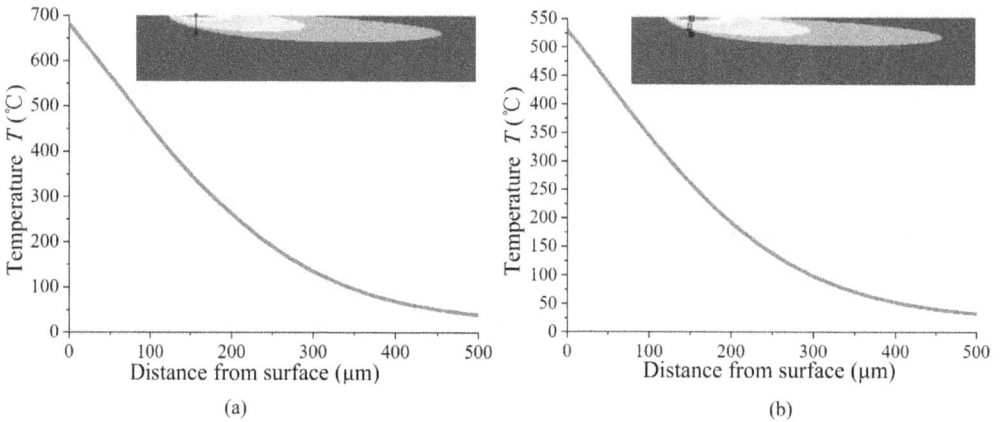

(a)   (b)

FIGURE 15.10 Temperature distribution beneath the workpiece surface along a vertical direction: (a) electroplated CBN wheel, (b) brazed CBN wheel.

consistency, as displayed in Figure 15.6. A high grinding temperature is always produced within the tool-workpiece contact zone when the thermal source moves along the workpiece surface. The heat conducted into the workpiece is transferred away to the surroundings during grinding, which results in a temperature field tail, as shown in Figure 15.9.

Figure 15.10 displays the temperature distribution along a vertical direction from the ground surface to the workpiece interior part within the tool-workpiece contact zone. The variation of the temperature gradient along the grinding depth direction (i.e., a depth of $200\,\mu m$ beneath the ground surface) could be observed. The grinding temperature in the tool-workpiece contact zone drops from the maximum temperature (i.e., 682°C) on the surface to the lower temperature (i.e., 190°C) at a depth of $250\,\mu m$ when the electroplated CBN wheel is applied, as shown in Figure 15.10a; meanwhile, the corresponding grinding temperature drops from 530°C on the surface to 192°C at a depth of $200\,\mu m$ when using the brazed wheel (Figure 15.10b).

Figure 15.11 displays the measured and simulated grinding temperature as functions of workpiece infeed speed and depth of cut. All the simulated temperatures using the triangle

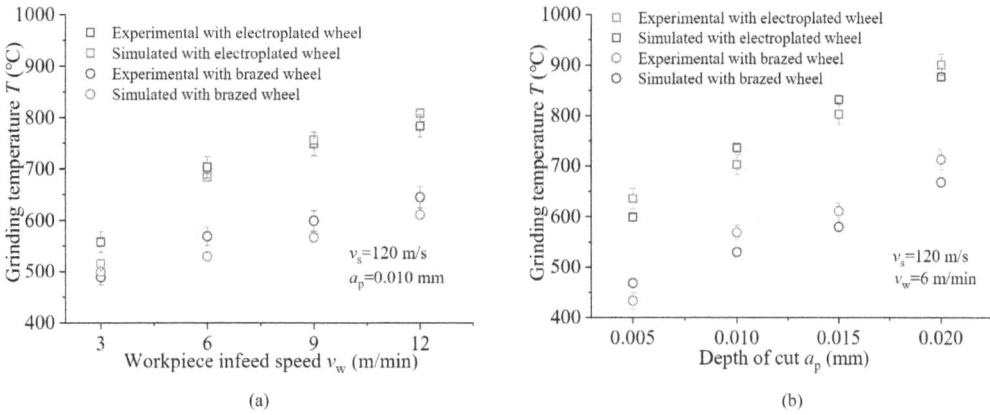

FIGURE 15.11 Effects of grinding parameters on grinding temperature: (a) workpiece infeed speed, (b) depth of cut.

heat source model show good agreement with those measured using the semi-artificial thermocouple method in the current high-speed grinding experiments. In general, the grinding temperature is significantly increased with an increase in the workpiece infeed speed and depth of cut. As seen in Figure 15.11a, when the depth of cut is 0.010 mm and the workpiece infeed speed is increased from 3 to 12 m/min, the grinding temperature is increased from 558°C to 784°C (by 40%) with the electroplated CBN wheel and from 489°C to 645°C (by 32%) with the brazed CBN wheel, respectively. When the depth of cut is increased from 0.005 to 0.020 mm and the workpiece infeed speed is fixed at 6 m/min, the grinding temperature is raised rapidly from 636°C to 901°C (by 42%) with the electroplated wheel (Figure 15.11b). For the single-layer brazed CBN wheel, the grinding temperature is raised from 433°C to 713°C (by 65%). The variation mentioned above is due to the fact that a higher workpiece infeed speed or a larger depth of cut always results in a larger tangential force. As a consequence, more heat flux enters the grinding zone, which produces a higher grinding temperature within the workpiece to a certain extent.

According to Figure 15.11, it is also known that the grinding temperature obtained with the brazed CBN wheel is always lower than that with the electroplated wheel under identical grinding parameters. Especially when the larger workpiece infeed speed or the higher depth of cut is utilized, the grinding temperature difference between the two CBN wheels is more significant, which is attributed to the different cooling conditions in the tool-workpiece contact zone formed with different CBN wheels in grinding. Klocke and Baus (Klocke et al. 2000) have reported that the effective flow of coolants was proportional to the low grinding temperature. In the current investigation, the difference in detailed cooling conditions formed with the electroplated CBN wheel and brazed counterpart, respectively, are schematically demonstrated in Figure 15.12. As for the electroplated CBN wheel, the grain protrusion is merely about 20%–30% of the whole grain particles; however, it could reach approximately 50%–70% of the whole grains for the brazed wheel. Here, the grain protrusion refers to the particular part that is exposed beyond the connecting layer of the grinding wheel. Under such conditions, when the

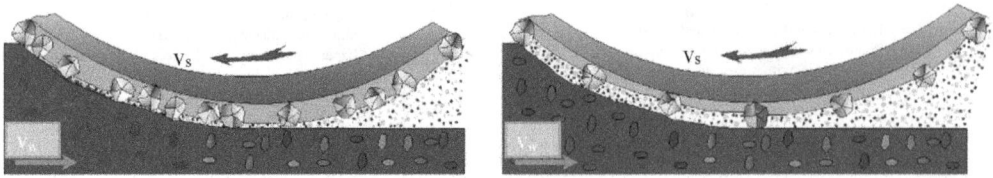

FIGURE 15.12 Comparison of the detailed cooling conditions within the tool-workpiece contact zone: (a) electroplated CBN wheel, (b) brazed CBN wheel.

electroplated CBN wheel is applied, the coolants could not have good effects due to the small space in the tool-workpiece contact zone during high-speed grinding. However, as for the brazed CBN wheel, the abundant storage space for coolants is provided owing to high grain protrusion, which ensures enough flow of the coolants and therefore produces excellent cooling effects in high-speed grinding. For this reason, lower grinding temperatures of PTMCs are always obtained with the single-layer brazed CBN wheel compared with the electroplated wheel. Quantitative analysis of the cooling effects in the tool-workpiece contact zone with the different single-layer CBN wheels will be an important topic in further investigation.

Furthermore, because the burnout phenomenon of the PTMCs surface would take place once the grinding temperature exceeded 700°C in the current experiments, the maximum material removal rate (MRR) could merely arrive at $1\,mm^3/(mms)$ when using the single-layer electroplated CBN wheel. Here, the material removal rate (MRR) is the product of the workpiece infeed speed and depth of cut. However, when the brazed CBN wheel is applied, even though the MRR has reached $2\,mm^3/(mm·s)$, the highest grinding temperature is still below 700°C, as displayed in Figure 15.11. That is to say, the maximum material removal rate in high-speed grinding of PTMCs with the brazed CBN wheel is much larger than that with the electroplated wheel.

15.5 GRINDING-INDUCED SURFACE FEATURES OF TITANIUM MATRIX COMPOSITES

The typical morphology of the ground PTMC surface is shown in Figure 15.13. Here, two grinding parameters are chosen: (i) the workpiece infeed speed of 6 m/min and the depth of cut of 0.005 mm; (ii) the workpiece infeed speed of 12 m/min and the depth of cut of 0.010 mm. The grinding-induced surface features, which are indicated by the number of arrows from 1 to 6 in Figure 15.13, include the grinding traces, fractured zone, voids, microcracks, debris adhere, and smooth zone. The voids and microcracks are always the dominant surface defects.

On the one hand, in the high-speed grinding process of PTMCs, the voids are produced due to the fracture and pullout of the TiC reinforcing particles under the grinding forces, as demonstrated schematically in Figure 15.14. Meanwhile, it is found from Figure 15.13 that the void size is distinctly increased with increasing workpiece infeed speed and depth of cut for the two types of CBN wheels. This is because the undeformed chip thickness increases with an increase in workpiece infeed speed

Ground with electroplated CBN wheel	Ground with brazed CBN wheel

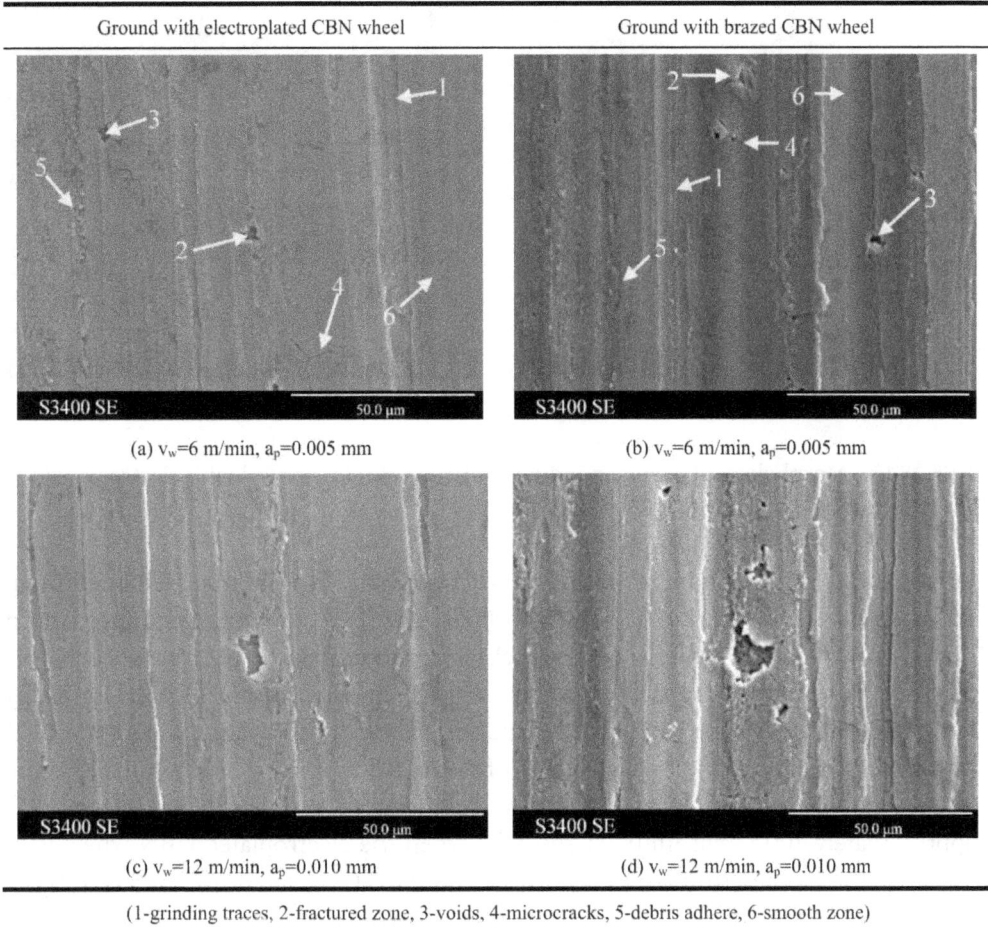

(a) v_w=6 m/min, a_p=0.005 mm

(b) v_w=6 m/min, a_p=0.005 mm

(c) v_w=12 m/min, a_p=0.010 mm

(d) v_w=12 m/min, a_p=0.010 mm

(1-grinding traces, 2-fractured zone, 3-voids, 4-microcracks, 5-debris adhere, 6-smooth zone)

FIGURE 15.13 Morphology of the surface of PTMCs produced in high-speed grinding.

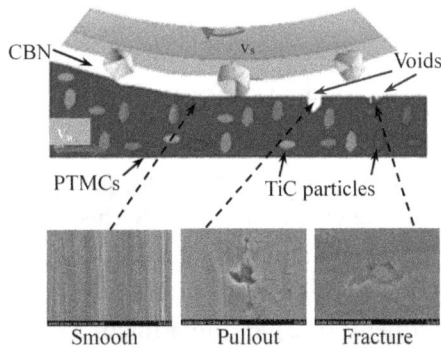

FIGURE 15.14 Schematics of the PTMCs surface features produced in high-speed grinding.

and depth of cut. In other words, the effective CBN grains taking part in the grinding practice are increased, thereby increasing the grinding force, which results in a great increase in the void size in high-speed grinding.

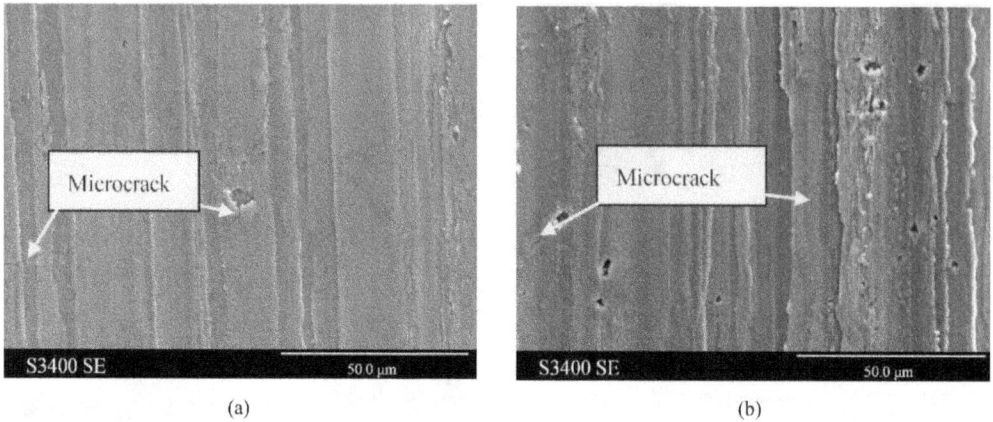

(a)   (b)

FIGURE 15.15   Morphology of grinding-induced microcracks: (a) ground with electroplated CBN wheel, (b) ground with a brazed CBN wheel.

On the other hand, grinding-induced microcracks are observed in the current investigation, as displayed in Figure 15.15, which is obtained at the workpiece infeed speed of 6 m/min and depth of cut of 0.010 mm. Microcracks are produced due to the fracture behaviour of the Ti-6Al-4V matrix materials under too large mechanical-thermal loads in grinding. For example, in the current high-speed grinding process, the surface temperature in the tool-workpiece contact zone rises at a rate of 2.85×10^{4}°C/s (from 20°C to 703°C within 0.024 seconds) and drops down to the lower temperature at 5.48×10^{4}°C/s (from 703°C to approximately 100°C within 0.011 seconds) when the electroplated CBN wheel is utilized, while the temperature rises abruptly at 2.50×10^{4}°C/s (from 20°C to 569°C within 0.022 seconds) and falls at 3.91×10^{4}°C/s (from 569°C to approximately 100°C within 0.012 seconds) when using the brazed CBN wheel. High grinding temperatures and abrupt temperature variation could result in large residual tensile stresses within the surface/subsurface layer of the ground PTMC specimens, which have a key influence on the formation of grinding-induced microcracks.

15.6 MATERIALS REMOVAL MECHANISM DURING GRINDING OF TITANIUM MATRIX COMPOSITES

A 2D finite element simulation is conducted using ABAQUS/Explicit 6.13.1 to reveal the formation mechanism of the ground surface, with special concern for the material removal mode of the reinforcements (Xi et al. 2017). Because the particles in PTMCs are more like an ellipse than a circle or a polygon, the particle is modelled as an ellipse with a minor axis semi-diameter of 2 μm and a major axis semi-diameter of 2.75 μm, as displayed in Figure 15.16. The particles are assumed to be perfectly bonded with the matrix, and the interface nodes of the matrix and particles are tied together. The contact arc length l_c is considered the square root of the product of the wheel diameter and the depth of cut, which is 0.9–2.8 mm in this work and 90–280 fold higher than the

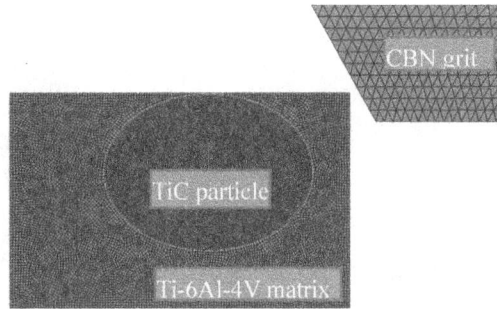

FIGURE 15.16   Finite element model of material removal behaviour of TiC particles in PTMCs during grinding.

length of the simulated workpiece. Therefore, the kinematic trajectory of the grains is assumed to be straight during the simulation. The undeformed chip thickness a_{gmax} in this work is 0.2–1.2 μm. Three values of undeformed chip thickness, such as 0.2, 0.7, and 1.2 μm, are selected to investigate the different grinding conditions.

The Johnson–Cook constitutive equation and shear failure mode are employed to characterize the matrix material, Ti-6Al-4V alloy, while the TiC particles are modelled as an isotropic, perfectly elastic material following the generalized Hook's law. Particle fracture simulation is performed using the brittle cracking model. The material constants of Ti-6Al-4V matrix, TiC reinforcements, and CBN grain are obtained from Refs. (Chen et al. 2011; Zhang et al. 2011; Dai et al. 2015). The friction during grinding is described by the Coulomb's Law. The critical point where chip separation occurs for Ti-6Al-4V matrix and TiC reinforcements is determined according to Refs. (Chen et al. 2011; Zhou et al. 2014), respectively.

In this work, interactions between the CBN grains and the TiC reinforcing particles are discussed. As the undeformed chip thickness of the CBN grain changes, different removal patterns of particles could occur when the grain approaches them, as displayed in Figure 15.17.

Figure 15.17a shows the simulation result of material removal at a large value of undeformed chip thickness, such as 1.2 μm, which is obtained in the case of the workpiece infeed speed of 12 m/min, depth of cut of 0.020 mm, and wheel speed of 80 m/s. The cutting path of CBN grain passes through about a quarter of the particle. Under such conditions, a rude fracture can occur in the upper part of the TiC particle due to the initiation and development of cracks. Then the particle debris could be washed away by the cooling fluids in grinding, and accordingly, a large and deep cavity was produced on the ground surface, as experimentally illustrated in Figure 15.17b. For this reason, the ground workpiece strength may seriously deteriorate due to surface defects.

In the case of a moderate value of undeformed chip thickness, such as 0.7 μm (Figure 15.17c), which is obtained at the workpiece infeed speed of 9 m/min, depth of cut of 0.010 mm, and wheel speed of 120 m/s, gentle particle fracture behaviour can be generated.

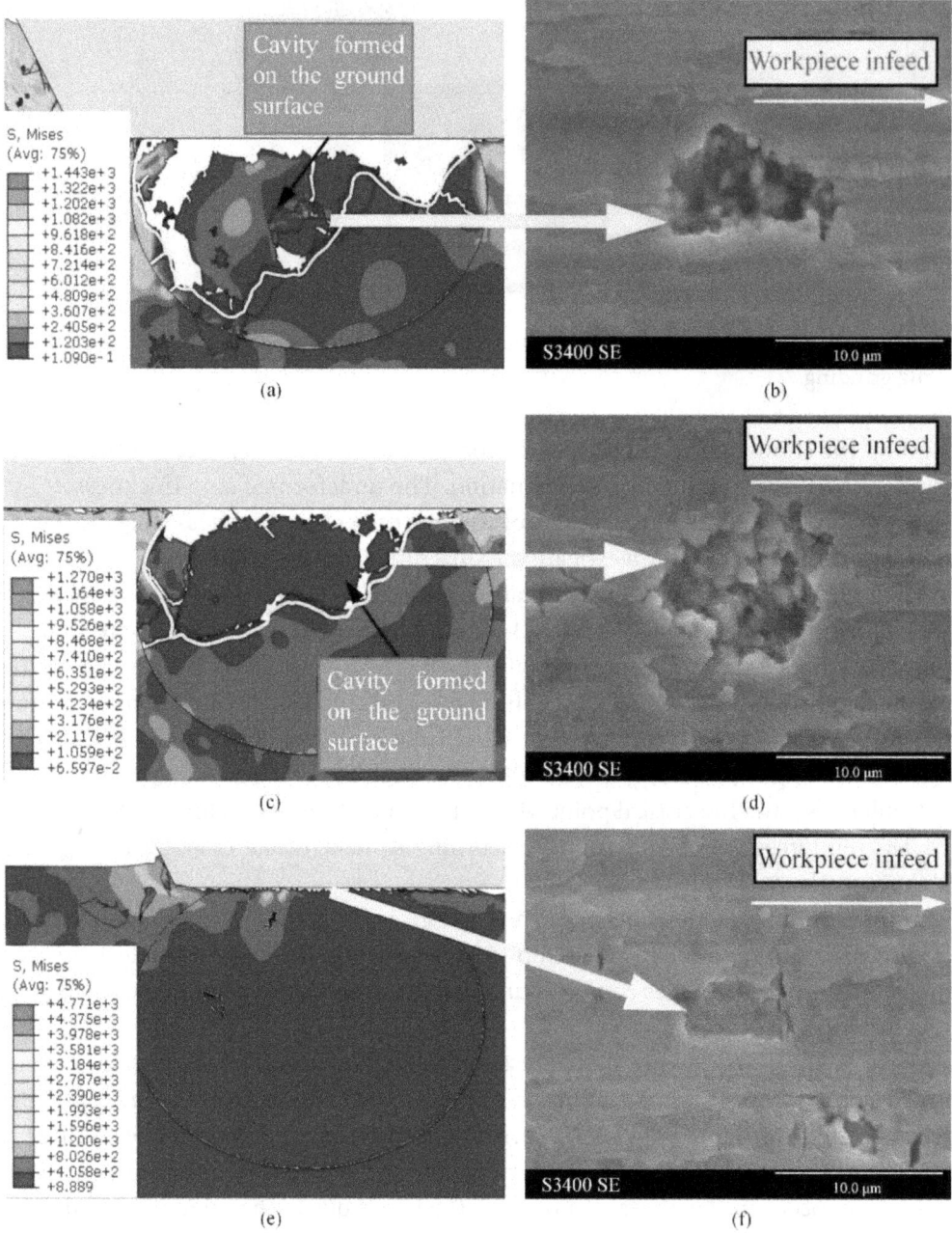

FIGURE 15.17 Simulation and experimental results of material removal of PTMCs in grinding: (a) simulation result at undeform chip thickness of 1.2 μm, (b) v_w = 12 m/min, a_p = 0.020 mm, v_s = 80 m/s, (c) simulation result undeform chip thickness of 0.7 μm, (d) v_w = 9 m/min, a_p = 0.010 mm, v_s = 120 m/s, (e) simulation result at undeform chip thickness of 0.2 μm, (f) v_w = 1 m/min, a_p = 0.002 mm, v_s = 140 m/s.

As such, a small and shallow cavity could be formed on the ground surface, as experimentally displayed in Figure 15.17d.

As for a small value of undeformed chip thickness, such as 0.2 µm, which is obtained at the workpiece infeed speed of 1 m/min, depth of cut of 0.002 mm, and wheel speed of 140 m/s, the cutting path of CBN grit passes through the top of TiC particles (Figure 15.17e). As such, the particle fracture is invisible, producing a good ground surface without significant defects, as experimentally shown in Figure 15.17f. This phenomenon can also be understood from the viewpoint of the ductile removal mechanism of brittle materials (such as Si_3N_4 and SiC ceramics) during grinding when an undeformed chip thickness below the critical value is applied (Liu et al. 2018).

REFERENCES

Aramesh, M., Y. Shaban, M. Balazinski, et al. 2014. Survival life analysis of the cutting tools during turning titanium metal matrix momposites (Ti-MMCs). *Procedia CIRP*, 14: 605–609.

Bejjani, R., B. Shi, H. Attia, et al. 2011. Laser assisted turning of titanium metal matrix composite. *CIRP Annals-Manufacturing Technology*, 60(1): 61–64.

Blau, P. J., and B. C. Jolly. 2009. Relationships between abrasive wear, hardness, and grinding characteristics of titanium-based metal-matrix composites. *Journal of Materials Engineering and Performance*, 18(4): 424–432.

Chen, G., C. Z. Ren, X. Y. Yang, et al. 2011. Finite element simulation of high-speed machining of titanium alloy (Ti-6Al-4V) based on ductile failure model. *International Journal of Advanced Manufacturing Technology*, 56(9): 1027–1038.

Choi, B. J., I. Y. Kim, Y. Z. Lee, et al. 2014. Microstructure and friction/wear behavior of (TiB+TiC) particulate-reinforced titanium matrix composites. *Wear*, 318: 68–77.

Dai, J. B., W. F. Ding, L. C. Zhang, et al. 2015. Understanding the effects of grinding speed and undeformed chip thickness on the chip formation in high-speed grinding. *International Journal of Advanced Manufacturing Technology*, 81(5): 995–1005.

Dong, G., S. Gao, L. Wang. 2022. Three dimensional shape model of TiBw mesh reinforced titanium matrix composites in rotary ultrasonic grinding. *Journal of Manufacturing Processes*, 75: 682–692.

Klocke, F., A. Baus, and T. Beck. 2000. Coolant induced forces in CBN high speed grinding with shoe nozzles. *CIRP Annals-Manufacturing Technology*, 49(1): 241–244.

Klocke, F., S. L. Soo, B. Karpuschewski, et al. 2015. Abrasive machining of advanced aerospace alloys and composites. *CIRP Annals-Manufacturing Technology*, 64(2): 581–604.

Li, Z., W. F. Ding, L. Shen, et al. 2016. Comparative investigation on high-speed grinding of TiCp/Ti-6Al-4V particulate reinforced titanium matrix composites with single-layer electroplated and brazed CBN wheels. *Chinese Journal of Aeronautics*, 29(5): 1414–1424.

Liu, C., W. Ding, T. Yu, et al. 2018. Materials removal mechanism in high-speed grinding of particulate reinforced titanium matrix composites. *Precision Engineering*, 51: 68–77.

Xi, X. X., W. F. Ding, Z. Li, et al. 2017. High speed grinding of particulate reinforced titanium matrix composites using a monolayer brazed cubic boron nitride wheel. *International Journal of Advanced Manufacturing Technology*, 90: 1529–1538.

Zhang, Y., H. L. Zhang, J. H. Wu, et al. 2011. Enhanced thermal conductivity in copper matrix composites reinforced with titanium-coated diamond particles. *Scripta Materialia*, 65(12): 1097–1100.

Zhou, L., Y. Wang, Z. Y. Ma, et al. 2014. Finite element and experimental studies of the formation mechanism of edge defects during machining of SiCp/Al composites. *International Journal of Machine Tools and Manufacture*, 84: 9–16.

CHAPTER 16

Speed Effect on Materials Removal during Grinding

16.1 INTRODUCTION

High-speed grinding with a cubic boron nitride (CBN) abrasive wheel is a highly efficient precision machining technique for difficult-to-cut materials, such as nickel-based superalloys and titanium alloys. However, the chip formation mechanism, which is a key issue for understanding the material removal mechanisms and optimizing the machining parameters to improve the workpiece performance, is not clear due to the difficulty in experimentally detecting the deformation of a chip in-situ (Dai et al. 2015). Apart from the high grinding speed, the simultaneous interactions between multiple abrasive grains and the workpiece material make the investigation more challenging.

Recently, researchers have tried to use single-grain grinding processes to explore chip formation mechanisms. For example, Xu et al. (2012) performed such tests on nickel-based superalloy Inconel 718 to study the chip morphologies at grinding speeds of 20–150 m/s. Chen et al. (2014) also investigated the grinding of EN24T steel, respectively, utilizing single-grain scratching tests. They reported that the grain was cut at the inlet side of the scratch but ploughed more at the outlet side.

The groove characteristics and chip formation behaviours were discussed (Rao et al. 2021; Zhao et al. 2023). Tawakoli et al. (2013) and Aurich and Steffes (2011) have carried out research to understand plastic flow in single-grain scratching. Also, the mechanisms of ductile material removal during the scratching of brittle materials (e.g., monocrystal sapphire and 3C-SiC) using a single-grained diamond (Liang et al. 2013; Yang et al. 2023; Wang et al. 2023). It was reported that the activation of sufficient dislocation and slipping systems in the material is a major factor in the ductile-regime cutting of alumina, whereas phase transformation under mechanical stresses dominates the ductile cutting of monocrystalline silicon. Also, Chen et al. (2014) utilized the finite element (FE) method to investigate the residual stress fields during ultra-high-speed grinding of 40Cr steel. The authors found that surface tensile residual stress decreases when the grinding speed is higher.

 DOI: 10.1201/9781032678047-19

However, some core issues, such as the critical grinding speed, have not yet been understood. As is commonly known, chip formation behaviour in grinding is mainly governed by the mechanical properties of workpiece materials. Sima and Özel (2010), Guo and Yen (2004), and Calamaz et al. (2008) postulated that both grinding speed and undeformed chip thickness have great influences on chip formation due to the combined effect of strain hardening, strain-rate hardening, and thermal softening.

At present, it is difficult to measure the dynamic characteristics of chip formation behaviour as a result of plastic strain, strain-rate, and stresses in-situ during the grinding process. Furthermore, the dynamic force signals in single-grain grinding are not measured very precisely. This is because the grain-workpiece interaction frequency, as a result of the high rotational speed of the grinding wheel, is much greater than the response frequency of the force measurement systems available. For example, grain-workpiece interaction frequencies tend to be 10 kHz for a wheel speed of 20 m/s, 40 kHz at 80 m/s, and 60 kHz at 120 m/s, but the highest natural frequency of a dynamometer tends to be far lower (e.g., that of a Kistler 4-Component Dynamometer 9272 is 6.3 kHz). A numerical analysis, therefore, is often a helpful technique to characterize chip formation behaviour and to explore the critical grinding speed.

In cutting or grinding, workpiece materials tend to experience elastic-plastic deformation under the coupled effects of high temperatures, strains, and strain rates. It is necessary to know the flow stress of workpiece material in such conditions to better understand the mechanisms of chip formation, grain wear, damage, etc. The FE method is, therefore, an effective tool to analyse machining processes such as cutting and grinding, as it can provide a deep understanding of the variations in force, temperature, stress, and strain. This can be used to model changes to the system, such as tool wear and surface integrity.

The work will use Abaqus/Explicit to simulate the process of grinding the Inconel 718 alloy under single-grain grinding and to investigate the effects of the grinding speed and undeformed chip thickness on chip formation to determine the critical grinding speed.

16.2 FINITE ELEMENT MODELLING

A 2D FE model for the single-grain grinding process of Inconel 718 alloy has been established using the FE commercial software Abaqus/Explicit v6.12. The explicit method was utilized mainly due to the advantages of computational efficiency for highly nonlinear problems in most mechanical manufacturing processes and its suitability for modelling brief and transient dynamic events. The FE model is composed of the workpiece and grain. Then the single-grain grinding process is performed by a two-dimensional plane-strain-coupled thermo-mechanical analysis using orthogonal assumptions.

16.2.1 Materials Model

The grain material is CBN. Compared with workpiece material, the hardness of grain is much higher. Therefore, the grain is modelled as an isothermal, rigid body. Some material properties of the grain are listed in Table 16.1.

The behaviour of the workpiece material, Inconel 718 alloy, was described by the Johnson–Cook (J-C) constitutive model, including the effects of deformation hardening, strain rate, and temperature-dependent factors. The equivalent plastic flow stress (σ), namely the instantaneous yield strength, was taken as a function of the von Mises equivalent plastic strain $\dot{\varepsilon}$, the dimensionless plastic strain rate ($\dot{\varepsilon}/\dot{\varepsilon}_0$) ($\dot{\varepsilon}_0$=0.001s^{-1} is a quasi-static reference), and the homologous temperature T^*=$(T-T_{room})/(T_{melt}-T_{room})$. Thus,

$$\sigma=\left(A+B\varepsilon^n\right)\left(1+C\ln\dot{\varepsilon}^*\right)\left(1-\left(T^*\right)^m\right) \tag{16.1}$$

where A is the yield strength, B is the hardening modulus, C is the strain rate stringing coefficient, m is the thermal softening coefficient, and n is the strain-hardening coefficient; T_{room} and T_{melt} refer to the use of ambient temperature and material melting temperature, respectively, and T is the current grinding temperature. The J-C material model parameters used for Inconel 718 alloy are summarized in Table 16.1.

TABLE 16.1 Material Properties of Inconel 718 Alloy Workpiece and CBN Grain

Material	E (GPa)	μ	ρ (kg/m³)	A (MPa)	B (MPa)	C	n	m	T_melt (°C)	T_room (°C)
						Johnson–Cook Model Parameters				
Inconel 718	220	0.3	8,420	450	170	0.017	1.3	0.65	1,320	20
CBN	909	0.12	3,120							

16.2.2 Criterion of Material Fracture

Shear fracture, as a consequence of shear band localization, is one of the main mechanisms in the fracture of ductile metallic materials, including the nickel-based superalloy Inconel 718 to be investigated in this study. Therefore, in the present FE analysis, chip formation was modelled to be achieved by shear failure, with the equivalent plastic strain taken as a measure of the failure. In doing so, the failure occurs when the damage parameter w_s exceeds a unit. The damage parameter w_s was defined by

$$w_s=\sum\left(\frac{\Delta\varepsilon^{pl}}{\varepsilon_f^{pl}}\right) \tag{16.2}$$

where $\Delta\varepsilon^{pl}$ is an increment of the equivalent plastic strain and ε_f^{pl} is the strain at failure, in which the summation is performed over all increments.

16.2.3 Contact Law

Whilst grinding metallic materials, deformation and friction produce heat. If the induced temperature rise is high enough, a structural change in a workpiece material can happen during grinding. In the present FE analysis, friction between the grain and the workpiece material was described by the Coulomb's Law $\tau_f=\mu\sigma_n$, where τ_f is the frictional stress, σ_n is the normal contact stress, and μ is the coefficient of friction (taken to be 0.3). *Abaqus/Explicit* allows the introduction of an inelastic heat fraction, called the Taylor-Quinney empirical constant (often assigned as 90%), to reflect the percentage of plastic deformation and friction work converted to heat in grinding.

16.2.4 Two-dimensional Adaptive FE Model

Figure 16.1 schematically displays a single-grain grinding model. The size of the workpiece was 5 mm and 1 mm in the x- and y-directions, respectively, as displayed in Figure 16.2. Arbitrary Lagrangian-Eulerian (ALE) adaptive meshing was used to maintain a high-quality mesh to prevent possible errors due to severe mesh distortion. The workpiece was discretized by 30,000 bilinear, four-noded quadrilateral elements with reduced integration (CPE4R). As shown in Figure 16.1, the CBN grain shape is more like a regular hexagon in two dimensions. For simplicity, a half-regular hexagon with a rake angle of $-30°$ is assumed for the CBN geometry in the FE model, as displayed in Figure 16.2. The CBN grain size is 620 μm. In the simulations, the bottom edge of the workpiece was constrained in both the x- and y-directions, and its left edge was fixed in the horizontal direction.

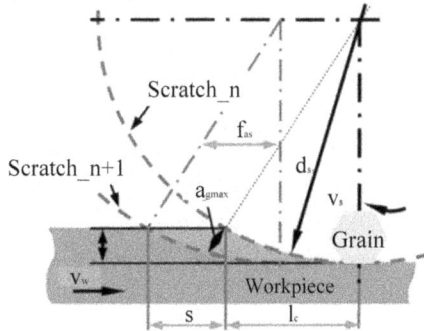

FIGURE 16.1   A single-grain grinding process.

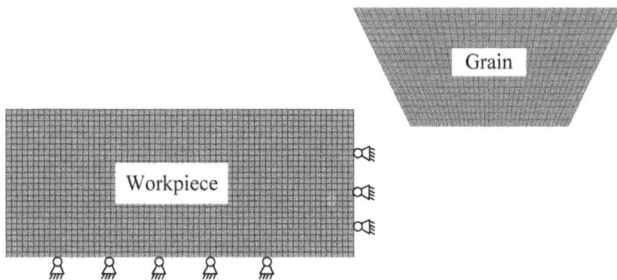

FIGURE 16.2   The initial FE mesh.

To investigate the effects of grinding speed (20–400 m/s) and undeformed chip thickness ($a_{gmax} = 1, 2, 4, 6,$ and 8 μm) on the chip formation behaviour, the grinding parameters listed in Table 16.2 were chosen. The undeformed chip thickness (a_{gmax}) can then be calculated by

$$a_{g\max} \approx 2 \cdot \pi \cdot d_{eq} \cdot \left(\frac{v_w}{v_s}\right)\sqrt{\frac{a_p}{d_{eq}}} \qquad (16.3)$$

where a_p is the depth of the cut, d_{eq} is the grinding wheel diameter (400 mm), v_w is the workpiece feed-in speed, and v_s is the grinding wheel speed.

TABLE 16.2   Grinding Parameters

No.	Grinding Wheel Speed v_s (m/s)	Workpiece Infeed Speed v_w (m/min)	Dept of Cut a_p (μm)	Maximum Undeformed Chip Thickness $a_{g\max}$ (μm)
1	20	0.055, 0.111, 0.222, 0.330, 0.443	30	1, 2, 4, 6, 8
2	40	0.111, 0.222, 0.443, 0.665, 0.886		
3	60	0.166, 0.332, 0.665, 0.997, 1.329		
4	80	0.222, 0.443, 0.887, 1.329, 1.773		
5	100	0.277, 0.554, 1.108, 1.661, 2.215		
6	120	0.333, 0.665, 1.330, 1.995, 2.660		
7	150	0.415, 0.831, 1.662, 2.492, 3.323		
8	180	0.498, 0.997, 1.994, 2.990, 3.987		
9	200	0.554, 1.108, 2.215, 3.323, 4.430		
10	300	0.831, 1.661, 3.323, 4.984, 6.645		
11	400	1.108, 2.215, 4.430, 6.645, 8.860		

16.3 CHIPPING STAGES IN SURFACE GRINDING

Chip formation during the grinding process of Inconel 718 with a single CBN grain can be separated into three stages: tool cut-in, chip generation, and chip formation, as shown in Figure 16.3. When compared to the element shape before cutting (Figure 16.1), significant shear deformation can be seen in the cut-in stage (Figure 16.3a). Meanwhile, a plastic flow region is observed to grow towards the free surface of the workpiece. As shown in Figure 16.3b, when the chip is initiated, the plastic flow region is narrowed down to a shear band that extends from the grain tip to the free surface, a typical characteristic of cutting before chip curling. Once chip curling occurs, the plastic flow region bends and covers the chip-grain contact area (Figure 16.3c). A chip segment is formed accordingly.

The duration of each segmented chip generation is extremely short and varies with the grinding speed. For example, when the undeformed chip thickness ($a_{g\max}$) is fixed to be 8 μm, the total duration of the chip formation stage decreases with increasing the grinding speed, which is 5.7 μs at 20 m/s, 1.25 μs at 80 m/s, 0.8 μs at 120 m/s, 0.45 μs at 180 m/s, 0.42 μs at 200 m/s, 0.3 μs at 300 m/s, and 0.255 μs at 400 m/s.

Figure 16.4 shows the effects of grinding speed and undeformed chip thickness on segmentation frequency according to the results of FE simulations. It can be seen that the frequency increases linearly with grinding speed, which correlates well with

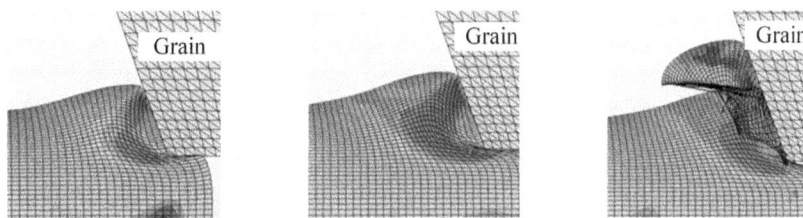

FIGURE 16.3 Three stages of chip formation during grinding (grinding speed of 150 m/s): (a) tool cut-in, (b) chip generation, (c) chip formation.

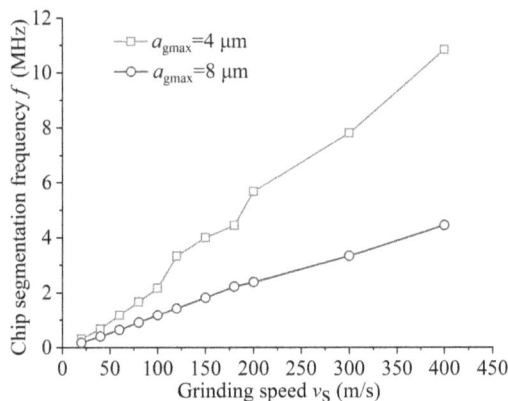

FIGURE 16.4 Chip segmentation frequency.

findings by other researchers. When the grinding speed is given, the chip segmentation frequency changes with the undeformed chip thickness. For example, undeformed chip thickness $a_{gmax} = 4\,\mu m$ gives a remarkable lager frequency compared with that of $a_{gmax} = 8\,\mu m$. It has been pointed out that the chip segmentation frequency is comprehensively affected by thermal softening and strain and strain-rate hardening of the adiabatic shear banding in chip formation.

Figure 16.5 shows the equivalent plastic strain contours in the three chip formation stages where a grinding speed of 150 m/s and an undeformed chip thickness of 8 μm were used. In the cut-in stage, a deformation zone forms in front of the tool tip, and the maximum plastic strain (3.071) of the deformation zone is relatively low. As the grinding proceeds, the strain increases significantly and spreads to a narrow zone where shear banding takes place in the chip generation stage. As observed in Figure 16.5c, deformation is significant in the adjacent shear bands but little in the material between the shear bands. This may be due to the low heat conductance of the ground Inconel 718 alloy.

Figure 16.6 shows the corresponding temperature contours at the different chip formation stages. As can be seen in Figs. 16.6a and b, the highest temperature (up to 1,350°C) appears in the primary deformation zone and close to the grain tip in the tool cut-in and chip generation stages. However, Figure 16.6c shows that the highest temperature

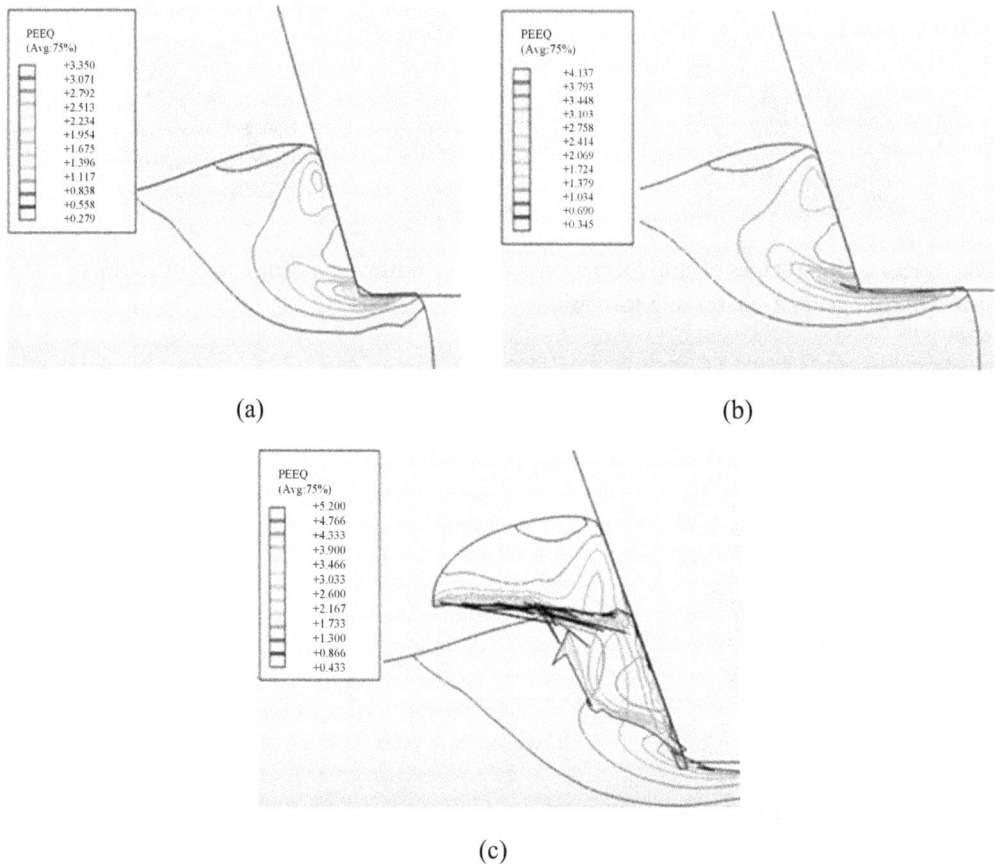

(a)

(b)

(c)

FIGURE 16.5   Equivalent plastic strain contour (grinding speed of 150 m/s): (a) tool cut-in, (b) chip generation, (c) chip formation.

(1,553°C) is in the contact area of the primary shear zone and along with the grain–chip interface in the chip formation stage. In this stage, the temperature is far higher than those in the tool cut-in and chip generation stages, indicating that a mass of heat has been generated due to the contact friction between the workpiece and grain tool. By comparing Figures 16.5 and 16.6, it can be observed that the shape and distribution of the contours are similar for the equivalent plastic strain and the grinding temperature in the corresponding chip formation stages. The smaller deformation taking place in the regions between the shear bands, as displayed in Figure 16.5c, leads to a lower temperature increase; in the FE simulations, the highest temperature in the shear bands reaches 1,129°C or more, whereas the less deformed regions remain at a temperature below 423°C, as displayed in Figure 16.6c.

This phenomenon can be explained as follows: The higher equivalent plastic strain reflects drastic plastic deformation of the material under grinding. The severer the plastic deformation, the more the energy consumed, leading to more heating and therefore higher temperatures.

(a)

(b)

(c)

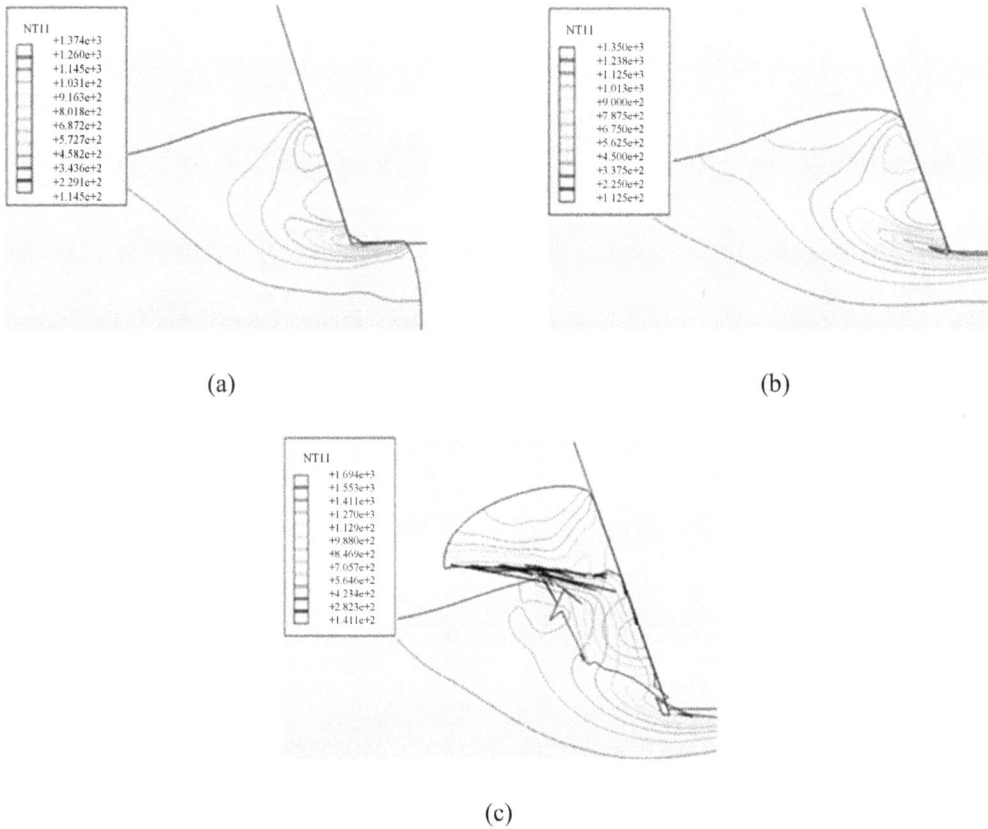

FIGURE 16.6 Temperature distribution contour (grinding speed of 150 m/s): (a) tool cut-in, (b) chip generation, (c) chip formation.

16.4 EFFECT OF GRINDING SPEED ON CHIP FORMATION

Figures 16.7 and 16.8 illustrate the strain and stress distribution contours during chip formation at various typical grinding speeds (60–400 m/s). Obviously, the stress and equivalent plastic strain are greatly influenced by grinding speed. For example, the maximum equivalent plastic strain is 5.107 at a wheel speed of 60 m/s, 5.271 at 150 m/s, 4.956 at 200 m/s, 4.589 at 300 m/s, and 4.328 at 400 m/s. The rate of variation of the maximum equivalent plastic strain is 17.9% at 150 m/s when compared to that at 60 m/s. The maximum von Mises stress is found to be 1.772 GPa at 60 m/s, 1.927 GPa at 150 m/s, 1.825 GPa at 200 m/s, 1.56 GPa at 300 m/s, and 1.465 GPa at 400 m/s. This variation rate in von Mises stress can reach 24% at 150 m/s when compared with that at 60 m/s.

Figure 16.9 illustrates the changes in the maximum equivalent plastic strain and maximum von Mises stress under different grinding speeds. Here, the undeformed chip thickness (a_{gmax}) is fixed (at 4 or 8 μm). The curves of the maximum von Mises stress and the maximum equivalent plastic strain exhibit an identical variation trend, and

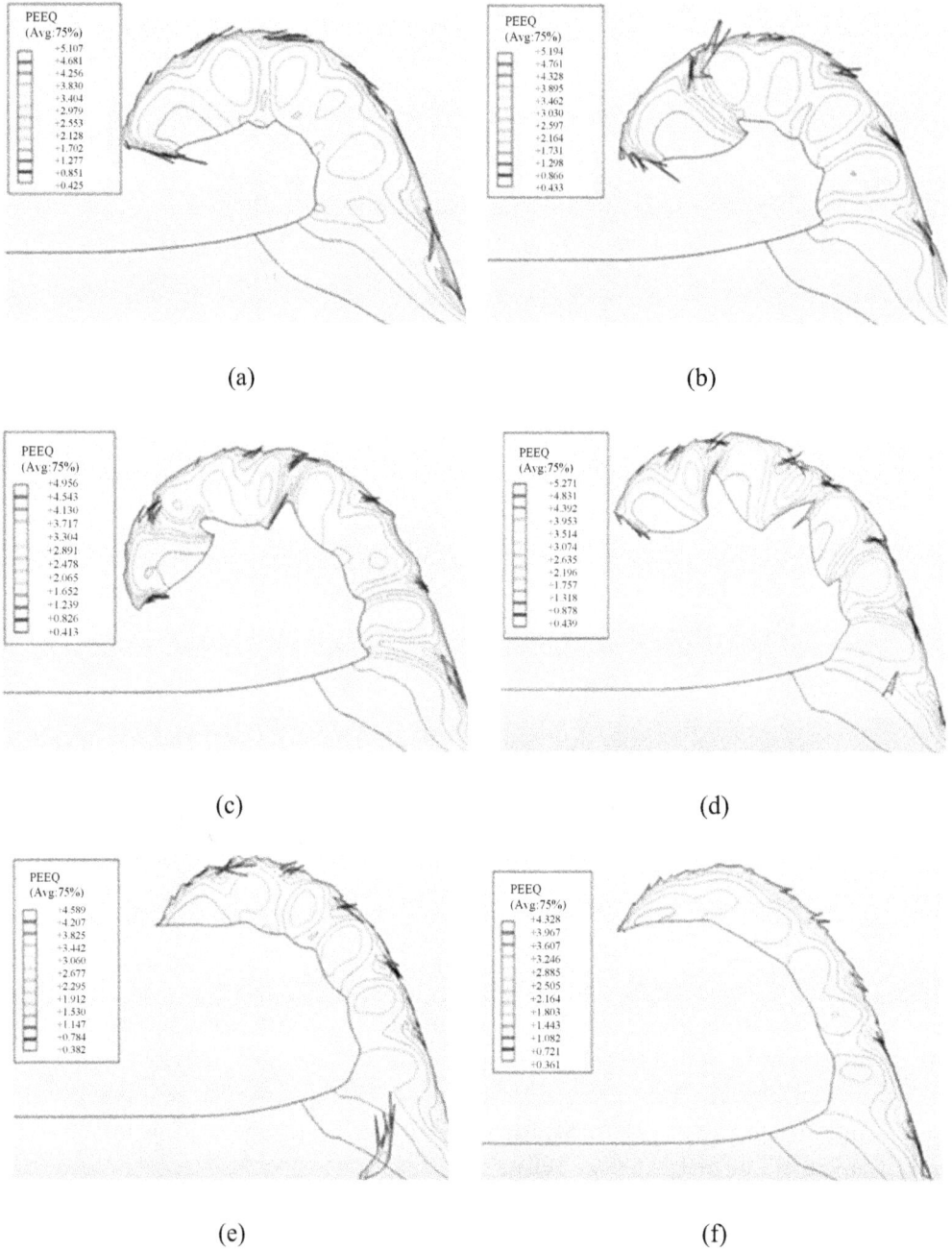

(a)

(b)

(c)

(d)

(e)

(f)

FIGURE 16.7 Effect of the grinding speed on the equivalent plastic strain distributions: (a) $v_s = 60$ m/s, (b) $v_s = 120$ m/s, (c) $v_s = 150$ m/s, (d) $v_s = 200$ m/s, (e) $v_s = 300$ m/s, (f) $v_s = 400$ m/s.

they increase with increasing grinding speed between 60 and 150 m/s but then decrease between speeds of 150 and 400 m/s. The highest values are both at the speed of 150 m/s, which is likely to be due to the coupled effect of strain and strain-rate hardening and thermal softening.

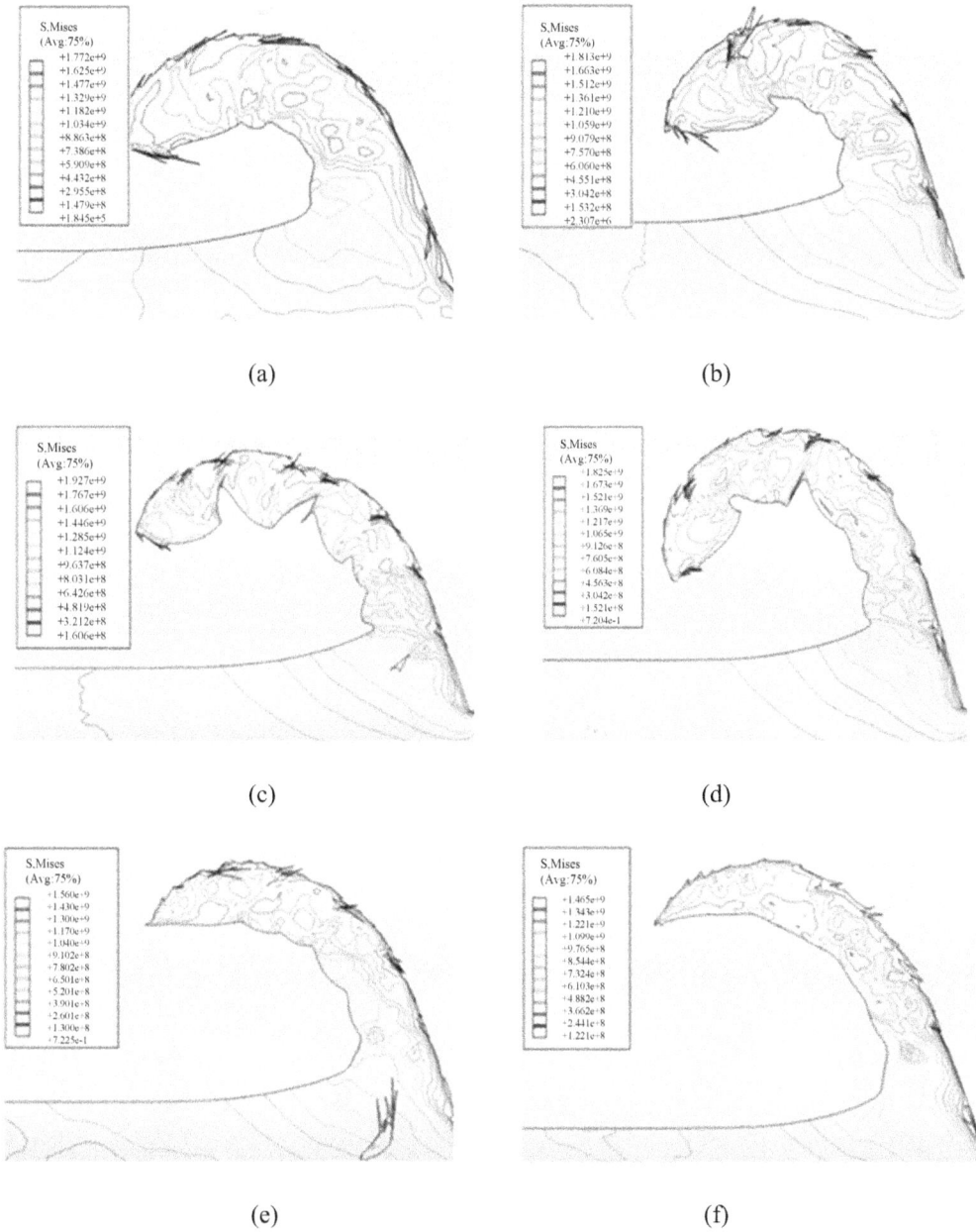

(a)

(b)

(c)

(d)

(e)

(f)

FIGURE 16.8 Effect of grinding speed on von Mises stress distributions: (a) v_s=60 m/s, (b) v_s=120 m/s, (c) v_s=150 m/s, (d) v_s=200 m/s, (e) v_s=300 m/s, (f) v_s=400 m/s.

16.5 GRINDING FORCES DURING CHIP FORMATION

Figure 16.10a shows the simulated grinding forces at the grinding speed of 150 m/s, showing that the forces fluctuate during the steady grinding stage. Although the force curves are complex, certain dynamic details of the instantaneous forces can be observed. The force fluctuations from 0 to 15 μs, as magnified in Figure 16.10b, correspond to shear bands

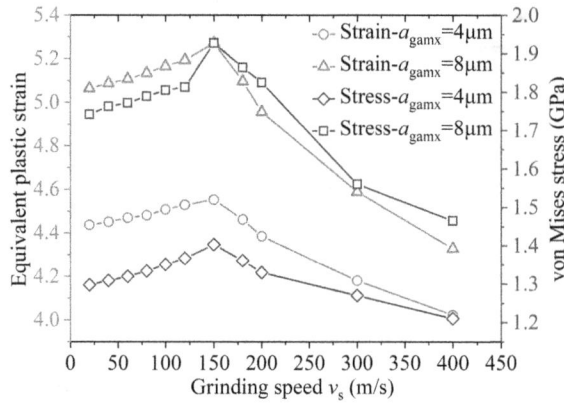

FIGURE 16.9 Effect of grinding speeds on the maximum values of equivalent plastic strain and von Mises stresses.

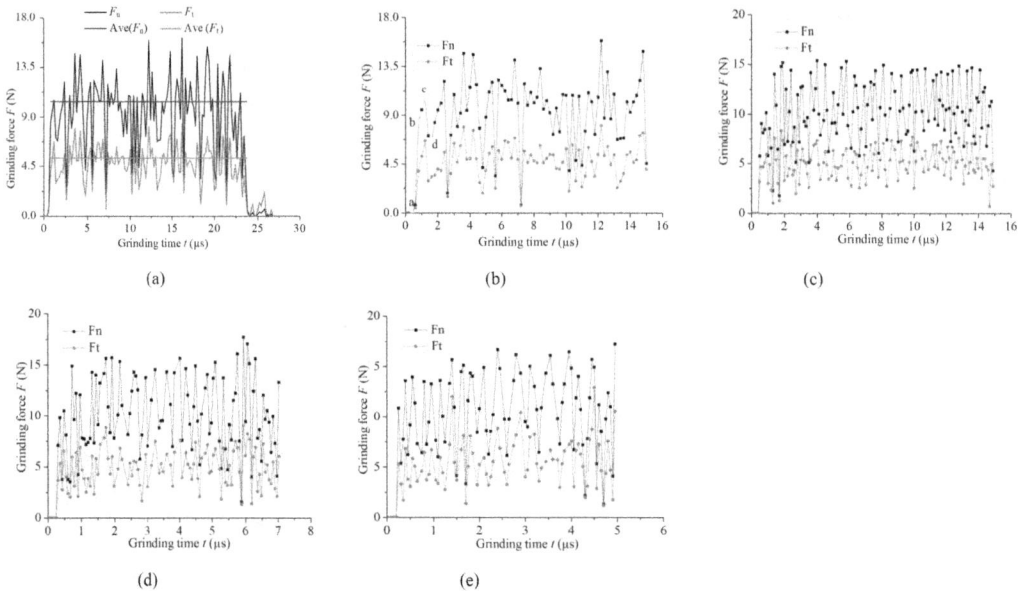

FIGURE 16.10 Grinding force variation versus grinding speed: (a) force curve (grinding speed of 150 m/s), (b) magnified curve (grinding speed = 150 m/s), (c) force curve (grinding speed of 200 m/s), (d) force curve (grinding speed of 300 m/s), (e) force curve (grinding speed of 400 m/s).

during grinding. For example, line segment a–b corresponds to the tool cut-in stage, where elastic deformation occurs and the normal grinding force increases linearly from 0 to 8.4 N. The chip generation begins at Point b, with the grinding force again increasing, thought to be a result of plastic deformation taking place once the flow stress reaches the yield stress point of the workpiece material. The workpiece material then continuously accumulates on the grain surface. The normal grinding force reaches a peak value (10.5 N) at the end of the chip-generation stage. The chip formation stage begins at Point c. The normal grinding force falls to a trough (6.8 N) while the lower edge forms a sawtooth, accompanied by shear sliding of the adiabatic shear band. Finally, the segmented chip formation is complete. According to Figs. 16.10b–e, there are a total of 17 force peaks in the normal force curve during the grinding period from 4 to 14 μs at the grinding speed of 150 m/s, 24 peaks during the grinding period of 4 to 14 μs at a grinding speed of 200 m/s, 14 peaks during the grinding period of 2 to 6 μs at a grinding speed of 300 m/s and 11 peaks during the grinding period of 1 to 4 μs at 400 m/s. Hence, the fluctuation frequency of the grinding force is 1.7 MHz for a grinding speed of 150 m/s, 2.4 MHz for a grinding speed of 200 m/s, 3.5 MHz for 300 m/s, and 3.67 MHz for 400 m/s, respectively. In contrast, the chip segmentation frequency obtained is 1.818 MHz for a grinding speed of 150 m/s, 2.38 MHz for 200 m/s, 3.3 MHz for 300 m/s, and 3.67 MHz for 400 m/s. The deviation in the fluctuation frequency of the simulated force and the chip segmentation frequency is merely within a variation tolerance of 7%, indicating that the fluctuation frequency of the grinding force agrees well with the segmentation frequency of the saw-tooth chips. It can, therefore, be considered that the grinding force fluctuation is predominantly caused by the saw-tooth chip formation in the grinding process. This phenomenon agrees with our understanding of the correlation between chip segmentation frequency and force fluctuation frequency.

Figure 16.11 provides the details of the average grinding force under different grinding speeds (20–400 m/s) and different undeformed chip thicknesses (1–8 μm). The average force was obtained based on the arithmetic mean of the simulated force curve in the steady stage, as shown in Figure 16.10. In the case of the undeformed chip thickness of 8 μm, the average normal force increases significantly from 8.4 to 10.4 N for grinding speeds ranging from 20 to 150 m/s and then decreases rapidly by 34.6% to 6.8 N when the speed changes from 150 to 400 m/s. However, when the undeformed chip thickness equals 4 or 6 μm, the variations in grinding forces are similar to those of $a_{gmax} = 8$ μm, with the magnitude of the average normal grinding force changing only slightly by 14.45% and 21.1% for the 4 and 6 μm thicknesses, respectively. Further, when $a_{gmax} = 1$ or 2 μm, the variation is even less than 10%. This indicates that the undeformed chip thickness significantly affects the grinding force variation during high-speed grinding and

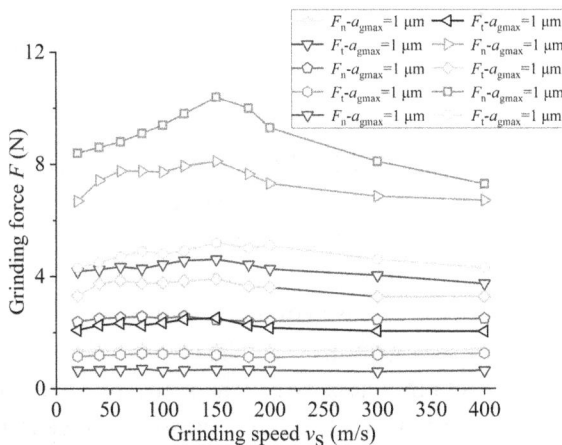

FIGURE 16.11 Average grinding force versus grinding speed and undeformed chip thickness.

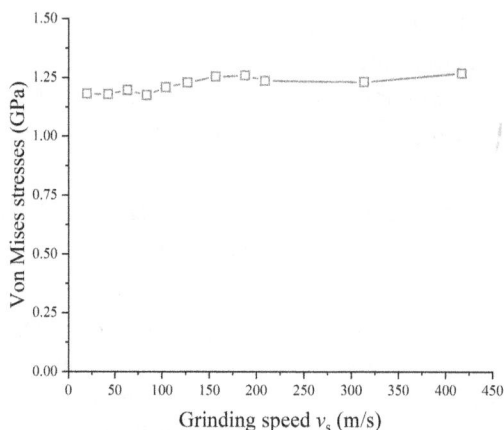

FIGURE 16.12   von Mises stress variation versus grinding speed ($a_{gmax} = 2\,\mu m$).

that the grinding speed effect on the force becomes insignificant when the undeformed chip thickness is less than $2\,\mu m$. Figure 16.12 displays the von Mises stresses versus the grinding speeds for samples with $a_{gmax} = 2\,\mu m$. The von Mises stresses at particular points $6\,\mu m$ above the cutting plane in the primary deformation zone are 1.176, 1.162, 1.215, 1.243, 1.224, 1.218, and 1.256 GPa for grinding speeds of 20, 80, 120, 150, 200, 300, and 400 m/s, respectively. This shows that the stress changes within a narrow range of about 5%, further confirming that the grinding speed effect becomes insignificant when the undeformed chip thickness is less than $2\,\mu m$. This trend can be explained as follows: the initial increase in the grinding force with increasing the grinding speed (below 150 m/s) is largely a result of strain and strain-rate hardening, and the subsequent decrease in the grinding force with the further grinding speed increase (above 150 m/s) can be attributed to the temperature rise and the consequent thermal softening.

16.6 RESULTANT STRESSES UNDER DIFFERENT GRINDING SPEEDS

Figure 16.13 shows the simulated stress-strain history experienced by a material point at a position $13\,\mu m$ above the cutting plane (element code is 6152). The stress-strain curves clearly show the three stages: elastic deformation (strain ranging from 0 to 0.1), plastic deformation (strain ranging from 0.1 to 1.1), and failure (strains beyond 1.1). Yield stress is dependent on strain rate. As can be seen in Figure 16.13, the yield stress of Inconel 718 is 596 MPa at a grinding speed of 20 m/s, 623 MPa at a grinding speed of 80 m/s, 672 MPa at a grinding speed of 120 m/s, 695 MPa at a grinding speed of 150 m/s, 609 MPa at a grinding speed of 200 m/s, 570 MPa at a grinding speed of 300 m/s, and 522 MPa at a grinding speed of 400 m/s.

In Figure 16.14, the yield stress versus grinding speed curve is fitted based on the yield stresses of Inconel 718 at different grinding speeds, obtained based on the data exhibited in Figure 16.13. The yield stress increases when the grinding speed increases from 20 to 150 m/s, but decreases afterwards when the grinding speed exceeds 150 m/s. The highest yield stress reaches 695 MPa when a grinding speed of 150 m/s is used. This is likely to be

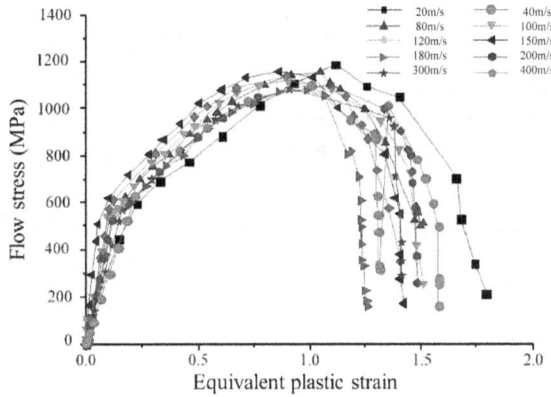

FIGURE 16.13 Effect of grinding speeds on the effective stress-strain relationship ($a_{\text{gmax}} = 8\,\mu\text{m}$).

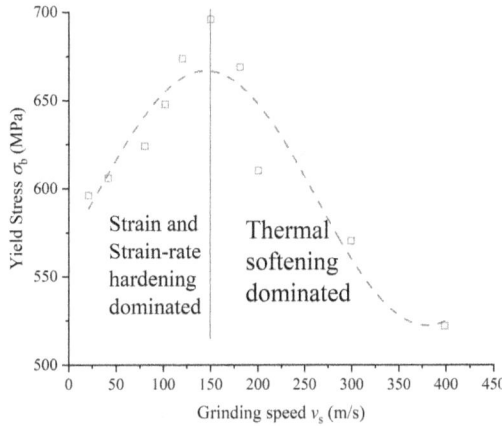

FIGURE 16.14 Effect of the grinding speeds on the yield stresses of the ground materials ($a_{\text{gmax}} = 8\,\mu\text{m}$).

because dynamic strain hardening and strain rate hardening become more competitive than thermal softening, leading to a significant yield stress rise with increasing the grinding speed from 20 to 150 m/s. Thermal softening starts to contribute more after that.

To further understand the effect of grinding speed, the dependence of the equivalent stresses on the strain, strain rate, and temperature are considered. As Inconel 718 was described by the J-C plasticity model, when the strain rate and temperature are constant, the equivalent stress increment $\Delta\sigma_1$ due to a strain increment can be expressed as

$$\Delta\sigma_1 = \left(\varepsilon_2^n - \varepsilon_1^n\right)\left(1 + C\ln\left(\frac{\dot{\varepsilon}}{\dot{\varepsilon}_0}\right)\right)\left(1 - \left(\frac{T - T_{\text{room}}}{T_{\text{melt}} - T_{\text{room}}}\right)^m\right) \tag{16.4}$$

where ε_1 and ε_2 represent two different strains of consequent strains before and after the strain increment, respectively.

By keeping the strain and temperature constant, the equivalent stress increment $\Delta\sigma_2$ caused by a strain rate increment can then be expressed as

$$\Delta\sigma_2 = \left(A + B\varepsilon^n\right)\left(C\ln\left(\frac{\dot{\varepsilon}_2 + \dot{\varepsilon}_0}{\dot{\varepsilon}_1 + \dot{\varepsilon}_0}\right)\right)\left(1 - \left(\frac{T - T_{\text{room}}}{T_{\text{melt}} - T_{\text{room}}}\right)^m\right) \tag{16.5}$$

where $\dot{\varepsilon}_1$ and $\dot{\varepsilon}_2$ represent two consequent strain rates before and after the strain rate increment, respectively.

Further, by keeping the strain and strain rate constant, the equivalent stress increment $\Delta\sigma_3$ caused by a temperature increment can be written as

$$\Delta\sigma_3 = \left(A + B\varepsilon^n\right)\left(1 + C\ln\left(\frac{\dot{\varepsilon}}{\dot{\varepsilon}_0}\right)\right)\left(\frac{\left(T_1 - T_{\text{room}}\right)^m - \left(T_2 - T_{\text{room}}\right)^m}{\left(T_{\text{melt}} - T_{\text{room}}\right)^m}\right) \tag{16.6}$$

where T_1 and T_2 present two consequent temperatures before and after the temperature increment, respectively.

Given an undeformed chip thickness of 8 μm, the equivalent plastic strain, strain rate, and temperature values at the same point above under different grinding speeds can be obtained from the simulation results. According to Eqs. (16.4) to (16.6), the variation in the equivalent stresses due to increasing strain, strain rate, and temperature can also be obtained. These are listed in Table 16.3.

Figure 16.15 shows the magnitude of equivalent stress variation from Table 16.3, compared to the quasi-stable state ($\varepsilon = 0$, $\dot{\varepsilon} = 0.001$, $T = 20°C$) under different grinding speeds. According to Table 16.3, the equivalent plastic strain values are ~ 1 at different grinding speeds, and the maximum increase in equivalent stress caused by strain rate hardening is only 400 MPa, as shown in Figure 16.15. However, the increase in equivalent stress arising from strain hardening reaches a value of 1,000 MPa for various grinding speeds. Thus, the strain-rate hardening effect is not as obvious as that of strain hardening with an equivalent plastic strain of 1. From Table 16.3, the grinding temperature increases dramatically with

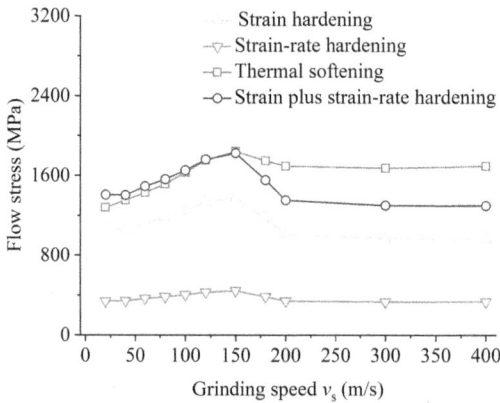

FIGURE 16.15 Variation of flow stress caused by strain, strain rate, and temperature at a different grinding speed.

TABLE 16.3   The Calculation Results

v_s (m/s)	Strain	Strain rate (s⁻¹)	Temperature (°C)	$\Delta\sigma_1$ (MPa)	$\Delta\sigma_2$ (MPa)	$\Delta\sigma_3$ (MPa)
20	0.924644	5.08×10^5	523	1,068	339	1,281
40	0.936317	9.89×10^5	550	1,060	344	1,353
60	0.979173	1.51×10^6	553	1,125	366	1,429
80	1.02021	2.09×10^6	572	1,178	383	1,513
100	1.05698	2.46×10^6	597	1,248	404	1,631
120	1.09954	3.12×10^6	612	1,332	428	1,754
150	1.14954	3.75×10^6	636	1,381	443	1,841
180	1.07758	4.16×10^6	662	1,170	384	1,747
200	1.00208	4.90×10^6	672	1,012	341	1,693
300	0.97684	6.81×10^6	683	968	332	1,674
400	0.97688	9.65×10^6	690	963	334	1,693

an increase in grinding speed. On the other hand, as shown in Figure 16.15, the variation in equivalent stress induced by the temperature increase is not monotonic, which rises first but drops when grinding speeds exceed 150 m/s, clearly indicating the coupled effect of strain, strain rate, and thermal softening. Comparing the curves of strain hardening and strain-rate hardening with those of thermal softening, it can be seen that the total effect of strain and strain-rate hardening is slightly greater than that of thermal softening (for grinding speeds ranging from 20 to 150 m/s). However, the effect of thermal softening becomes a dominant factor when the speed is beyond 150 m/s. This means that the critical grinding speed is 150 m/s.

REFERENCES

Aurich, J. C., and M. Steffes. 2011. Single-grain scratch tests to determine elastic and plastic material behavior in grinding. *Advanced Materials Research*, 325: 48–53.

Calamaz, M., D. Coupard, and F. Girot. 2008. A new material model for 2D numerical simulation of serrated chip formation when machining titanium alloy Ti-6Al-4V. *International Journal of Machine Tools & Manufacture*, 48: 275–288.

Chen, Z. Z., L. Tian, Y. C. Fu, et al. 2012. Chip formation of nickel-based superalloy in high speed grinding with single diamond grain. *International Journal of Abrasive Technology*, 5(2): 93–106.

Chen, J. B., Q. H. Fang, and L. C. Zhang. 2014. Investigation on distribution and scatter of surface residual stress in ultra-high speed grinding. *International Journal of Advanced Manufacture Technology*, 75: 615–627.

Dai, J. B., W. F. Ding, L. C. Zhang, et al. 2015. Understanding the effects of grinding speed and unde-formed chip thickness on the chip formation in high-speed grinding. *International Journal of Advanced Manufacturing Technology*, 81: 995–1005

Guo, Y. B., and D. W. Yen. 2004. A FEM study on mechanisms of discontinuous chip formation in hard machining. *Journal of Materials Processing Technology*, 155–156: 1350–1356.

Liang, Z., X. Wang, Y. Wu, et al. 2013. Experimental study on brittle-ductile transition in elliptical ultrasonic assisted grinding (EUAG) of monocrystal sapphire using single diamond abrasive grain. *International Journal of Machine Tools and Manufacture*, 71: 41–51.

Rao, Z., G. Xiao, B. Zhao, et al. 2021. Effect of wear behaviour of single mono- and poly-crystalline cBN grains on the grinding performance of Inconel 718. *Ceramics International*, 47(12): 17049–17056.

Sima, M., and T. Özel. 2010. Modified material constitutive models for serrated chip formation simulations and experimental validation in machining of titanium alloy Ti-6Al-4V. *International Journal of Machine Tools & Manufacture*, 50: 943–960.

Tawakoli, T., H. Kitzig, and R. D. Lohner. 2013. Experimental investigation of material removal mechanism in grinding of alumina by single-grain scratch test. *Advanced Materials Research*, 797: 96–102.

Wang, S., G. Sun, Q. Zhao, et al. 2023. Monitoring of ductile-brittle transition mechanisms in sapphire ultra-precision grinding used small grit size grinding wheel through force and acoustic emission signals. *Measurement*, 210: 112557.

Yang, M., C. Liu, B. Guo, et al. 2023. Understanding of highly-oriented 3C-SiC ductile-brittle transition mechanism in ELID ultra-precision grinding. *Materials Characterization*, 203: 113136.

Zhao, C., J. Li, and Y. Liu. 2023. Study on the grinding force of single grain in rail grinding based on open-type belt grinding. *Journal of Manufacturing Processes*, 99: 794–811.

CHAPTER 17

Formation and Affecting Factors of Burrs during Grinding

17.1 INTRODUCTION

The exit-direction burrs are always produced in the grinding operations of difficult-to-cut metallic materials, i.e., Ti-6Al-4V titanium alloy, and Inconel 718 nickel-based superalloy. The machining quality of workpiece components is always greatly affected negatively by the presence of exit-direction burrs, the removal of which is one of the key factors in decreasing the production efficiency and increasing the machining cost. For this reason, how to decrease the exit-direction burrs in the grinding of difficult-to-cut metallic materials has been an important issue in the present day (Fu et al. 2017).

Formation of the exit-direction burrs is a common phenomenon in metal cutting and grinding, which is because the cutting/grinding layer material is separated from the workpiece, a part of which remains at the workpiece (Qu, 2008). According to the previous literature (Chen, 2014), three types of burrs could be defined: positive burr, negative burr, and residual burr; furthermore, it was found that the positive burr was produced due to plastic bending of the material, while the negative burr was formed due to the crack appearing in the negative shear zone and further extended tearing, and the residual burr was the chip left at the workpiece exit-direction. Gillespie and Boltt (1976) and Romon (2007) discovered that the main reason for burr formation during cutting was the friction between the tip radius, flank, and machined surface, which resulted in the plastic deformation of materials within the cutting layer. Based on the grinding experiments, descriptive models and the mechanism of burr formation in grinding operations were studied (Wang et al. 2023b). Based on the shear strain theory, Ko and Dornfeld (1996) analysed the formation mechanism of positive/negative burr and established the boundary rule of negative burr formation. Furthermore, Toropov et al. (2006) established the theoretical calculation formula of the burr size using the theory of slip line and equilibrium equation; Segonds et al.

DOI: 10.1201/9781032678047-20

(2013) proposed the analytical models of burr formation and predicted the burr properties. Generally, because the exit-direction burrs cannot be avoided completely during the actual cutting and grinding operations, it is usually expected to produce the positive burrs (due to their easy removal) and restrain the negative burrs and residual burrs.

It is known from the previously reported research work about cutting and grinding operations that it is usually rather difficult and even impossible to observe the grinding wheel-workpiece contact and the removal behaviour under the effect of millions of abrasive grains. Under such conditions, the formation mechanism of exit-direction burrs in grinding, due to the complex, could not be well analysed using conventional mathematical and physical methods. However, the finite element method (FEM) could be effectively utilized to simulate burr formation and even detect the cutting/grinding mechanism. For example, Shet and Deng (2000) and Yan et al. (2012), respectively, analysed the cutting and grinding mechanisms using the FEM. Lu et al. (2016) investigated the fracture surfaces in the cutting exit direction of the workpiece based on FEM, in which the minimum energy principle and strain hardening theory are also applied to explain the relationship between the fracture surface and the hardened layer thickness during cutting. Additionally, Qu et al. (2007) simulated the formation behaviour of burrs in cutting. They pointed out that the different fracture extensions resulted in two types of burrs, that is, positive burrs and negative burrs.

In this work, an investigation was made on the formation mechanism and geometry characteristics of the exit-direction burrs in single-grain surface grinding of Ti-6Al-4V titanium alloy based on Abaqus/Explicit Dynamics finite element simulation. A discussion is accordingly made on the effects of the critical grinding parameters (i.e., grinding speed, uncut chip thickness, and negative rake angle of abrasive grains) on the size and types of exit-direction burrs produced in surface grinding. All are expected to provide theoretical guidance for optimizing the grinding parameters to obtain a favourable burr type and minimize the burr size during the grinding operation.

17.2 FINITE ELEMENT MODELLING

17.2.1 Geometric Model of Single-Grain Surface Grinding

Figure 17.1 schematically displays the surface grinding model with a single cubic boron nitride (CBN) grain. The initial two-dimensional (2D) finite element model of the surface grinding operations with a single CBN grain is provided in Figure 17.2. The model of a single CBN grain is simplified to a hexagon. The size of the workpiece sample applied is 30 μm in the x-direction and 10 μm in the y-direction, respectively, which is discretized by 53,000 bilinear, four-node plane strain thermally coupled quadrilateral elements with reduced integration and enhanced hourglass control (CPE4RT). Arbitrary Lagrangian-Eulerian (ALE) adaptive meshing is utilized to maintain a high-quality mesh to prevent possible errors due to severe mesh distortion. To guarantee the efficiency and accuracy of the finite element calculation work, the division of meshes in the workpiece grinding layer is more intensive. In the present simulation work, the bottom edge of the workpiece sample is constrained in both the x- and y-direction, while the CBN grain grinds the workpiece sample from right to left at a certain speed.

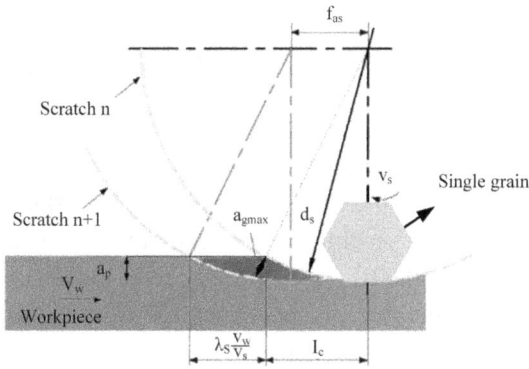

FIGURE 17.1 Surface grinding model with single CBN grain.

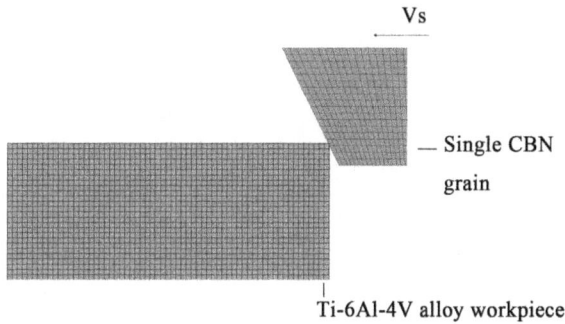

FIGURE 17.2 The initial two-dimensional finite element model for single grain surface grinding.

17.2.2 Materials Model

Some key material properties of CBN grain and Ti-6Al-4V titanium alloy are listed in Table 17.1. Compared with the workpiece material of Ti-6Al-4V titanium alloy, the hardness of the CBN grain is much larger; as such, the grain could be regarded as an isothermal rigid body. Particularly, Tables 17.2–17.4 list the thermal conductivity, heat capacity, and thermal expansion coefficient of the Ti-6Al-4V alloy, which vary with the temperature produced in grinding and have been taken into account in the present work.

During the conventional cutting and grinding process, the workpiece materials always tend to experience an extremely elastic-plastic deformation under the coupled effects of high temperatures, high strains, and high strain rates. Under such conditions, the Johnson–Cook constitutive model (J–C model for short) (Johnson et al. 1985) is applied to describe the behaviour of the workpiece material of the Ti-6Al-4V titanium alloy, including the effects of deformation hardening, strain rate, and temperature-dependent factors. The J–C model is described as follows:

$$\sigma = \left(A + B\varepsilon^n\right)\left[1 + C\left(\frac{\dot{\varepsilon}}{\dot{\varepsilon}_0}\right)\right]\left[1 - \left(\frac{T - T_r}{T_m - T_r}\right)^m\right] \qquad (17.1)$$

where A is the yield strength, B is the strain hardening modulus, C is the strain rate hardening coefficient, m is the thermal softening coefficient, and n is the strain-hardening coefficient; ε is the equivalent plastic strain; $\dot{\varepsilon}$ is the equivalent plastic strain rate; $\dot{\varepsilon}_0$ is the reference plastic strain rate (i.e., 0.001); T_{room} and T_{melt} refer to the use of ambient temperature and material melting temperature, respectively. The initial temperature of workpiece materials and CBN grain is 20°C. The parameters in the J–C model of Ti-6Al-4V titanium alloy are summarized in Table 17.5.

TABLE 17.1 Materials Properties of Ti-6Al-4V Titanium Alloy and CBN Grain

Material	E (GPa)	μ	P (kg/m3)
Ti-6Al-4V	108	0.33	4,440
CBN	909	0.12	3,120

TABLE 17.2 Thermal Conductivity of Ti-6Al-4V Titanium Alloy

T (°C)	20	100	200	300	400	500
λ, W/(m°C)	6.8	7.4	8.7	9.8	10.3	11.8

TABLE 17.3 Heat Capacity of Ti-6Al-4V Titanium Alloy

T (°C)	20	100	200	300	400	500
C (J/(kg·°C))	611	624	653	674	691	703

TABLE 17.4 Thermal Expansion Coefficient of Ti-6Al-4V Titanium Alloy

T (°C)	20	100	200	300	400	500
α (10^{-6}°C)	9.1	9.2	9.3	9.5	9.7	1

TABLE 17.5 Johnson–Cook Model Parameters of Ti-6Al-4V Titanium Alloy

A (MPa)	B (MPa)	C	n	m	T_{melt} (°C)	T_{room} (°C)
875	793	0.01	0.386	0.71	1,560	20

17.2.3 Criterion of Material Fracture in Surface Grinding

The material fracture criterion proposed by J–C considers the effects of strain, strain rate, temperature, and stresses. In the present work, the J–C shear failure (shear fracture) criterion is used as the initial damage criterion of the Ti-6Al-4V titanium alloy. Based on the equivalent plastic strain of the element integral point, failure occurs when the damage parameter w_s exceeds a unit. The damage parameter w_s is defined by

$$\omega_s = \sum \frac{\Delta \varepsilon^{pl}}{\varepsilon_f^{pl}} \tag{17.2}$$

where $\Delta \varepsilon^{pl}$ is an increment of the equivalent plastic strain; ε_f^{pl} is the strain at failure and defined as

$$\varepsilon_f^{pl} = \left(D_1 + D_2 e^{D_3 \sigma^*} \right) \left(1 + D_4 \ln \frac{\dot{\varepsilon}^{pl}}{\dot{\varepsilon}_0} \right) \left[1 - D_5 \left(\frac{T - T_r}{T_m - T_r} \right) \right] \tag{17.3}$$

where the strain at failure is determined by the variable σ^*, $\dot{\varepsilon}_0$, T.

The J–C shear failure parameters $(D_1–D_5)$ of Ti-6Al-4V titanium alloy are taken from the literature (Lesuer, 2000) and summarized in Table 17.6.

TABLE 17.6   J-C shear Failure Parameters of Ti-6Al-4V Titanium Alloy

Material	D_1	D_2	D_3	D_4	D_5
Ti-6Al-4V	−0.09	0.25	−0.5	0.014	3.87

17.2.4 Contact Law

When grinding metallic materials (i.e., Ti-6Al-4V titanium alloy), the grinding heat, causing the workpiece temperature to rise rapidly, is generated not only by the elastic-plastic deformation of machined materials but also by the friction behaviour of the grain-chip interaction and grain-machined surface interaction. In the present finite element analysis, friction between the grain and the workpiece material could be described by the Coulomb's Law, as follows:

$$\tau_f = \mu \sigma_n \tag{17.4}$$

where τ_f is the frictional stress, σ_n is the normal contact stress, and μ is the coefficient of friction (taken to be 0.3 in the present work).

Particularly, Abaqus/Explicit allows the introduction of an inelastic heat fraction, called the Taylor-Quinney empirical constant (often assigned as 90%), to reflect the percentage of plastic deformation and friction work converted to heat during surface grinding.

17.2.5 Grinding Parameters

In the present work, the critical surface grinding parameters (i.e., grinding speed, uncut chip thickness, and negative rake angle of abrasive grains) are chosen to investigate their effects on the exit-direction burrs in single-grain surface grinding, as listed in Table 17.7. It is noted that the grinding parameters listed here have been broadly applied in the practical conventional-speed grinding and high-speed grinding of Ti-6Al-4V titanium alloy.

TABLE 17.7   Surface Grinding Parameter Applied in the Present Finite Element Simulation Work

No.	Grinding Speed v_s (m/s)	Workpiece Infeed Speed v_w (m/min)	Dept of Cut a_p (µm)	Uncut Chip Thickness a_{gmax} (µm)	Negative Rake Angle of CBN Grain γ_g (°)
1	15	0.043, 0.050, 0.057, 0.064, 0.072, 0.079	10	0.6, 0.7, 0.8, 0.9, 1.0, 1.1	−15°, −23°, −30°, −38°, −45°, −60°
2	20	0.057, 0.067, 0.076, 0.086, 0.095, 0.105			
3	60	0.172, 0.201, 0.229, 0.258, 0.286, 0.315			
4	80	0.229, 0.267, 0.306, 0.344, 0.382, 0.420			
5	100	0.286, 0.334, 0.382, 0.430, 0.477, 0.525			
6	120	0.344, 0.401, 0.458, 0.516, 0.573, 0.630			

Particularly in surface grinding operations with single grains, the uncut chip thickness (a_{gmax}), which has a great influence on grinding resultants, could be calculated by Tian et al. (2015)

$$a_{g\max} = 2 \cdot \lambda_s \cdot \left(\frac{v_w}{v_s} \right) \cdot \sqrt{\frac{a_p}{d_{eq}}} \tag{17.5}$$

where a_p is the depth of cut, d_{eq} is the grinding wheel diameter (400 mm in the experimental work), v_w is the workpiece infeed speed, v_s is the grinding speed (that is, the circumferential speed of the grinding wheel), and λ_s is the circumference length of the grinding wheel when considering the single CBN grain on the wheel surface.

17.3 FORMATION STAGES OF EXIT-DIRECTION BURRS IN SURFACE GRINDING

17.3.1 Deformation and Geometry Characteristics of Exit-Direction Burrs

In the traditional steady stage of grinding operations, the grinding region, similar to the cutting region, is usually classified into three zones, as schematically displayed in Figure 17.3, which include the first deformation zone (also called the primary shear slip zone), the second deformation zone (mainly existing extrusion and friction behaviour at the interface between chip and rake face), and the third deformation zone (mainly existing extrusion and friction between the machined surface and the tip radius or flank). Furthermore, in the formation process of exit-direction burrs, there perhaps exits a

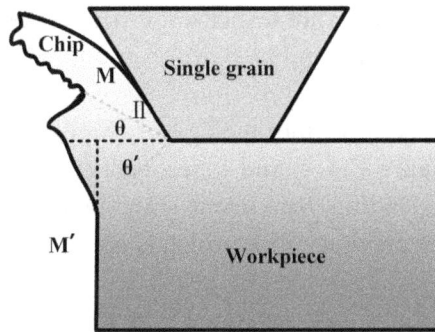

FIGURE 17.3   Formation models of exit-direction burrs in grinding.

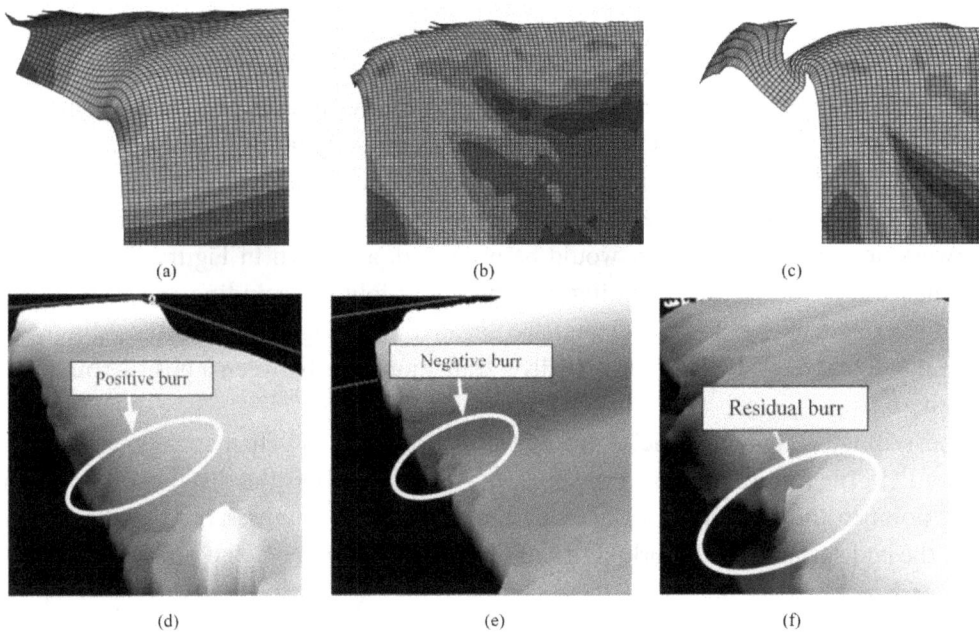

FIGURE 17.4   Simulated and experimental results of three types of exit-direction burrs produced in surface grinding of Ti-6Al-4V titanium alloy: (a) positive burr, (b) negative burr, (c) residual burr, (d) positive burr, (e) negative burr, (f) residual burr.

different deformation zone, which is defined as the fourth deformation zone or the negative shear zone, in which the ground material produces a complicated flow behaviour.

Cutting-direction burr could always be divided into positive burr, negative burr, and residual burr. Figure 17.3 also provides the formation model of exit-direction burrs in grinding. OM is a shear slip line (Figure 17.3), and the angle between the grinding forward direction and OM is θ; on the contrary, OM' is a negative shear line, and the angle between the grinding forward direction and OM' is θ'. When the CBN grain approaches the end part of the workpiece sample, a flexure deformation could take place under the grinding forces; as such, the negative shear slip line (OM') is produced in the negative shear zone (that is, the fourth deformation zone).

Figure 17.4 provides the typical morphology of three types of exit-direction burrs obtained by simulation (Figure 17.4–c) and experiments (Figure 17.4d–f). It is noted that the details of the surface grinding experiment with single-brazed CBN/diamond grain have been introduced in the previous work, which is not provided any more in this work. Generally, in the surface grinding process of the particular region adjacent to the end part of the workpiece sample, if the chip is separated from the shear slip line (OM) of the first deformation zone, the positive burr would be generated, as shown in Figure 17.4a, which is obtained under the surface grinding parameters as follows: grinding speed of 100 m/s, uncut chip thickness of 1 μm, and negative rake angle of CBN grain of −30°. Generally, positive burr is comparatively removed easily.

However, if the chip shear slip occurs along the negative shear slip line (OM') in the fourth deformation zone, in which the crack would take place, and the extension causes the chip to be separated from the workpiece, the negative burr is generated, as shown in Figure 17.4b, which is obtained under the surface grinding parameters as follows: grinding speed of 100 m/s, uncut chip thickness of 1 μm, and abrasive negative rake angle of −15°.

During the formation process of the negative burr, if the chip is not separated from the workpiece, the residual burr would be produced, as shown in Figure 17.4c, which is obtained under the surface grinding parameters as follows: grinding speed of 100 m/s, uncut chip thickness of 1 μm, and abrasive negative rake angle of −23°. Accordingly, it is also known that a residual burr is actually a transition state between a positive burr and a negative burr.

The geometry characteristic of an exit-direction burr is schematically displayed in Figure 17.5. The root thickness of the burr, B, is represented as the distance from the pivoting point to the machined surface; the burr height, H, is represented as the distance from the end surface of the workpiece sample to the burr outer side. Using the parameter

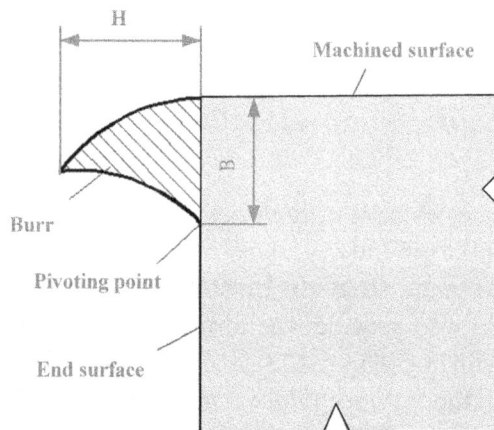

FIGURE 17.5   Cross-section view of exit-direction burrs produced in surface grinding.

combination of B and H, the size and morphology of the burr cross-section could be characterized quantitatively. Particularly, the burr height directly affects the dimensional and shape accuracy of the workpiece component, while the burr root thickness influences the deburring behaviour.

17.3.2 Formation Stages of Positive Burrs

Figure 17.6 shows the simulation results of deformation behaviour and equivalent stresses distribution at the different formation stages of positive burrs, where the surface grinding

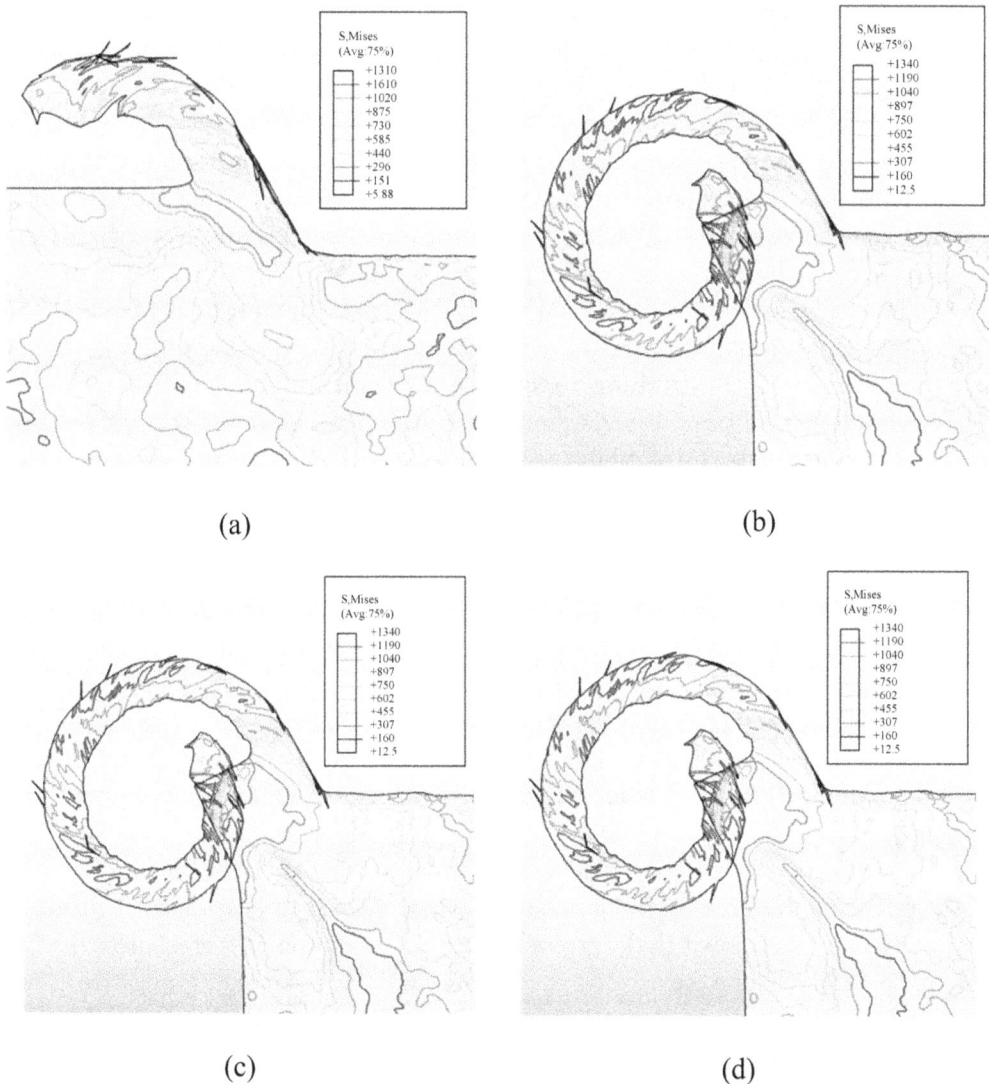

(a)

(b)

(c)

(d)

FIGURE 17.6   Four stages of positive burr formation in the surface grinding of Ti-6Al-4V titanium alloy: (a) steady grinding stage, (b) pivoting point stage, (c) successive grinding stage, (d) positive burr formation stage.

speed is 100 m/s, the uncut chip thickness is 1 μm, and the negative rake angle of CBN grain is −30°. In general, the formation process of positive burrs could be divided into four stages: the steady grinding stage (Figure 17.6a), the pivoting point stage (Figure 17.6b), the successive grinding stage (Figure 17.6c), and the positive burr formation stage (Figure 17.6d).

At the steady grinding stage, the shear angle of the material deformation region is positive, as shown in Figure 17.6a, and the maximum von Mises stress (1.31 GPa) is the largest one in the whole shear zone. The chip is formed along the shear slip surface. This stage is thought of as adiabatic shear, and an adiabatic shear band is formed in the first deformation zone. As demonstrated in Figure 17.6b, at the pivoting point stage, the grain goes near the workpiece end part, which results in decreased support stiffness of the workpiece sample. Due to the grinding force, the workpiece end part rotates around the terminal point, and the plastic deformation takes place; accordingly, the negative shear zone is produced and extended to the end part surface. At the same time, a small slip deformation occurs as the grinding layer material rotates. Furthermore, due to the rotation of the grinding layer material, the actual depth of cut is greater than the theoretical value, which increases the maximum von Mises stress (i.e., 1.34 GPa). At the successive grinding stage, as shown in Figure 17.6c, the negative shear zone disappears, and the material within the grinding layer does not rotate any longer. With the continuous removal of the material, the crack is induced at the primary shear zone along the grinding direction, and the maximum von Mises stress (i.e., 1.31 GPa) decreases continuously. As shown in Figure 17.6d, the crack expands unceasingly at the positive burr formation stage, and finally the burr is cut out; accordingly, the positive burr is formed.

17.3.3 Formation Stages of Negative Burrs

Figure 17.7 displays the simulation results of deformation and von Mises stress distribution at the different formation stages of negative burrs, which are obtained at a grinding speed of 100 m/s, an uncut chip thickness of 1 μm, and a negative rake angle of CBN grain of −15°. According to the variation characteristics, the formation process of negative burrs could also be divided into four stages, which include the steady grinding stage, pivoting point stage, crack formation and propagation stage, and the negative burr formation stage.

The formation of negative burrs is generally similar to that of positive burrs at the steady grinding stage and the pivoting point stage. The difference is mainly at the crack formation and propagation stage, in which the ground material would produce a crack near the CBN grain tip and extend along the negative shear plane, and the shear angle is negative (∠AOB in Figure 17.7b), as shown in Figure 17.7c. Finally, as demonstrated in Figure 17.7d, at the negative burr formation stage, the crack extends along the negative shear surface with the rotation of the grinding layer material. Under such conditions, the chip is separated from the workpiece material, and a negative burr is therefore generated.

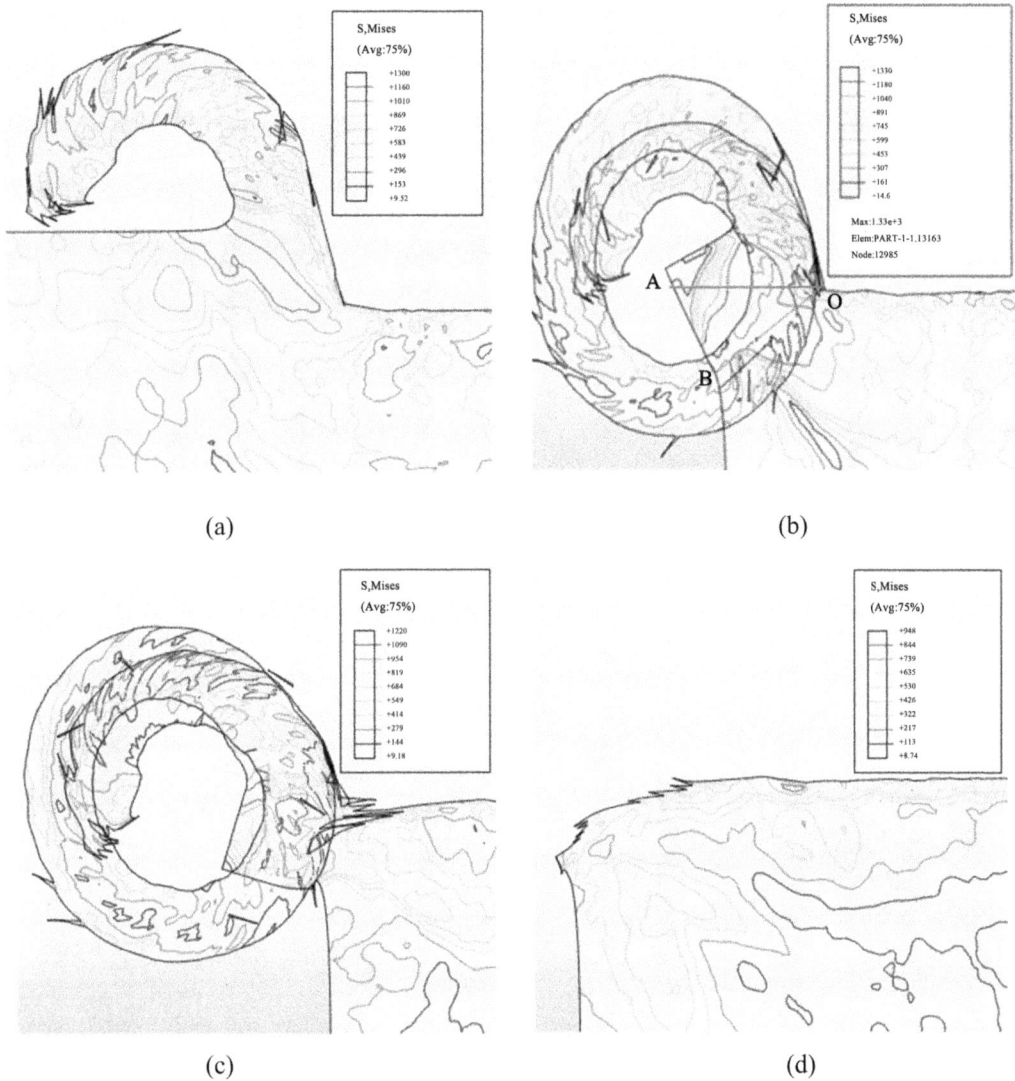

FIGURE 17.7 Four stages of negative burr formation in surface grinding of Ti-6Al-4V titanium alloy: (a) steady grinding stage, (b) pivoting point stage, (c) crack formation and propagation stage, (d) negative burr formation stage.

17.3.4 Formation Stages of Residual Burrs

The simulation results of deformation and von Mises stress in the formation stages of residual burrs are demonstrated in Figure 17.8, where the grinding speed is 100 m/s, the uncut chip thickness is 1 μm, and the negative rake angle of CBN grain is −23°. There are also four stages in the formation of residual burrs: a steady grinding stage, a pivoting point stage, crack formation and propagation stage, and a residual burr formation stage.

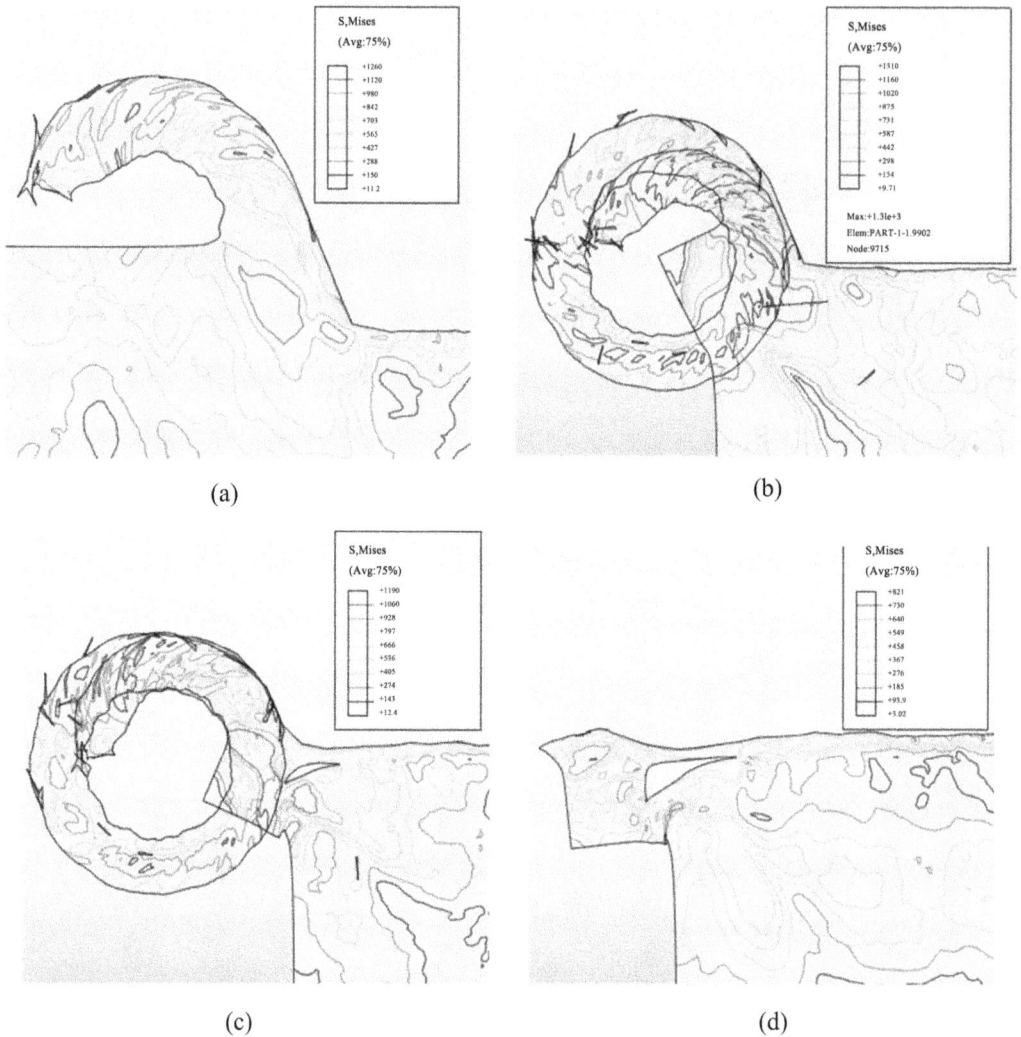

FIGURE 17.8   Four stages of residual burr formation in surface grinding of Ti-6Al-4V titanium alloy: (a) steady grinding stage, (b) pivoting point stage, (c) crack formation and propagation stage, (d) residual burr formation stage.

The formation of residual burrs is generally the same as that of the negative burrs; in particular, the major difference is that the chip does not break but remains at the workpiece end part at the final stage.

17.4 GRINDING SPEED EFFECT ON EXIT-DIRECTION BURRS IN SURFACE GRINDING

Figure 17.9 shows the simulation results of the exit-direction burr types and resultant equivalent plastic strain contour at various grinding speeds (i.e., 15–120 m/s), where the uncut chip thickness is fixed at 0.6 μm and the negative rake angle of CBN grain is fixed at −30°.

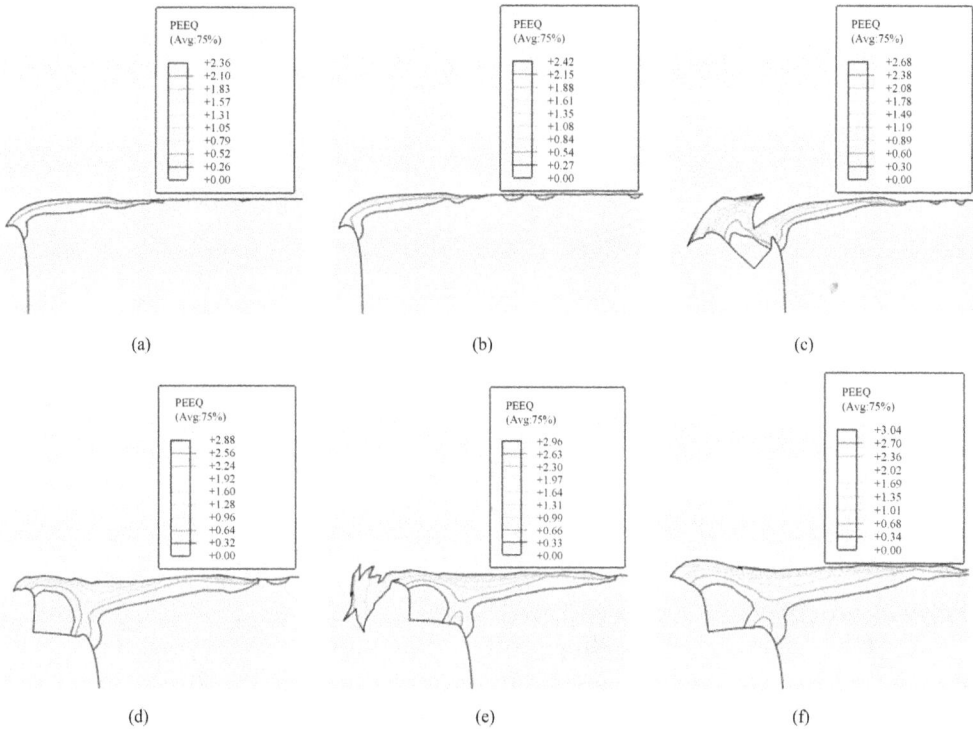

FIGURE 17.9 Effect of grinding speed on burr types and equivalent plastic strain in surface grinding of Ti-6Al-4V titanium alloy: (a) v_s = 15 m/s, (b) v_s = 20 m/s, (c) v_s = 60 m/s, (d) v_s = 80 m/s, (e) v_s = 100 m/s, (f) v_s = 120 m/s.

As seen in Figure 17.9, the grinding speeds generally have a great influence on the burr types and the maximum equivalent plastic strain. The maximum equivalent plastic strain increases with increasing grinding speeds, ranging from 15 to 120 m/s. For example, the maximum values of equivalent plastic strain are 2.36 at a grinding speed of 15 m/s, 2.42 at 20 m/s, 2.68 at 60 m/s, 2.88 at 80 m/s, 2.96 at 100 m/s, and 3.04 at 120 m/s, respectively. That is, the maximum equivalent plastic strain increases by 28.8% at the grinding speed of 120 m/s in comparison to that at the grinding speed of 15 m/s, which corresponds to the growing plastic slip deformation.

The morphology of exit-direction burrs also changes significantly at different grinding speeds. With the increase in grinding speeds from 15 to 20 m/s, the negative burr is produced; afterwards, it is transformed into residual burr at the grinding speed of 60 m/s and further changed into positive burr at the grinding speed of 80 m/s. Based on the J-C model of Ti-6Al-4V titanium alloy, strain rate, and temperature are the two key factors for the plastic deformation behaviours (Lee et al. 1998). A high strain rate leads to work-hardening of the workpiece material, while a high temperature results in work-softening, both of which take effect simultaneously in a different way under the conditions of different surface grinding parameters. Due to the rather small value of uncut chip thickness during single-grain surface grinding, the resultant hardening and

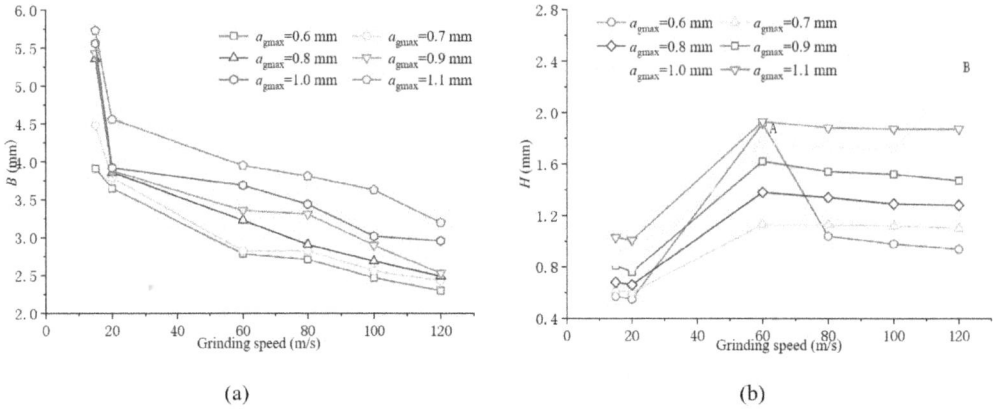

FIGURE 17.10 Burr sizes versus grinding speeds: (a) burr root thickness B, (b) burr height H.

softening effects are very complicated. When the grinding speed is low (i.e., at 15 m/s), the thickness of the plastic hardening layer can be very thin; as such, the chip is more likely to be torn to produce a negative burr.

On the other hand, in the grinding process, the strain hardening plays a dominating role in the material removal behaviour at the grinding speed of 15–120 m/s. Based on the previous literature (Velasquez et al. 2010), it is known that the thickness of the plastic hardening layer increases with the cutting and grinding speeds. Therefore, compared with the primary deformation zone, the machined surface forms a thicker plastic hardening layer and higher strength with an increase in grinding speeds, which makes the maximum equivalent plastic strain gradually increase. According to the minimum energy principle, the chips are more easily cut off in the primary deformation zone to produce a positive burr in grinding.

Figure 17.10 provides the size details of exit-direction burrs obtained at different grinding speeds (i.e., 15–120 m/s) and different uncut chip thicknesses (i.e., 0.6–1.1 μm). Here, the negative rake angle of the CBN grain is fixed at −30°.

As demonstrated in Figure 17.10a, the burr root thickness decreases with increasing grinding speeds. Comparing the burr root thickness obtained at the grinding speed of 15 and 120 m/s, it is decreased by 41.2% from 3.91 to 2.3 μm ($a_{gmax} = 0.6$ μm), by 45.8% from 4.48 to 2.43 μm ($a_{gmax} = 0.7$ μm), by 53.5% from 5.36 to 2.49 μm ($a_{gmax} = 0.8$ μm), by 53.4% from 5.43 to 2.53 μm ($a_{gmax} = 0.9$ μm), by 46.8% from 5.56 to 2.96 μm ($a_{gmax} = 1.0$ μm), and by 44.2% from 5.73 to 3.2 μm ($a_{gmax} = 1.1$ μm). As displayed in Figure 17.10b, when the grinding speed is below 20 m/s, all the burrs are negative; however, the burrs become positive when the grinding speed is beyond 60 m/s. Except for the two mutation points (i.e., Point A and Point B) in Figure 17.10b, a transitional residual burr could be produced. Furthermore, the burr height decreases with increasing grinding speed, and meanwhile, the smaller the uncut chip thickness, the smaller the burr height.

When the negative burr is changed to the positive burr, the burr height increases rapidly, while the burr root thickness is less affected. This is because the burr root thickness directly reflects the position of the pivoting point; however, it does nothing

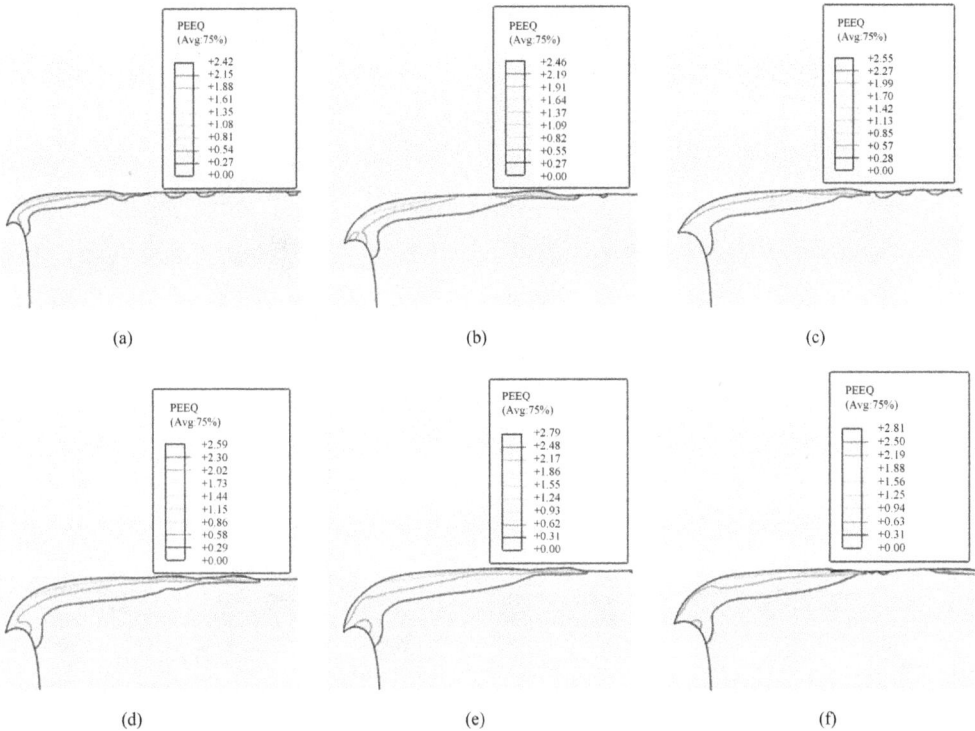

FIGURE 17.11 Effect of uncut chip thickness on exit-direction burr type and equivalent plastic strain ($v_s = 20$ m/s): (a) $a_{gmax} = 0.6$ μm, (b) $a_{gmax} = 0.7$ μm, (c) $a_{gmax} = 0.8$ μm, (d) $a_{gmax} = 0.9$ μm, (e) $a_{gmax} = 1.0$ μm, (f) $a_{gmax} = 1.1$ μm.

for the burr type. With increasing grinding speeds, the thickness of the plastic-hardening layer increases gradually; at this time, the strain hardening plays a main role in the material removal behaviour, so the deformation is reduced and the burr size decreases. Also, with an increase in grinding speed, the thermal softening effect is strengthened, and the distance from the pivoting point to the machined surface becomes smaller. The depth of the pivoting point directly determines the burr root thickness, so the burr root thickness gets smaller and smaller.

It can significantly reduce the burr size with increasing grinding speeds. However, the increasing grinding speed also results in a rising grinding temperature, which may enhance the surface burning behaviour of the machined components in grinding.

17.5 EFFECT OF UNCUT CHIP THICKNESS ON EXIT-DIRECTION BURRS IN SURFACE GRINDING

Figures 17.11 and 17.12 provide the simulation results of the exit-direction burr type and equivalent plastic strain contour at various uncut chip thicknesses (0.6–1.1 μm), where the negative rake angle of the CBN grain is −30°, and the grinding speed is 20 and 60 m/s, respectively.

According to Figs. 17.11 and 17.12, when the uncut chip thickness is in the range of 0.6–1.1 μm, the negative burr is always generated at the grinding speed of 20 m/s, while the

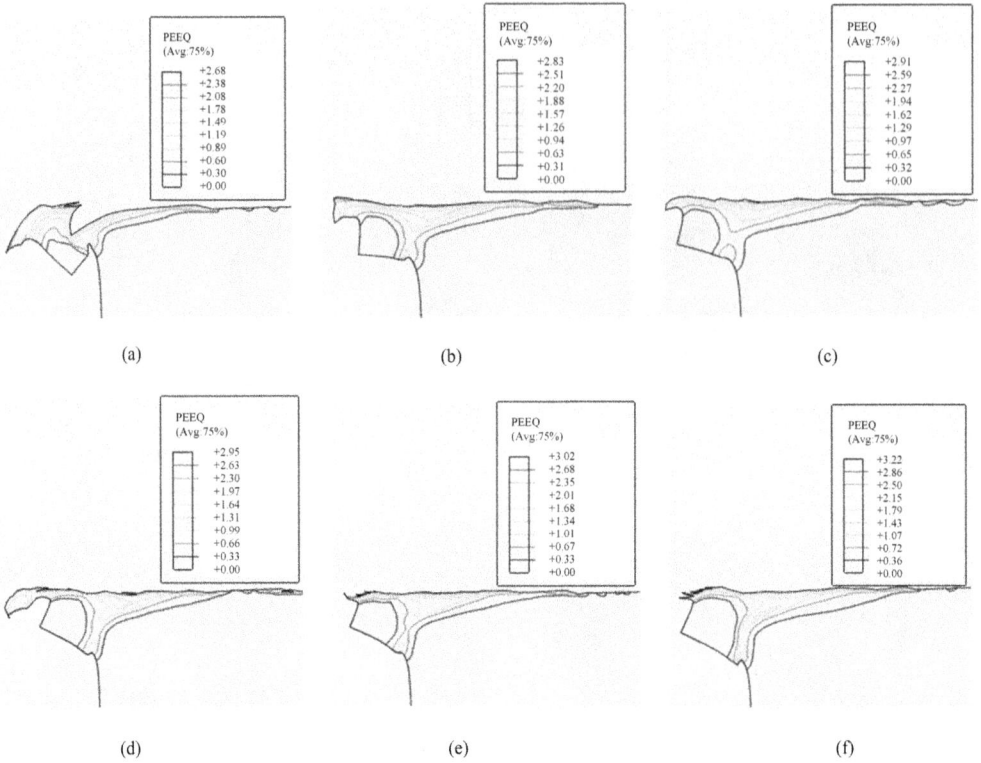

FIGURE 17.12 Effect of uncut chip thickness on exit-direction burr type and equivalent plastic strain ($v_s = 60$ m/s): (a) $a_{gmax} = 0.6\,\mu$m, (b) $a_{gmax} = 0.7\,\mu$m, (c) $a_{gmax} = 0.8\,\mu$m, (d) $a_{gmax} = 0.9\,\mu$m, (e) $a_{gmax} = 1.0\,\mu$m, (f) $a_{gmax} = 1.1\,\mu$m.

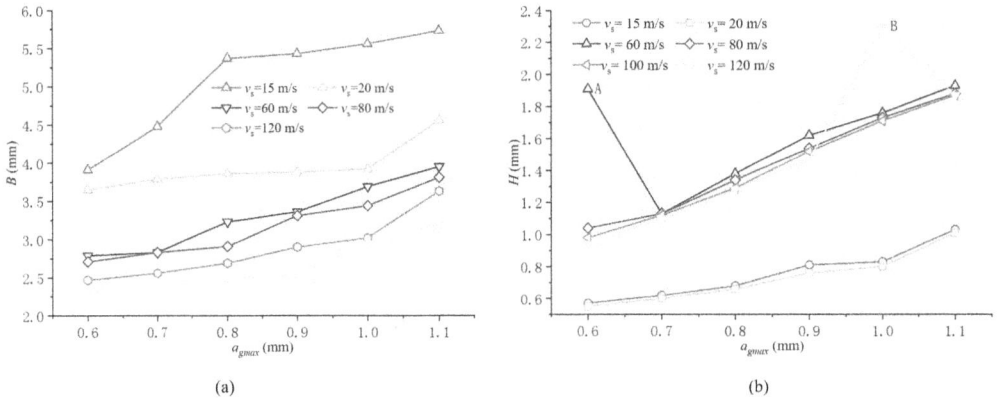

FIGURE 17.13 Burr sizes versus uncut chip thickness: (a) burr root thickness B, (b) burr height H.

residual and positive burrs are produced at the grinding speed of 60 m/s. The value of the maximum equivalent plastic strain increases when the uncut chip thickness increases from 0.6 to 1.1 μm. For example, at a grinding speed of 60 m/s, the maximum equivalent plastic strains are 2.68, 2.83, 2.91, 2.95, 3.02, and 3.33 in the case of the uncut chip thickness of

0.6, 0.7, 0.8, 0.9, 1.0, and 1.1 μm, respectively. The variation rate of the maximum equivalent plastic strain is 20.1% at the uncut chip thickness of 1.1 μm when compared with that at the uncut chip thickness of 0.6 μm. Lekkala et al. (2011) pointed out that, with the increasing depth of cut, the workpiece component stiffness along the grinding direction decreases due to the increasing grinding force and high-frequency vibration. The material deformation increases; meanwhile, the equivalent plastic strain and burr size also increase. Under such conditions, the equivalent plastic strain increases with an increase in the uncut chip thickness.

Figure 17.13 provides the size details of burrs obtained with different uncut chip thicknesses (0.6–1.1 μm) and at different grinding speeds (15–120 m/s). Here, the negative rake angle of CBN grains is fixed at −30°.

As shown in Figure 17.13a, with an increase in uncut chip thickness from 0.6 to 1.1 μm, the burr root thickness increases. For example, when the uncut chip thickness is increased from 0.6 to 1.1 μm, the burr root thickness increases from 3.91 to 5.73 μm at a grinding speed of 15 m/s, from 3.65 to 4.56 μm at a grinding speed of 20 m/s, from 2.79 to 3.95 μm at a grinding speed of 60 m/s, from 2.71 to 3.81 μm at a grinding speed of 80 m/s, from 2.47 to 3.63 μm at a grinding speed of 100 m/s, and from 2.30 to 3.20 μm at a grinding speed of 120 m/s, respectively. That is to say, the variation rate reaches 41.2%, 45.8%, 53.5%, 53.4%, 46.8%, and 44.2%, respectively, at the different grinding speeds, which is because the larger the uncut chip thickness, the thicker the grinding layer material and the more of the deformation material volume. The farther the distance of a pivoting point from the machined surface is, the greater the burr thickness is.

As shown in Figure 17.13b, the burr height increases linearly with increasing uncut chip thickness. The uncut chip thickness is 0.57 μm in case of the uncut chip thickness of 0.6 and 1.03 μm in case of the uncut chip thickness of 1.1 μm ($v_s = 15$ m/s), 0.55 μm with the uncut chip thickness of 0.6 and 1.01 μm with the uncut chip thickness of 1.1 μm ($v_s = 20$ m/s), 1.91 μm with the uncut chip thickness of 0.6 and 1.93 μm with the uncut chip thickness of 1.1 μm ($v_s = 60$ m/s), 1.04 μm with the uncut chip thickness of 0.6 and 1.88 μm with the uncut chip thickness of 1.1 μm ($v_s = 80$ m/s), 0.98 μm with the uncut chip thickness of 0.6 and 1.87 μm with the uncut chip thickness of 1.1 μm ($v_s = 100$ m/s), 0.94 μm with the uncut chip thickness of 0.6 and 1.87 μm with the uncut chip thickness of 1.1 μm ($v_s = 120$ m/s), respectively. That is, the burr height increases by 80.7%, 83.6%, 1.0% (due to the formation of residual burrs at Point A), 80.8%, 90.8%, and 98.9%, respectively, at the varied grinding speed. The burr height obtained at the grinding speeds of 15 and 20 m/s, respectively, is smaller than that at other grinding speeds, which is produced due to the different burr types; for instance, the height of the negative burr is far less than that of the positive and residual ones. In the case of grinding speeds of 15 and 20 m/s, the negative burrs are produced; however, at the grinding speed range of 60 to 120 m/s, the positive burrs are generated. In addition to the two mutation points A and B in Figure 17.13b, a transitional residual burr is also produced. The reason for burr sizes increasing is that the stiffness of the workpiece decreases with the increase in the uncut chip thickness and the enhancement of the thermal softening effect.

17.6 EFFECT OF GRAIN RAKE ANGLE ON EXIT-DIRECTION BURRS IN SURFACE GRINDING

Figure 17.14 shows the simulation results of the exit-direction burr types and equivalent strain contour at various negative rake angles of CBN grains ($-15°$ – $-60°$), where the grinding speed is 100 m/s and the uncut chip thickness is 1.0 μm.

As shown in Figure 17.14, with increasing of the magnitude of the negative rake angle of CBN grains from $-15°$ to $-60°$, the burr type changes from the negative burr with the negative rake angle of $-15°$ to the residual burr with the negative rake angle of $-23°$, and then transforms to a positive burr with the negative rake angle beyond $-30°$. At the same time, the maximum equivalent plastic strain increases with the increasing of the magnitude of the negative rake angle of CBN grains. For example, the maximum equivalent plastic strain reaches 3.23, 3.29, 3.32, 3.33, 3.35, and 3.39, respectively, when the corresponding negative rake angle of CBN grains is $-15°$, $-23°$, $-30°$, $-38°$, $-45°$, and $-60°$, respectively. This phenomenon can be explained as follows: the larger magnitude of rake angle of CBN grains, the greater of contact area and friction between the chip and abrasive grains, leading to a higher temperature and grinding force and therefore a larger chip deformation.

For this reason, the maximum equivalent plastic strain increases. Usually, the effect of the tool rake angle on the plastic hardening layer is small, and the hardened layer is less affected by the burr types. When the magnitude of the negative rake angle of a CBN grain is small, the stress concentration is more likely to form at the grain tip. The grinding layer material is more

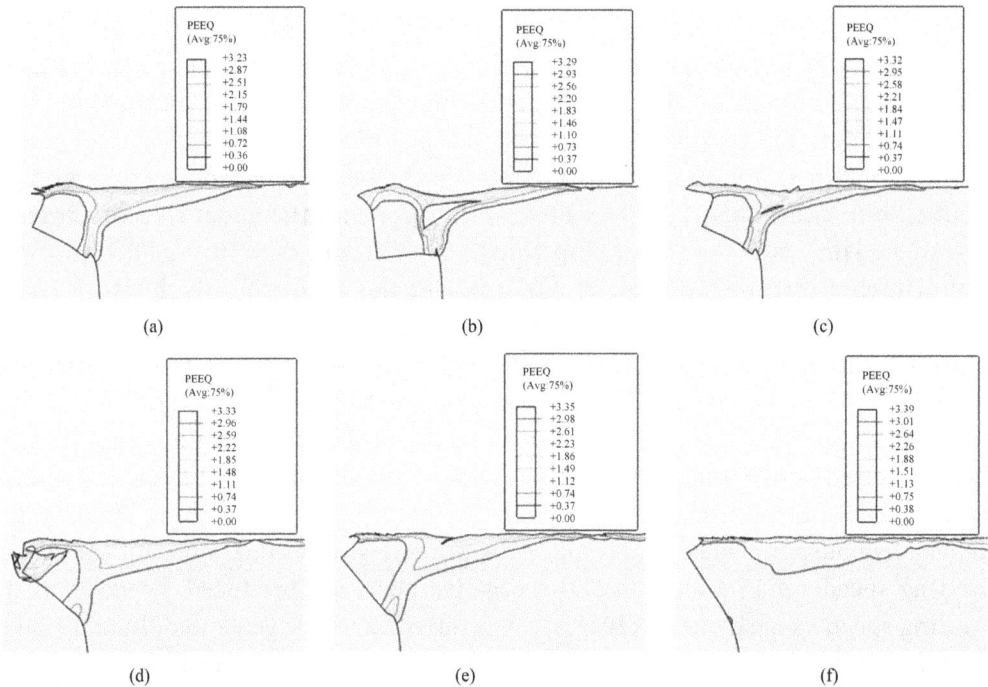

FIGURE 17.14 Effect of negative rake angle of CBN grains on exit-direction burr type and equivalent plastic strain: (a) $\gamma_g = -15°$; (b) $\gamma_g = -23°$; (c) $\gamma_g = -30°$; (d) $\gamma_g = -38°$; (e) $\gamma_g = -45°$; (f) $\gamma_g = -60°$.

easily torn along the negative shear section; as such, a negative burr is formed. The larger the magnitude of the negative rake angle of CBN grains, the larger the grinding force; moreover, the direction of the grinding force also changes, resulting in a variation of the burr morphology.

Figure 17.15 provides the size details of burrs obtained under different conditions of negative rake angle of CBN grains ($-15°$ to $-60°$) and at different grinding speeds (15–120 m/s). Here, the uncut chip thickness is fixed at 1 μm.

As shown in Figure 17.15a, the burr root thickness increases linearly with an increase in the magnitude of negative rake angle of CBN grains, which ranges from $-15°$ to $-60°$. For example, when the grinding speed is fixed at 15 m/s, the burr root thickness increases with increasing magnitude of the negative rake angle of CBN grains, which is 4.11 μm at $-15°$, 4.16 μm at $-23°$, 4.72 μm at $-30°$, 4.93 μm at $-38°$, 4.99 μm at $-45°$, and 7.04 μm at $-60°$. The curves of the burr root thickness at the grinding speeds of 20, 60, 80, 100, and 120 m/s, respectively, exhibit a nearly identical variation trend with that at the grinding speed of 15 m/s. For example, when the grinding speed is 20 m/s, the burr root thickness is 3.32 μm at $-15°$, 3.86 μm at $-23°$, 3.92 μm at $-30°$, 4.84 μm at $-38°$, 4.94 μm at $-45°$, 6.89 μm at $-60°$, respectively; when the grinding speed reaches 60 m/s, the burr root thickness is 2.93 μm at $-15°$, 3.38 μm at $-23°$, 3.69 μm at $-30°$, 4.24 μm at $-38°$, 4.34 μm at $-45°$, 6.83 μm at $-60°$, respectively; when the grinding speed is finally increased to 120 m/s, the root thickness is 2.37 μm at $-15°$, 2.95 μm at $-23°$, 2.96 μm at $-30°$, 3.17 μm at $-38°$, 3.51 μm at $-45°$, 5.35 μm at $-60°$, respectively. Comparing the burr root thickness obtained at the negative rake angle of CBN grains of $-15°$ and $-60°$, respectively, it is increased by 71.3% ($v_s = 15$ m/s), 107.5% ($v_s = 20$ m/s), 133.1% ($v_s = 60$ m/s), 140.2% ($v_s = 80$ m/s), 122.4% ($v_s = 100$ m/s), and 125.7% ($v_s = 120$ m/s), respectively.

As seen in Figure 17.15b, the burr height is greatly influenced by the negative rake angle of CBN grains. For example, when the grinding speed is fixed at 15 m/s, the burr height increases with increasing magnitude of the negative rake angle of CBN grains, which is 0.74 μm at $-15°$, 0.77 μm at $-23°$, 0.80 μm at $-30°$, 0.80 μm at $-38°$, 1.94 μm at $-45°$, and 2.26 μm at $-60°$, respectively. The curves of the burr height obtained at the other grinding

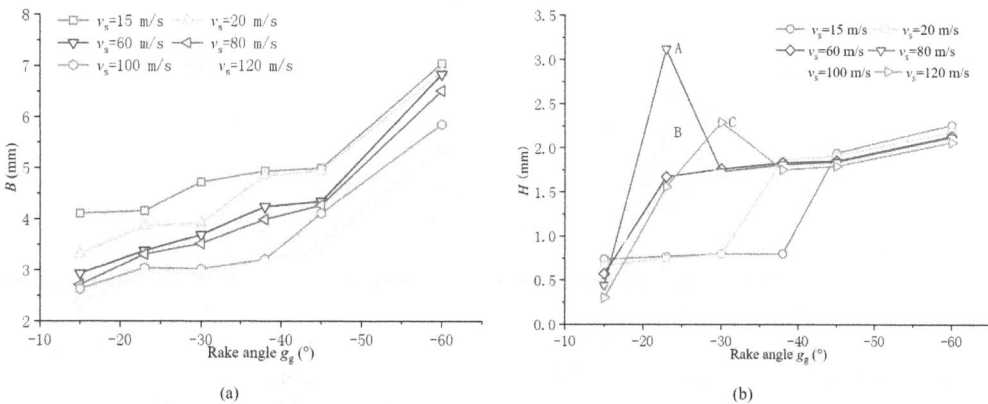

FIGURE 17.15 Burr sizes versus negative rake angle of CBN grains: (a) burr root thickness B, (b) burr height H.

speeds, i.e., 20–120 m/s, also have the almost identical variation trend obtained at the grinding speed of 15 m/s. As an example, the burr height is 0.67 μm at −15°, 0.75 μm at −23°, 0.80 μm at −30°, 1.86 μm at −38°, 1.91 μm at −45°, 2.18 μm at −60°, respectively, when the grinding speed is 20m/s; it is 0.57 μm at −15°, 1.67 μm at −23°, 1.76 μm at −30°, 1.83 μm at −38°, 1.85 μm at −45°, 2.13 μm at −60°, respectively, when the grinding speed is 60 m/s; it is 0.30 μm at −15°, 1.56 μm at −23°, 2.29 μm at −30°, 1.75 μm at −38°, 1.79 μm at −45°, 2.06 μm at −60°, respectively, when the grinding speed reaches 120 m/s. Furthermore, comparing the burr height obtained at the negative rake angle of CBN grains of −15° and −60°, respectively, it is increased by 205.4% (v_s = 15 m/s), 225.4% (v_s = 20 m/s), 273.7% (v_s = 60 m/s), 381.8% (v_s = 80 m/s), 533.3% (v_s = 100 m/s), and 586.7% (v_s = 120 m/s), respectively.

The increase in burr sizes can be explained by the larger magnitude of the negative rake angle of CBN grains, the higher of grinding temperature and grinding force, causing the enhanced thermal softening effects. Therefore, the chip deformation is increased, and the burr sizes are also enlarged.

As shown in Figure 17.15b, a negative burr is produced at a negative rake angle of CBN grains of −15°; in addition to the three mutation points A, B, and C in Figure 17.15b, a transitional residual burr could be generated. However, a positive burr starts to be produced in the case of the negative rake angle of −23° and the grinding speed of 120 m/s, or the negative rake angle of −30° and the grinding speed of 60–120 m/s, or the negative rake angle of −38° and the grinding speed of 20 m/s, or the negative rake angle of −45° and the grinding speed of 15 m/s. This indicates that a positive burr could be generated at the parameter combination (i.e., higher grinding speed and a smaller magnitude of negative rake angle of CBN grains).

REFERENCES

Chen, Y. L. 2014. *Modeling and experimental study on burr formation process fly-cutting*. Ph.D. thesis, Harbin: Harbin Institute of Technology, China.

Fu, D. K., W. F. Ding, S. B. Yang, et al. 2017. Formation mechanism and geometry characteristics of exit-direction burrs generated in surface grinding of Ti-6Al-4V titanium alloy. *International Journal of Advanced Manufacturing Technology*, 89: 2299–2313.

GillesPie, L. K., and P. T. Boltt. 1976. The formation and properties of machining burr. *Journal of Manufacturing Science & Engineering*, 98(1): 66–74.

Ko, S. L., and D. A. Dornfeld. 1996. Burr formation and fracture in oblique cutting. *Journal of Materials Processing Technology*, 62(1): 24–36.

Lu, J. P., J. B. Chen, Q. H. Fang, et al. 2016. Finite element simulation for Ti-6Al-4V alloy deformation near the exit of orthogonal cutting. *International Journal of Advanced Manufacturing Technology*, 85: 2377–2388.

Qu, H. J. 2008. *Burr formation mechanism and simulation of right angle cutting*. Ph.D. thesis, Zhenjiang: Jiangsu University, China.

Qu, H. J., G. C. Wang, H. J. Pei, et al. 2007. Formation and simulation of cutting-direction burr in orthogonal cutting. *Advanced Materials Research*, 24–25: 249–254.

Romon. 2007. *Study on formation mechanism and control method of burr formation in metal cutting process*. Ph.D. thesis, Shanghai: Shanghai Jiaotong University, China.

Shet, C., and X. M. Deng. 2000. Finite element analysis of the orthogonal metal cutting process. *Journal of Materials Processing Technology*, 105(1–2): 95–109.

Segonds, S., J. Masounave, V. Songmene, et al. 2013. A simple analytical model for burr type prediction in drilling of ductile materials. *Journal of Materials Processing Technology*, 213: 971–977.

Toropov, A. A., S. L. Ko, and J. M. Lee. 2006. A new burr formation model for orthogonal cutting of ductile materials. *CIRP Annals - Manufacturing Technology*, 55(1): 55–58.

Wang, H., C. Wang, J. Chen, et al. 2023a. Burr formation mechanism and morphological transformation in grinding of nickel-based superalloy honeycomb cores under ice freezing and MQL conditions. *Journal of Materials Processing Technology*, 318: 118005.

Wang, H., J. Chen, C. Wang, et al. 2023b. Burr formation mechanism and morphological transformation critical conditions in grinding of nickel-based superalloy honeycomb cores. *Chinese Journal of Aeronautics*. https://doi.org/10.1016/j.cja.2023.02.015.

Yan, L., F. Jiang, and Y. M. Rong. 2012. Grinding mechanism based on single grain cutting simulation. *Journal of Mechanical Engineering*, 48(11): 172–182.

For Product Safety Concerns and Information please contact our EU
representative GPSR@taylorandfrancis.com
Taylor & Francis Verlag GmbH, Kaufingerstraße 24, 80331 München, Germany

www.ingramcontent.com/pod-product-compliance
Lightning Source LLC
Chambersburg PA
CBHW080918220326
41598CB00034B/5613

9 781032 678054